Vol. 7

Frontiers in
Nanobiomedical
Research

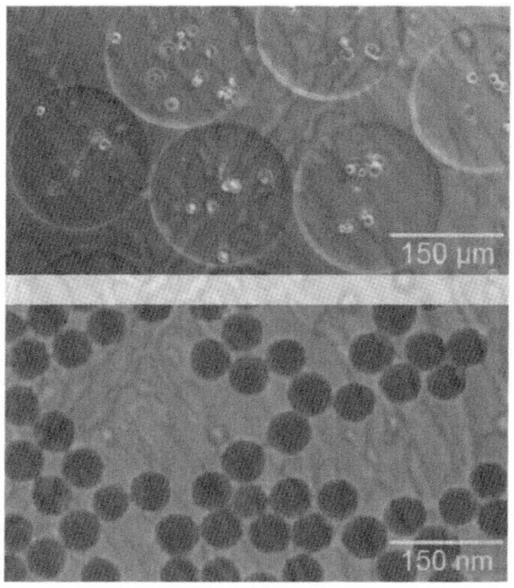

MULTISCALE TECHNOLOGIES FOR CRYOMEDICINE

Implementation from Nano to Macroscale

Frontiers in Nanobiomedical Research

ISSN: 2251-3965

Series Editors: Martin L. Yarmush *(Harvard Medical School, USA)*
Donglu Shi *(University of Cincinnati, USA)*

Published

Vol. 1: Handbook of Immunological Properties of Engineered Nanomaterials
edited by Marina A. Dobrovolskaia and Scott E. McNeil (SAIC-Frederick, Inc., USA)

Vol. 2: Tissue Regeneration: Where Nano-Structure Meets Biology
edited by Qing Liu (3D Biotek, USA and Tongji University, China) and Hongjun Wang (Stevens Institute of Technology, USA)

Vol. 3: Nanobiomedical Research: Fundamentals, Main Applications and Recent Developments (In 4 Volumes)
edited by Vladimir P. Torchilin (Northeastern University, USA)

Vol. 4: Bio-Inspired Nanomaterials and Applications: Nano Detection, Drug/Gene Delivery, Medical Diagnosis and Therapy
edited by Donglu Shi (Tongji University School of Medicine, China and University of Cincinnati, USA)

Vol. 5: The World Scientific Encyclopedia of Nanomedicine and Bioengineering I: Nanotechnology for Translational Medicine: Tissue Engineering, Biological Sensing, Medical Imaging, and Therapeutics (A 5-Volume Set)
edited by Yu Cheng (University of Chicago, USA), Yarong Liu (University of Southern California, USA), Jiang Chang (Chinese Academy of Sciences, China), Jia Huang and Bingbo Zhang (Tongji University School of Medicine, China)

Vol. 6: Handbook of Immunological Properties of Engineered Nanomaterials (In 3 Volumes)
edited by Marina A. Dobrovolskaia and Scott E. McNeil (Leidos Biomedical Research Inc., USA)

Vol. 7: Multiscale Technologies for Cryomedicine: Implementation from Nano to Macroscale
edited by Xiaoming He (The Ohio State University, USA) and John C. Bischof (University of Minnesota, USA)

Forthcoming title

The World Scientific Encyclopedia of Nanomedicine and Bioengineering II: Bioimplants, Regenerative Medicine, and Nano-Cancer Diagnosis and Phototherapy (A 3-Volume Set)
edited by Vibhor Chaswal (University of Cincinnati, USA) and Maoquan Chu (Tongji University, China)

Vol. 7

Frontiers in Nanobiomedical Research

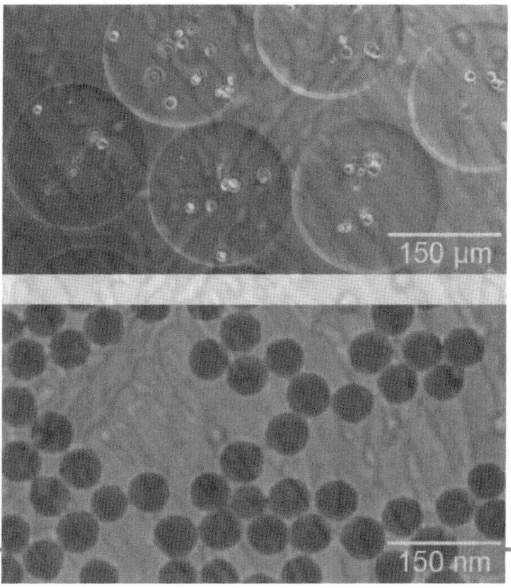

MULTISCALE TECHNOLOGIES FOR CRYOMEDICINE

Implementation from Nano to Macroscale

editors

Xiaoming He
The Ohio State University, USA

John C Bischof
University of Minnesota, USA

World Scientific

NEW JERSEY • LONDON • SINGAPORE • BEIJING • SHANGHAI • HONG KONG • TAIPEI • CHENNAI • TOKYO

Published by

World Scientific Publishing Co. Pte. Ltd.

5 Toh Tuck Link, Singapore 596224

USA office: 27 Warren Street, Suite 401-402, Hackensack, NJ 07601

UK office: 57 Shelton Street, Covent Garden, London WC2H 9HE

Library of Congress Cataloging-in-Publication Data
Names: He, Xiaoming (Of Ohio State University), editor. | Bischof, John C., editor.
Title: Multiscale technologies for cryomedicine : implementation from nano to macroscale /
edited by Xiaoming He & John C. Bischof.
Other titles: Frontiers in nanobiomedical research ; v. 7. 2251-3965
Description: New Jersey : World Scientific, 2016. | Series: Frontiers in nanobiomedical research ; volume 7 |
Includes bibliographical references and index.
Identifiers: LCCN 2015047284 | ISBN 9789814733182 (hardcover : alk. paper)
Subjects: | MESH: Cryopreservation | Cryosurgery | Preservation, Biological--methods |
Temperature | Cryoprotective Agents | Nanotechnology
Classification: LCC QH324.9.C7 | NLM QY 95 | DDC 570.75/2--dc23
LC record available at http://lccn.loc.gov/2015047284

British Library Cataloguing-in-Publication Data
A catalogue record for this book is available from the British Library.

Typeset by Stallion Press
Email: enquiries@stallionpress.com

Printed in Singapore

Contents

List of Contributors

Jason P. Acker
Canadian Blood Services, Centre for Innovation
8249 114th Street
Edmonton, AB, T6G 2R8, Canada

Department of Laboratory Medicine and Pathology
University of Alberta
Edmonton, AB, T6G 2R3, Canada

John G. Baust
Institute of Biomedical Technology
State University of New York at Binghamton
4400 Vestal Parkway East
Binghamton, NY 13902, USA

Department of Biological Sciences
Binghamton University
4400 Vestal Parkway East
Binghamton, NY 13902, USA

John M. Baust
CPSI Biotech
2 Court St.
Owego, NY 13827, USA

Institute of Biomedical Technology
State University of New York at Binghamton
4400 Vestal Parkway East
Binghamton, NY 13902, USA

Sankha Bhowmick
Department of Mechanical Engineering
University of Massachusetts Dartmouth
285 Old Westport Road,
North Dartmouth, MA 02747, USA

John Bischof
Department of Mechanical Engineering
University of Minnesota
111 Church Street SE
Minneapolis, MN 55455, USA

Jeunghwan Choi
Department of Mechanical Engineering
University of Minnesota
111 Church Street SE
Minneapolis, MN 55455, USA

Chris Dames
Department of Mechanical Engineering
University of California at Berkeley
4164 Etcheverry Hall
Berkeley, CA 94720-1740, USA

Luciana da S. Cavalcante
Canadian Blood Services, Centre for Innovation
8249 114th Street
Edmonton, AB, T6G 2R8, Canada

Department of Laboratory Medicine and Pathology
University of Alberta
Edmonton, AB, T6G 2R3, Canada

Ram V. Devireddy
Department of Mechanical Engineering
Louisiana State University
2502A Patrick F. Taylor Hall,
Baton Rouge, LA 70803, USA

Andrew A. Gage
State University of New York at Buffalo
Medical School

462 Grider Street, David K. Miller Building
Buffalo, NY 14214, USA

Dayong Gao
Department of Mechanical Engineering
University of Washington
3900 E Stevens Way NE
Seattle, WA 98195, USA

Soham Ghosh
School of Mechanical Engineering
Purdue University
585 Purdue Mall
West Lafayette, IN 47907, USA

Bumsoo Han
School of Mechanical Engineering
Weldon School of Biomedical Engineering
Birck Nanotechnology Center
Purdue University
585 Purdue Mall
West Lafayette, IN 47907, USA

Xiaoming He
Department of Biomedical Engineering
Davis Heart and Lung Research Institute
Comprehensive Cancer Center
The Ohio State University
1080 Carmack Road
Columbus, OH 43210, USA

Jelena L. Holovati
Canadian Blood Services, Centre for Innovation
8249 114th Street
Edmonton, AB, T6G 2R8, Canada

Department of Laboratory Medicine and Pathology
University of Alberta
Edmonton, AB, T6G 2R3, Canada

Haishui Huang
Department of Mechanical Engineering
Department of Biomedical Engineering
The Ohio State University
460 W. 12th Avenue
Columbus, OH 43210, USA

Zhongping Huang
Department of Biomedical Engineering
Department of Mechanical Engineering
Widener University
Kirkbride Hall 269A
802 E. 16th Street
Chester, PA 19013-5792, USA

Weijie Li
Institute of Biomedical Technology
University of Shanghai for Science and Technology
516 Jungong Road
Shanghai 200093, China

Baolin Liu
Institute of Biomedical Technology
University of Shanghai for Science and Technology
516 Jungong Road
Shanghai 200093, China

Sean Lubner
Department of Mechanical Engineering
University of California at Berkeley
4164 Etcheverry Hall
Berkeley, CA 94720-1740, USA

Fukou Lv
Institute of Biomedical Technology
University of Shanghai for Science and Technology
516 Jungong Road
Shanghai 200093, China

Harishankar Natesan
Department of Mechanical Engineering
University of Minnesota

111 Church Street SE
Minneapolis, MN 55455, USA

Harriëtte Oldenhof
Clinic for Horses — Unit for Reproductive Medicine
University of Veterinary Medicine Hannover
Buenteweg 15
30559 Hannover, Germany

Seungman Park
School of Mechanical Engineering
Purdue University
585 Purdue Mall
West Lafayette, IN 47907, USA

Dinesh Pinisetty
Department of Engineering Technology
The California State University Maritime Academy
200 Maritime Academy Drive
Vallejo, CA 94590, USA

Mehdi Raessi
Department of Mechanical Engineering
University of Massachusetts Dartmouth
285 Old Westport Road,
North Dartmouth, MA 02747, USA

Wei Rao
Department of Biomedical Engineering
Davis Heart and Lung Research Institute
The Ohio State University
460 W. 12th Avenue
Columbus, OH 43210, USA

Anthony T. Robilitto
CPSI Biotech
2 Court St. Owego, NY 13827, USA

Institute of Biomedical Technology
State University of New York at Binghamton
4400 Vestal Parkway East
Binghamton, NY 13902, USA

Alexander N. Sinkevich
Department of Mechanical Engineering
University of Massachusetts Dartmouth
285 Old Westport Road,
North Dartmouth, MA 02747, USA

Hai Wang
Department of Biomedical Engineering
Davis Heart and Lung Research Institute
Comprehensive Cancer Center
The Ohio State University
460 W. 12th Avenue
Columbus, OH 43210, USA

Willem F. Wolkers
Institute of Multiphase Processes
Leibniz Universität Hannover
Callinstrasse 36
30167 Hannover, Germany

Gang Zhao
Department of Electronic Science & Technology
University of Science and Technology of China
96 JinZhai Road
Hefei 230027, P. R. China

Preface

Recent developments in micro and nanotechnologies have led to advancements in many biomedical fields including cancer research, drug delivery, stem cell therapy, and regenerative medicine. In this text, we explore how these same multiscale technologies are now also helping to shape and change the field of cryomedicine at or below the scale of the individual cell.

Cryomedicine is a broad field at the intersection of fundamental sciences and engineering that implements low temperatures to both destroy and preserve biological materials. Cryoablation (or cryosurgery/cryotherapy) is used to destroy diseased tissues with low temperatures. On the other hand, cryopreservation is a method to bank viable and/or functional biospecimens at cryogenic temperatures. Finally, lyopreservation is a cryomedical application for drying and storing biospecimens at hypothermic or even ambient temperatures.

The importance of preservation of living cells in healthcare exists in cell-based medicine including stem cell therapy, assisted reproduction, tissue engineering, and beyond. In addition, cryoablation is increasingly being used to treat cancer and dermatological and cardiovascular diseases. This volume presents current cryomedical research that incorporates micro, nano, and classical macrotechnologies and examines their impact in cellular and tissue systems.

We have collected a total of eleven chapters from leading international researchers in three areas of cryomedicine: cryoablation, cryopreservation, and lyopreservation. Each section is divided into three or four chapters based on the scale of the problems that the technologies address: macroscale, microscale, and nanoscale. This is the first book that is focused on both an introductory and in-depth account of the recent progress in implementing multiscale technologies in cryomedicine particularly for cellular and tissue systems.

At the macroscale: Chapter 1 gives an overview of the technologies to enhance cryoablation; Chapter 2 covers the delicate cell-fluid-matrix interactions during cryopreservation of engineered tissues; and Chapter 3 presents an in-depth numerical analysis of potential strategies to overcome the transport limitation during lyopreservation applications.

At the microscale: Chapter 4 reviews the miniaturized materials and devices used to reduce the cytotoxic cryoprotective agents (CPAs) needed for cell cryopreservation by vitrification; Chapter 5 discusses the theories and techniques for thermal conductivity measurement at the micro and nano length scales; Chapter 6 outlines the modeling, fabrication, and characterization of microscale array of thermal sensors and actuators for use in biological systems; and Chapter 7 summarizes the theory and measurement of cell membrane permeability with a microperfusion system.

At the nanoscale: Chapter 8 surveys the application of polymeric nanoparticles for drug delivery to enhance the outcome of cryoablation and achieve biopreservation with trehalose as a nontoxic CPA; Chapter 9 is focused on an overview of the recent application of liposomal delivery systems for cryopreservation and hypothermic storage of cells; Chapter 10 explores the effect of hydroxyapatite (HA) nanoparticles on the thermal physical properties of CPA solutions; and Chapter 11 presents the theory and implementation of FTIR analysis of lipids, proteins, and CPA diffusion kinetics in cells and tissues for cyromedical applications.

We hope this book offers both introductory and in-depth knowledge for existing and future researchers to further advance the implementation of micro and nanoscale technologies, so as to improve cell and tissue based methods in cryomedicine.

Section 1

Macroscale

Chapter 1

Enhanced Cryoablative Methodologies

John M. Baust, Anthony T. Robilitto, Andrew A. Gage and John G. Baust

1. Introduction

Cryoablation (cryosurgery) (CA) is an energy-deprivation therapy that relies on the application of a heat-sink to targeted tissues with the expectation of freeze destruction. The origin of modern CA is commonly attributed to the development of a closed-end cryoprobe in which liquid nitrogen (LN_2) was circulated while placed within a tissue target [Cooper and Lee, 1961]. Cooper taught that 1–2 minute tissue exposure to –20°C was adequate to form a necrotic lesion in the brains of patients suffering from motion disorders (i.e. Parkinson disease). Future studies identified the variability of tissue responses to freezing and extended the lethal temperature to a <–40°C nadir [see review: Gage and Baust, 1998]. More recently, Baust *et al.* [2014a] reviewed the diversity of CA mechanisms critical to clinical efficacy. Three methodological approaches with an emphasis on cancer ablation are discussed in this chapter. First, elucidation of the mechanisms of cancer cell death following a CA insult will be expanded beyond the initial understanding of necrosis as the path of cell death. Second, the engineering advancements in cryosurgical device technologies are reviewed. These include subcooled LN_2, supercritical nitrogen and Joule–Thompson cooling. Third, the use of cryosensitizers, when applied in concert with freezing to optimize a tissues ablative response, is discussed.

On the surface CA appears to be a rather simple method to ablate unwanted tissue. With an appropriate cryogen (Table 1) tissue temperature is "targeted" to a pre-determined nadir in an effort to assure destructive ice

Table 1. Overview of common cryogens used in cryosurgery.

Cryogen type	Cryogen temperature	Approximate device operation pressure
Carbon dioxide (CO_2)	–78.5°C	800 psi
Nitrous oxide (N_2O)	–89.5°C	800 psi
Argon (Ar)	–185.7°C	3,200 psi
Nitrogen (N2)		
Boiling liguid (LN_2	–195.8°C	
Subcooled LN_2	–210°C (@ 0.119 atm)	75 psi
Critical (CN)	–147.2°C (@ 33.4 atm)	492 psi
Supercritical (SCN)	–190.2°C(@ 68 atm)	1,000 psi

formation. In practice, cells of diverse tissues demonstrate different susceptibilities to freezing [Gage and Baust, 1998] with lethality ranging between –2°C to –70°C and further influenced by cooling rate, hold time at nadir temperature, rates of thaw and tumor cell biology. With the exceptions of radical surgery and CA, all other therapies rely on "energy additive" strategies (i.e. radiation, radio frequency, photo dynamic, laser and diverse chemo therapies), which if heat-based, often extend the ablative consequences for up to one week post-treatment [Chu and Dupuy, 2014]. When applied as monotherapies with curative intent, these treatments often yield the suppression of cancer growth for varying periods related to the biology of the disease (i.e. stage of cell cycle, focal vs. extent of dispersion and cancer cell-based defensive responses).

A cancer's continuous mutation and recruitment of stromal and immune cells to support the growth of a tumor microenvironment presents a "moving target" to treatment which commonly defeats clinically accepted "gold standard" monotherapies [Ghisolfi *et al.*, 2012]. Cancer's defensive strategies represent evolved capabilities characterized as well-established protective adaptations to anti-tumor efforts [Dimond, 2013; Shackleton *et al.*, 2009]. To successfully manage cancer the activation of multiple cell-type-specific "kill switches" is necessary. Without "kill switch" activation the combined defensive capabilities of the tumor microenvironment provide a strategy to defeat the curative intent of monotherapies [Ghisolfi *et al.* 2012; Sun *et al.* 2012]. As an example, radiation therapy for prostate cancer yields >25% positive biopsies in up to one-third of patients requiring secondary radiation treatment following cancer recurrence. A common outcome is radioresistant prostate cancer, the fourth most common genitourinary malignancy [Jone, 2011].

The biology of cancer supports resistance to both radiation and chemotherapy [Sun *et al.*, 2012]. Both therapeutic strategies primarily attack only actively dividing cancer cells in specific stages of the cell division cycle. Non-dividing cancer cells are weakly affected. Further, select tumor cell populations, such as cancer stem cells, have several molecular adaptations in which this sub-population can actively enter or halt the cell cycle, secrete protective proteins, induce extended dormancy, etc. to avoid targeting. Treatment plans must then be repetitive with an aim of inducing cell death in each of the tumor's sub-populations including dividing, non-dividing, cancer stem cells, dormant cells, etc. [Baust *et al.*, 2015; Moltzahn and Thalmann, 2013]. An unintended consequence of repetitive bouts of treatment is the acceleration of protective (defensive) mutations which progressively result in a treatment-resistant tumor cell population. In the case of cancer stem cells, these protective adaptations can be amplified due to increased molecular plasticity of this population [Dimond, 2013; Shackleton *et al.*, 2009; Sun *et al.*, 2012]. These cells become refractory to secondary re-treatment or alternative attempts at therapeutic intervention [Corn, 2012].

Hence, the development of new therapeutic approaches requires a therapy that is (a) non-repetitive to deny the cells of the tumor microenvironment the ability to activate defensive mutations; (b) capable of simultaneous activation of multiple cell death cascades to limit cancer's ability to activate survival mechanisms; (c) cell cycle independent; (d) damaging to tumor vasculature and (e) uniform in the destruction of all cells within the tumor microenvironment to eliminate both cancer cell/cancer stem cell survival through dormancy mechanisms. This also allows cryoablation to be a repeatable and even salvage therapy. A critical consequence of cryoablation is the disruption of most of the hallmarks of cancer within tumor cells [Hanahan and Weinberg, 2011] resulting in complete cell death, provided a cancer-specific nadir temperature is reached throughout the tumor microenvironment [Baust *et al.*, 2014a, 2015].

2. Mechanisms of Cell Death

From an historical perspective, cell death following cryosurgery was attributed to two mechanisms, freeze rupture which led to vascular stasis and tissue necrosis, together referred to as "coagulative necrosis" [Bischof *et al.*, 1992, 1993; Cooper, 1965; Gage, 1978; Gage and Baust 1998, 2002; Ikekawa *et al.*, 1985; Smith *et al.*, 1999]. Missing from this concept of "physical damage-related lethality" was recognition of cellular stress mechanisms that may be launched to either support cell recovery or to initiate multiple cell death cascades when repair was not possible [Baust and Gage, 2005; Gage and Baust, 1998; Hoffmann and Bischof, 2002]. Hollister *et al.* [1998] first observed late stage apoptosis in prostate cancer cells following exposure to elevated sub-freezing temperatures equivalent to those measured in the freeze zone periphery. More recently, Robilotto *et al.* [2013] demonstrated that not only was necrosis observed throughout the ablative freeze zone but also apoptotic activation was differentiated between membrane- and mitochondrial-mediated cell death pathways based on the intensity of the freeze. While the contribution and importance of the molecular response of cancer cells to the freeze process is now recognized to play a critical impact on outcome, the mechanisms of action are still fairly unknown. This due in part to the lack of detailed molecular-based studies and even more so as a result of the multi-pronged assault freezing places on cancer cells. This molecular-level assault includes alterations in cellular pH, ionic concentrations and gradients (hypertonic during freezing, hypotonic during thawing), membrane fluidity, cellular dehydration, uncoupling of metabolic pathways, waste accumulation, protein unfolding/misfolding, and cytoskeletal disruption among others [Baust *et al.*, 2013, 2014a, 2015; Baust, 2014]. Each of these individually has been shown to induce molecular-based cell death, whether apoptosis, secondary necrosis, autophagy, or some other form of cell death.

Further complicating this analysis is the fact that in addition to numerous initiation stresses, freezing has been shown to activate at least three individual apoptotic pathways including membrane, mitochondrial and nuclear induced apoptosis [Baust *et al.*, 2010a; Clarke *et al.*, 2004, 2007a, 2007b; Robilotto *et al.*, 2013]. Many of the induction pathway studies have gone hand in hand with cryoadjunctive or cryosensitization research (detailed below). For instance, Clarke *et al* [2004] found a significant alteration in Bcl-2 and Bax mitochondrial protein levels following freezing resulting in a shift in mitochondrial signaling. This study further demonstrated that the utilization of Taxotere resulted in a further pro-death shift in mitochondrial signaling when combined with mild freezing resulting in enhanced prostate cancer cell death. Studies by Goel *et al.* [2007, 2009], Han *et al.* [2007], Jiang *et al.* [2008] and Baust *et al.* [2010a] also found the involvement of membrane mediated apoptosis in cell death following freezing. Mechanistically, Clarke *et al.* [2007b] showed the activation of caspase 8 (membrane mediated apoptosis) following freezing which could be amplified by pretreatment with TRAIL. A study by Klossner *et al.* [2008] reported on the differential sensitivity of varying forms of prostate cancer (early and late stage) to freezing injury via the activation of apoptosis. Specifically, this study found that early stage androgen sensitive prostate cancer cells (LNCaP) were more sensitive to freezing injury and had increased levels of apoptosis compared to late stage androgen insensitive prostate cancer cells (PC3). This was further confirmed in a PC3 cell model in which the androgen receptor gene was transfected into PC3 cells resulting in the cells re-expressing the androgen receptor (PC3-AR). In this case, as with the LNCaP cells, the PC3-AR cells demonstrated a higher degree of sensitivity to freezing and increased apoptotic activity post-freeze. Similarly, when the LNCaP cell model was "aged" *in vitro* they lost AR expression which resulted in the LNCaP-HP population being more resistant to freezing injury with a reduction in apoptosis, similar to that of the PC3 cells. Additional mechanistic studies by Baust *et al.* [2010a] have shown the impact of the integrin attachment pathway on the induction of membrane-based apoptotic signaling following freezing. More recently, a report by Robilotto *et al.* [2013] demonstrated that the pathway and timing of apoptotic induction and progression was a result of the severity of freezing stress experienced. Specifically, this study found that lower sub-freezing temperatures (<−30°C) apoptotic induction was mediated by membrane-based pathway and initiated rapidly within ~30 minutes post-thaw progressing to complete cell death within 1–3 hours post-freeze. This was in contrast to cell death at temperatures >−20°C where apoptosis was found to progress through a mitochondrial mediated path initiating 6–12 hours post-thaw and was not completed until 24–36 hours post-freeze *in vitro*. This study demonstrated first-hand the complex and dynamic impact of freezing on cancer cell death. Interestingly, this study also began to provide insight into why many studies report only necrosis as the primary, or even sole, mode of cell death post cryo as the apoptotic process occurs and is complete within a few hours to few days window post-freeze resulting in most studies, especially *in vivo* studies, missing this critical window [Baust, 2014; Baust *et al.*, 2015]. In these cases, once apoptotic cell death is complete, it is typically seen as necrosis or more accurately secondary necrosis in the 1–7 day or greater assessment window. While complicated to analyze or often even distinguish/identify, it is without doubt that molecular-based cell death activation as a result of various cell stresses associated with freezing, beyond physical ice rupture, plays a critical role in therapeutic outcome. As discussed below, these molecular mechanisms and pathways are providing opportunities to tailor cryoablation to address specific disease states while having a minimal impact on non-cancerous cells within or surrounding structures of a given tissue or organ.

3. Technological Developments

The advent of modern cryoablative technologies is traced directly to the development of the closed end cryoprobe through which boiling liquid nitrogen was circulated [Cooper and Lee 1961]. This probe and its variations added a new surgical tool to treat difficult-to-excise tumors [Cooper, 1965; Gage, 1978; O'Donoghue *et al.* 1975]. Initially, ice growth around a single cryoprobe was monitored visually and with the aid of thermocouples. In the decades that followed, numerous other innovations were adopted including Joule–Thompson cooling, spray/pour freezing, cryogen laden wicks, surface contact metallic probes, and direct dry ice application to name but a few. Each approach to tissue freezing found success depending on physician ability to provide a nadir temperature exposure. None, however, provided a platform supportive of standardization that might have led to a "best practices" statement.

In the early 1990's three simultaneous developments led to the expanded use of cryoablative procedures. The development of the first multiprobe devices coupled with the use of intra-operative ultrasound provided a means of freezing deep tumors percutaneously and with the ability to sculpt the freeze zone to comport with tumor geometry [Onik *et al.*, 1991a, 1991b, 1993; Saliken *et al.*, 2002]. The third development, subcooled liquid nitrogen, provided a surgical tool that would bring the cryogenic liquid directly to the cryoprobe tip rather than less efficient cold nitrogen gas [Chang *et al.*, 1994]. For the first time, a single cryosurgical device with multiple (5), independently operated cryoprobes was available in the operating theatre.

Figure 1 is a schematic illustration of the liquid nitrogen subcooling system [Chang *et al.* 1994; Baust *et al.*, 1997]. Four chambers including a cryogen supply dewar (29), a cryogen return dewar (41), an evacuated nitrogen slush chamber (40) and a subcooled distribution chamber are pictured. The design of this system allowed for the subcooling of pressurized (~75 psi) liquid nitrogen so that it acquired an "excessive temperature" (differential between boiling point when pressurized and subcooled temperatures) which supported the transport of the cryogen through the delivery umbilical and cryoprobe with minimal phase change thereby supporting nucleate boiling in the cryoprobe tip. While subcooling allowed for cryogen delivery, immediate probe cooling was long (>5 mins) before freezing temperatures were reached. Once reached, probe tip temperatures decreased to ~−190°C within a minute. Despite noted advances, this technology required a large device (380 kg) and in surgery exhibited a "thermal inertia" due to

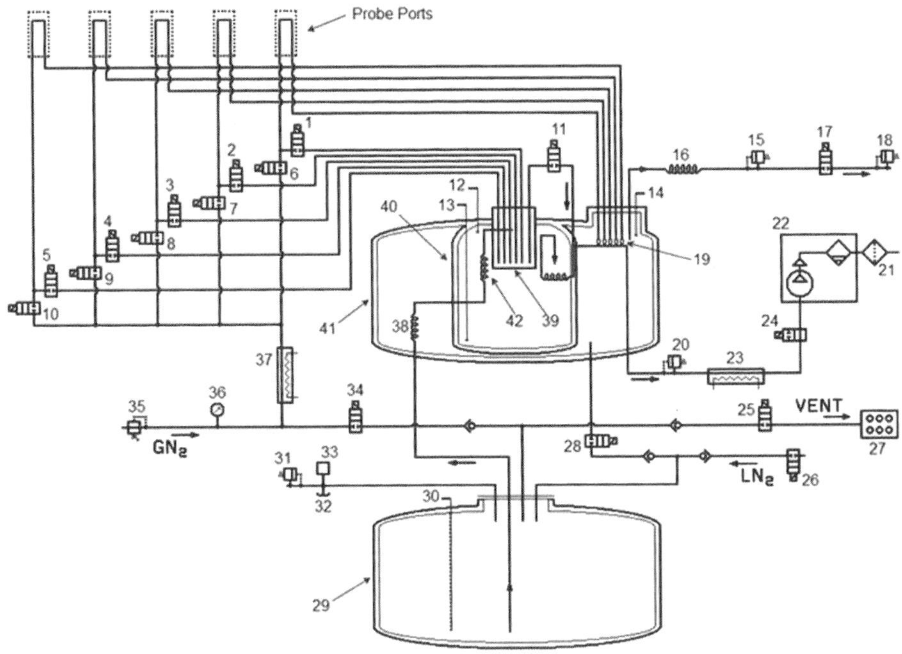

Figure 1. Schematic of the subcooled liquid nitrogen device. The subcooled liquid nitrogen device (AccuProbe) is designed to pass liquid nitrogen through a subcooling chamber containing a reservoir of LN2 upon which a vacuum is drawn, thereby reducing the boiling point of the LN2 to ~−210°C. When LN2 at atmospheric pressure passes through the subcoooling chamber the temperature is reduced from −196°C to ~210°C allowing for excess heat gain by the LN2 prior to boil off.

Diagram key: freezing solenoid (1–5); thawing solenoids (6–10); slush-fill solenoid (11); slush-level thermocouple (12); slush-temperature thermocouple (13); top dewar overfill indicator (14); pressure relief valves (15, 18, 25psi); top dewar heat exchanger (16); top dewar venting solenoid (17); phase separators (19); cryogenic pressure relief valve (20, 5psi); charcoal filter (21); vacuum pump (22); vacuum-line heater (23); vacuum-line solenoid (24); bottom dewar vent (25); LN2 fill (26); venting mufflers (27); LN2 regenerating solenoid (28); supply (bottom) dewar (29); bottom dewar level sensor (30); safety valve (31, 75psi); rupture disk (32); system pressure transducer (33); gaseous N2 pressurization solenoid (34); pressure regulator (35); pressure gauge (36); thawing heater (37); return (top) dewar heat exchanger (38); redistribution chamber (39); slush chamber (40); return (top) dewar (41); slush heat exchanger (42).

Source: Adapted from Baust *et al.* [1997]; Chang *et al.* [1994] and US Patent #5,520,682.

residual liquid nitrogen in the umbilical which provided continued freezing after cryoprobe inactivation. The latter feature required physicians to "anticipate" a freezing end-point by inactivating the freezing process prior to completion of the targeted freeze.

In the mid-1990s argon-based Joule–Thompson (JT) multiprobe devices were developed, one of which supported smaller diameter cryoprobes (2–2.5mm diameter) as well as cryoneedles (17 gauge) [Benoit *et al.*, 2000; Lee *et al.*, 2011; Saliken *et al.*, 2002]. These smaller cryoprobes could be passed through a guidance template integrated with an ultrasound probe. While cryoneedles cooled by JT gas expansion lack the cooling or "work" capacity of liquid nitrogen, they were more readily placed and controlled. A JT cryoprobe array could be assembled in the targeted tumor to accomplish more accurate sculpting. In addition to cryoprobe "miniaturization" JT cryoprobes initiate freezing within a few seconds of device activation whereas subcooled liquid nitrogen platforms required a pre-cool period of ~5 minutes. While attaining more rapid ice formation, JT systems yield smaller and "warmer" iceballs and thereby require 8–12 or more cryoprobes to accomplish what 3–5 LN_2 probes could in a similar total procedure freeze time (~10 mins).

Any cryogen that will provide the desired nadir temperature to a targeted tissue in a safe and timely manner may well support cryoablative therapy. As noted above, liquid nitrogen and argon-based JT devices have been used successfully in the clinic but each has limitations [Cohen *et al.*, 2008]. Liquid nitrogen platforms are typically slow to cool, continue to support (briefly) ice growth following probe inactivation and have not been compatible with miniaturization efforts. Argon-based JT systems cool less effectively, require the use of numerous cryoprobes to match freeze zone volume/geometry of liquid nitrogen, and their use is limited by cryogen availability contained in multiple high pressure cylinders.

In an effort to combine the strengths of both technologies, devices utilizing supercritical cryogens have been recently developed [Baust *et al.*, 2010b, 2014]. Supercritical fluids are characterized by low viscosity (similar to a gas) and a high density as in a liquid. Since these fluids are neither gas nor liquid, there is no gas/liquid phase boundary and therefore no surface tension. Ideally, a supercritical fluid is transported friction-free, and when subcooled, supercritical nitrogen (SCN) can be transported through a capillary tube near-instantaneously without boiling thereby delivering an ultracold cryogenic fluid to the cryoprobe tip boiling chamber in seconds. This transport occurs as rapidly as with a gas but yields the cooling effectiveness of a liquid [Baust *et al.*, 2014b]. In the probe tip boiling chamber, heat is extracted from the surrounding tissue resulting in the formation of an ultracold iceball. Cryogen flow returns to the console during which there is a pressure drop and further cooling of the cryogen. To this end, in SCN systems the measured temperature of the returning cryogen at the console is often colder than when leaving the console despite the heat gain along the fluid path.

From a cryosurgical device perspective, a supercritical fluid can be rapidly transported through capillary tubes (not possible with atmospheric liquid nitrogen) into a cryoneedle yielding the cooling capacity of a liquid. Supercritical cryoneedles initiate freezing in seconds as do argon JT probes but with the cooling capacity of liquid nitrogen. As described, when "tuned" appropriately, the supercritical nitrogen continues to have its temperature lowered while being transported through the return umbilical so that atmospheric liquid nitrogen is recovered. Reprocessing the returned liquid nitrogen to the supercritical state provides for longer device operation. Figure 2 provides a

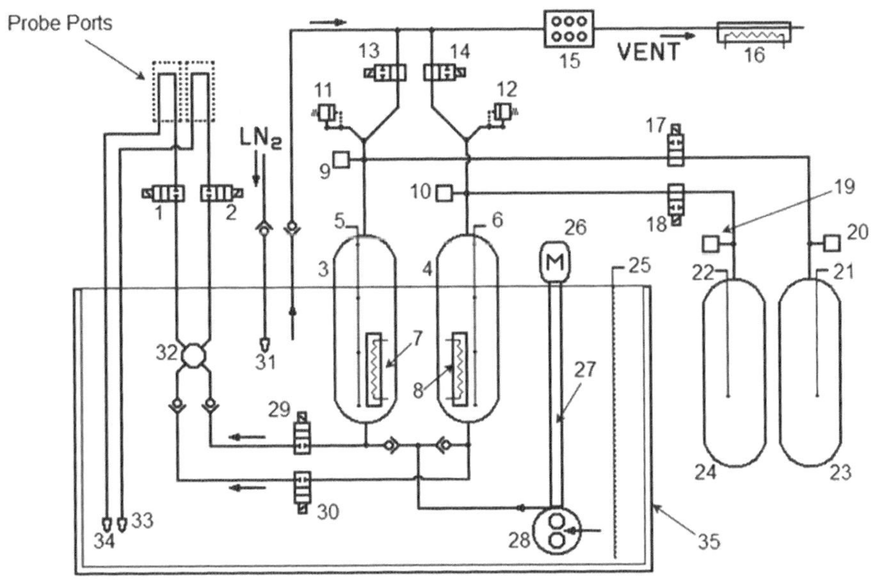

Figure 2. Schematic of the supercritical nitrogen cryoablation device. The SCN System is designed to convert nitrogen from a liquid to a supercritical state. Creation of the supercritical state allows for rapid and controlled delivery of a highly lethal ultra-cold cryogenic fluid to a target tissue.

Diagram key: injection solenoids (1, 2); supply cylinders (3, 4); cylinder level sensors (5, 6); cylinder heaters (7, 8); cylinder pressure transducers (9, 10); bleed valves (11, 12, 1500psi); cylinder fill/vent solenoids (13, 14); vent muffler (15); vent heat exchanger (16); pressure cylinder solenoids (17, 18); pressure cylinder pressure transducers (19, 20); pressure cylinder thermocouples (21, 22); pressure cylinders (23, 24); dewar level sensor (25); motor (26); drive shaft (27); LN2 gear pump (28); manifold solenoids (29, 30); diffusers (31, 33, 34); injection manifold (32); LN2 dewar (41).

Source: Adapted from Baust *et al.* [2014b]; US Patent 8,784,409, US Patent App. 2012/0059364 and PCT App. WO2014/144626A2.

schematic of a two channel (probe) supercritical nitrogen cryosurgical device. Four paired pressurization cylinders (7, 8, 23, 24) provide the LN_2 processing apparatus for the production of supercritical nitrogen. A cryogenic pump (30) fills the pressurization cylinders prior to processing. Cylinders 7 and 8 are partially submerged in atmospheric LN_2 to support subcooling of the supercritical nitrogen. Cylinders 25 and 26 provide pressure to support the transport of the supercritical nitrogen from the supply cylinders (7, 8) through the cryogen supply path to the cryoprobes (1, 2). LN_2 recovery from the cryoprobes to replenish dewar (41) levels occurs through 39 and 40. Beyond the benefits of the speed and controllability of a gas with the heat extraction capacity of a liquid, SCN systems are now supporting further probe miniaturization from the current 17 gauge limit in JT systems down to 21 gauge cryoprobes while still delivering a highly ablative cryogenic temperature to sounding tissues.

A key to the successful application of an ablative dose is an understanding of thermal gradients in the freeze zone and volume of tissue brought to the lethal nadir temperature [Baust *et al.*, 2007; Gage and Baust, 2007; Klossner *et al.*, 2007]. Independent of the cryogen circulating in the cryoprobe tip, a thermal gradient will extend from the probe surface to the outer edge of the freeze zone, the melting point of isotonic saline (–0.6°C). The position of the –40°C isotherm distal from the probe surface will identify the volume of tissue brought to a tissue-specific lethal nadir and also the volume of tissue experiencing damaging, partial-lethal freezing [Baust *et al.*, 2013, 2014a, 2014b]. Table 2 compares the differential position in time and space of –40°C isotherm in an argon-JT system vs. supercritical nitrogen. The more rapid freezing rate and lower probe surface temperatures of supercritical nitrogen probes results in the more distal positioning of –40°C isotherm. The enhanced heat extraction capacity also allows for a larger –40°C isotherm in shorter periods of time using smaller diameter cryoprobes [Baust *et al.*, 2014a, 2014b]

Ideally, the freeze zone should provide a uniform insult to the targeted tissue such that the freeze zone distal boundary is as lethal as the tissue proximal to the probe experiences. In practice one attempts to "move" the –40°C isotherm more distally with a second freeze and an extension of the freeze zone periphery beyond the targeted tissue boundary [Baust *et al.*, 2007; Gage *et al.*, 2009]. The latter, however, is problematic as the volume of partially freeze-damaged, non-targeted tissue can exceed the actual treatment target (Figure 3). There are three approaches to advancing the desired isotherm in the tissue target. The first involves multiple freeze-thaw cycles. Two freezes separated by either an active or passive thaw renders the tissue near 0°C when the second freeze is initiated. Cooler tissue allows freezing to progress distally. Also, target tissue vasculature is damaged following the first

Table 2. Comparison of SCN and argon cryogen average iceball diameter and distance (cm) of the critical isotherm penetration following a single and double freeze in a prostate cancer tissue engineered model.

Cryogen	Probe freeze zone size	Freeze interval	Average iceball diameter (cm)		Average lethal isotherm diameter (cm)			
			1st Freeze	2nd Freeze	1st Freeze −20°C	1st Freeze −40°C	2nd Freeze −20°C	2nd Freeze −40°C
SCN	1.8mm × 3cm	5 min	3.00	3.86	2.18	1.75	2.53	2.02
Argon	2.4mm × 3cm	5 min			1.93	1.40		
		10 min	3.65	4.27	2.28	1.60	2.63	1.87

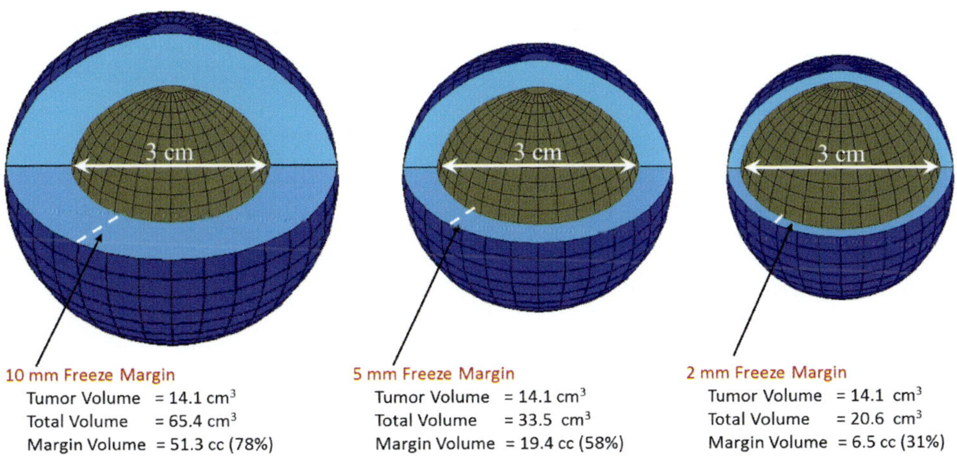

Figure 3. Impact of a positive freeze margin on overall frozen tissue volume. Diagrams illustrate the impact of a positive margin on overall frozen tissue volume when targeting a 14 cm^3 tumor tissue volume within a given lethal isotherm (i.e. –40°C). The diagram on the left illustrates a 1cm positive freeze margin which is recommended with typical Joule–Thompson cryodevices to deliver a lethal insult (<–40°C) to the target tissue. In this scenario, ~78% of the frozen mass is non-targeted tissue. Utilization of colder liquid nitrogen (center) or super-critical nitrogen (right) cryogens, a steeper isothermal gradient and deeper penetration of ultra-cold temperatures is achieved thereby reducing the necessary size of the positive freeze margin. In these cases, a larger percentage of the frozen mass is below the lethal temperature (–40°C) thereby resulting in a more lethal and targeted cryoleasion.

freeze which compromises blood flow, the vascular heat source. The second method of advancing the selected nadir temperature isotherm depends on faster freezing. Table 2 illustrates this outcome. A more rapid freezing rate projects the nadir isotherm more distally than will a slower cooling rate while maintaining a similar overall freeze zone geometry (i.e. similar sized but colder iceball) [Baust *et al.*, 2014a, 2014b]. The third method is dependent on the use of adjuvants, agents that sensitize tissue cells to a freezing stress through the elevation of the lethal nadir temperature [Baust *et al.*, 2009; Chu and Dupuy, 2014; Goel *et al.*, 2009].

4. Adjunctive Cryosensitization Strategies

The use of the modern monitoring techniques for freeze zone imaging (US, CT or MRI), while essential to ablative precision, are unable to predict the outcome of the cryosurgical procedure [Onik *et al.* 1991a, 1991b, 1993; Gilbert *et al.* 1993]. This is due to the uncertainty of a tissue target's lethal exposure to temperatures above the nadir and especially at the margin of the freeze zone where only extracellular freezing is anticipated. Accordingly, cryosurgery for cancer has two limitations: (1) tumor recurrence from cells surviving non-lethal freeze conditions and (2) collateral morbidity associated with freezing adjacent critical anatomical structures [Caso *et al.*, 2012a, 2012b; Pitman *et al.*, 2012; Saliken *et al.*, 2002]. To overcome these limitations, studies have more recently focused on the development of strategies to "make ice lethal at 0°C". The goal of these efforts is to yield complete cancer destruction with the frozen mass thereby allowing physicians to precisely visualize the zone of lethality in realtime under ultrasound or other imaging modality. One such strategy is the use of cryoadjunctive agents in combination with freezing. Adjuvants are designed to either enhance the physical effects of the freezing process and/or to activate cellular stress responses and attendant cell death cascades and/or inhibit cell survival/repair cascades. To date, numerous *in vitro*, *in vivo* and clinical studies have shown the potential benefit of this cryoadjunctive or cryosensitization strategy. For instance, independent studies by the Bischof and Baust laboratories have demonstrated the benefits of the activation of membrane receptors using TNF-α [Goel *et al.*, 2007, 2009; Han *et al.*, 2007; Jiang *et al.*, 2008] and TRAIL [Clarke *et al.*, 2007b] in combination with mild freezing (>−10°C) results in near complete ablation of prostate cancer in various models. Other studies have reported similar outcomes in a renal cancer cell model [Clarke *et al.*, 2007a]. A similar strategy using the combination of chemotherapeutic agents and cryo has also shown benefit. Le Pivert *et al.* [2004] demonstrated the enhancement of prostate cancer cell death through the combination of 5-fluorouracil (5-FU) pretreatment and cryoablation in a murine model. Similar results have also been reported using low-dose (subclinical) 5-FU pretreatment (1–2 days) in combination with mild freezing (–15°C) in an *in vitro* prostate cancer cell model [Clarke *et al.*, 2001]. Forest *et al.* [2005, 2006] have also demonstrated the enhanced cancer death caused by cryochemo combination strategies in *in vitro* and *in vivo* lung cancer models. Numerous other studies have reported similar outcomes with various classes of agents in several cancers including kidney, liver, skin, pancreas, lung and colorectal cancer, among others [Baust *et al.*, 2004, 2009; Gu *et al.*, 2011; Han and Bischof, 2004; Han *et al.*, 2004, 2007; Hanai *et al.*, 2001; Kuflik 2004; Redondo *et al.*, 2007; Rodriguez-Bigas *et al.*, 1996; Santucci *et al.*, 2011; Yang *et al.*, 2003]. To this end, five categories of cryoadjuvant have been identified to date and include (1) thermophysical adjuvants, (2) chemotherapeutics, (3) proinflammatory cytokines or vascular-based agents, (4) immunomodulators and (5) nutraceutical-based sensitization [Baust *et al.*, 2015; Goel *et al.*, 2007]. These categories serve as a general reference or classification as many agents are non-exclusive given that they may impact several different areas/pathways within a cell. For instance, TNF-α, can be considered a chemotherapeutics, a cytokine and an immunomodulator [Goel *et al.*, 2009]. The mechanism of action of these agents often overlap and/or amplify the molecular action of freezing thereby resulting in enhanced cell death.

4.1. *Thermophysical Adjuvants*

Thermophysical adjuvants are employed to affect the structure of ice crystals during freezing and therefore enhance the mechanical destruction [Koushafar *et al.*, 1997; Koushafar and Rubinsky, 1997; Muldrew *et al.*, 2001; Pham *et al.*, 1999]. Antifreeze proteins (AFP), salts (i.e. NaCl, KCl), and some amino acids (i.e. glycine) are examples. These adjuvants can enhance the threshold injury temperature up to –5°C [Han and Bischof, 2004; Han *et al.*, 2004; Pham *et al.*, 1999; Wang *et al.*, 2008] if delivered in high enough concentration to achieve the desired effect without causing toxicity.

4.2. *Chemotherapeutics*

The synergy between cryosurgery and chemotherapeutics was first reported in tumor model systems using peplomycin and adriamycin in combination with freezing [Ikekawa *et al.*, 1985]. Mechanistically, it was proposed that the frozen tumor preferentially trapped the anticancer drugs as compared to untreated tumor. With the identification of apoptosis as a primary mode of cell death at the periphery of the freeze zone in 1998 [Hollister *et al.*, 1998], studies began to focus on the activation and amplification of apoptotic activity with chemo agents and freezing [Baust *et al.*, 2009; Goel *et al.*, 2009]. These pro-apoptotic agents when combined with cryosurgery enhance cell destruction at elevated temperatures (–40 to –0.5°C). Pro-apoptotic agents include 5-fluorouracil, cisplatin, taxotere, doxorubicin, mitomycin and bleomycin, vinorelbine, navelbine and TRAIL (tumor necrosis factor-related apoptosis-inducing ligand) [Clarke *et al.*, 2001, 2007b; Forest *et al.*, 2005, 2006; LePivert *et al.*, 2004; Mir and Rubinsky, 2002; Yuan *et al.*, 2008]. In addition to combinations with traditional chemotherapeutic agents, cryo combination with radiation has also been shown to have benefit. Several reports by Xu *et al.* [2008a, 2008b] as well as others have reported improved outcomes using the combination of iodine seed implants in conjunction with cryo for the treatment of pancreatic cancer. As the majority of studies on adjuvants have been *in vitro, in vivo* studies are necessary to further establish clinical benefit [Forest *et al.*, 2005, 2006].

4.3. *Pro-inflammatory Cytokines and Vascular Agents*

The pro-inflammatory cytokine TNF-α in combination with cryosurgery has been shown to destroy tumors up to the freeze zone edge (–0.5°C) [Goel *et al.*, 2007; Jiang *et al.*, 2008; Shenoi *et al.*, 2013]. TNF-α is a multifunctional cytokine with complex actions on the tumor cells, their vasculature and the body's immune system (apoptosis vs. inflammation). The exact mechanisms of cell death enhancement caused by these and other pro-inflammatory agents following cryosurgery are still under investigation.

4.4. *Immunomodulators*

Several studies note the ability to enhance cryosurgical destruction by inducing and stimulating dendritic cells in the injury site utilizing agents such as imiquimod, CpG-oligodeoxynucleotides, BCG-CWS (bacillus calmette-guerin cell wall skeleton) and GM-CSF (macrophage-colony-stimulating factor) [Udagawa *et al.*, 2006; Gazzaniga *et al.*, 2001; Redondo *et al.*, 2007; den Brok *et al.*, 2006]. This topic has been recently reviewed [Sabel, 2009] and is being more fully explored using anti CTLA-4 antibody blocking [Waitz *et al.*, 2012]. Further, studies by Yuanying *et al.* [2013] have demonstrated the multiple combination of cryo, chemo, and immunomodulation can enhance treatment in a lung cancer model. The future use of immunomodulators appears promising but a better understand of the immunological mechanisms stimulated by cryosurgery are necessary.

4.5. *Nutraceutical — Small Molecule Adjuvants*

The success of cryosensitization strategies coupled with the desire to minimize side effects (both systemic and localized) support more recent efforts focused on the use of neutraceuticals (natural anti-cancer agents) such as vitamin D_3 (VD_3) and Resveratrol. Both *in vitro* and *in vivo* studies have demonstrated the benefits of the use of Calcitrol (VD_3 analog) in increasing the lethality of freezing in prostate cancer [Santucci *et al.*, 2011; Kimura *et al.*, 2010; Baust *et al.*, 2012]. These studies have shown that the exposure of prostate cancer to VD_3 prior to freezing results in complete cancer destruction at temperatures of ~−5°. Other studies utilizing the free radical scavenger Resveratrol have shown that pretreatment of renal and prostate cancer cells sensitizes them to freeze injury, yielding complete cell death at −10°C *in vitro* [Clarke *et al.*, 2007a]. Given the promise of these and other reports, numerous investigations and trials incorporating these as well as other agents are currently ongoing.

In addition to the above, several other categories and agents have shown promise in increasing cancer cell sensitivity to freezing. These include targeted activation of cell death pathways and inhibition of survival pathways [Tian *et al.*, 2011]. To this end, studies investigating the impact and ability to enhance the destructive effects of freezing targeting pathways such as cell attachment (integrin inhibition)[Baust *et al.*, 2010a], cell survival (AKT pathway)[Baust *et al.*, 2013], mitochondrial stress [Clarke *et al.*, 2001; Hanai *et al.*, 2001; Yang *et al.*, 2003], endoplasmic reticulum [Robilotto *et al.*, 2013], autophagy (or autophagocytosis), etc. have shown the potential of enhancing cryoablation of targeted structures while minimizing collateral damage to surrounding non-target tissues. As described above, these agents are designed to increase the lethality of mild freezing while having little or no negative side effects. This is especially true in prostate cancer CA where avoidance of damage to adjacent structures, including the neurovascular bundles and rectal wall, is critical. Accordingly, several studies have focused on tissue protective strategies in combination with CA to reduce comorbidities while enhancing ablative outcome [Baust *et al.*, 2010c; Mouraviev *et al.*, 2001a, 2001b; Rubinsky, 2000].

5. Future Directions

Cryoablation has a clear benefit and long established history of success in treating numerous cancers [Bischof *et al.*, 1992; Caso *et al.*, 2012a, 2012b; Cohen *et al.*, 2008; Gage and Baust, 2007; Gage *et al.*, 2009; Lee *et al.*, 2011;

Levy *et al.*, 2009; Littrup *et al.*, 2009; Mala, 2006]. As either a monotherapy or in combination with adjunctive agents, the therapeutic benefit of CA coupled with its minimally invasive nature results in both a molecular and physical insult on a targeted tissue, making cryoablation an ideal treatment paradigm for many cancers. Beyond cancer therapy, CA has also proven to be an effective treatment regime for treating a host of other disease states including cardiac arrhythmias, chronic pain, retinal detachment and neurodegenerative diseases, among others [Baust and Gage, 2005; Baust *et al.*, 2010a, 2010b; Berrada and Bischof, 2001; Gage *et al.*, 2009]. Further, the molecular nature of CA coupled with various adjunctive strategies is opening new doors in disease management, diagnostics, therapeutic development, and other areas. For example, recent work by Baust *et al.* [2014a, 2014b] has shown the potential to utilize the unique molecular response of cancer versus non-cancer cells to freezing and molecular sensitizers to selectively select for (protect) cancer cells within a tumor biopsy thereby providing for a more "purified" cancer cell population for which further diagnostics and therapeutic assessment can be conducted. In another case, manipulation of the molecular response to freezing can be utilized to identify the more plastic and robust cancer stem cell population both in *in vitro* and *in vivo* samples, allowing for downstream utilization in research, diagnostics and/or new therapy development. As such, even as we continue to dissect the molecular-based response and mechanisms of action associated with and/or activated by freezing, new avenues of utilization of CA beyond a therapeutic ablative strategy continue to emerge.

6. Summary

While cryoablation has a long and established history of therapeutic benefit, as studies over the last ~15 years have shown a critical molecular responses and the ability to couple this with new devices development, combination strategies etc., CA is well-positioned for resurgence as a primary approach to treat cancer. Studies have shown the involvement of numerous stressors and molecular cell death pathways in the activation and progression of cell death following freezing. While recognized as a critical factor in dictating therapeutic outcome, our understanding of the in-depth mechanisms and pathway progression remains in its infancy. What is clear, however, is that whether applied as a monotherapy or in conjunction with sensitization/adjunctive agents, CA is a highly effective means of treating cancer. As further studies emerge, CA should again enter mainstream therapeutics as a primary treatment strategy and may evolve into a gold standard personalized anti-cancer treatment option.

References

Baust, J. G., J. C. Bischof, J. Jiang-Hughes, T. J. Polascik, D. B. Rukstalis, A. A. Gage and J. M. Baust (2015). Re-purposing cryoablation: a combinatorial 'therapy' for the destruction of tissue. *Prostate Cancer and Prostatic Diseases*, online publication, Jan 27.

Baust, J. G. and A. A. Gage (2005). The molecular basis of cryosurgery. *BJU International*, 95, pp. 1187–1191.

Baust, J. G., A. A. Gage, T. E. Bjerklund Johansen and J. M. Baust (2014a). Mechanisms of cryoablation: Clinical consequences on malignant tumors. *Cryobiology*, 68(1), pp. 1–11.

Baust, J. G., A. A. Gage, D. Clarke, J. M. Baust and R. Van Buskirk (2004). Cryosurgery — a putative approach to molecular-based optimization. *Cryobiology*, 48, pp. 190–204.

Baust, J. G., A. A. Gage, D. P. Klossner, D. Clarke, R. Miller, J. Cohen *et al.* (2007). Issues critical to the successful application of cryosurgical ablation of the prostate. *Technology in Cancer Research and Treatment*, 6, pp. 97–109.

Baust, J. G., A. A. Gage, H. Ma and C. M. Zhang (1997). Minimally invasive cryosurgery — technological advances. *Cryobiology*, 34(4), pp. 373–384.

Baust, J. G., A. A. Gage, A. T. Robilotto and J. M. Baust (2009). The pathophysiology of thermoablation: optimizing cryoablation. *Current Opinion in Urology*, 19, pp. 127–132.

Baust, J. G., D. P. Klossner, R. G. Vanbuskirk, A. A. Gage, V. Mouraviev, T. J. Polascik *et al.* (2010a). Integrin involvement in freeze resistance of androgen-insensitive prostate cancer. *Prostate Cancer and Prostatic Diseases*, 13, pp. 151–161.

Baust, J. M. (2014). The molecular-based cell death continuum in hypothermic medicine: involvement in and impact on cryoablation and biopreservation outcomes. American College of Cryosurgery, *ACCryo2014*, Presentation — Jan 19.

Baust, J. M., D. Klossner, A. Gage, R. V. Buskirk and J. G. Baust (2013). Akt signaling mediates prostate cancer response to cryoablation. *Cryobiology*, 67, pp. 427–428.

Baust, J. M., D. P. Klossner, A. Robilotto, R. G. Vanbuskirk, A. A. Gage, V. Mouraviev *et al.* (2012). Vitamin D_3 cryosensitization increases prostate cancer susceptibility to cryoablation via mitochondrial-mediated apoptosis and necrosis. *BJU International*, 109, pp. 949–958.

Baust, J. M., A. T. Robilotto, C. Lueckge, P. Guerra, K. K. Snyder, R. Cheeks and J. G. Baust (2010b). Development of a novel cryoablation device for the treatment of cardiac tachyarrhythmias. *Heart Rhythm 2010*, Jun 12–15 Jun, Denver, CO.

Baust, J. M., K. Snyder, J. Katz, A. Gage, R. G. Van Buskirk and J. G. Baust (2010c). Investigation of neuronal cell stress response to transient freezing associated with cryosurgical ablation of the prostate. *Cryobiology*, 61(3), p. 364.

Baust, J. M., K. K. Snyder, K. L. Santucci, A. T. Robilotto, J. Smith, J. McKain *et al.* (2014b). Assessment of SCN and argon cryoablation devices in an in vivo-like 3-D tissue engineered prostate and renal cancer model. *Cryobiology*, 69, p. 198 (abstract).

Benoit, R. M., J. K. Cohen and R. J. Miller Jr. (2000). Cryosurgery for prostate cancer: new technology and indications. *Current Urology Reports*, 1(1), pp. 41–47.

Berrada, M. S. and J. C. Bischof (2001). Evaluation of freezing effects on human microvascular-endothelial cells (HMEC). *CryoLetters*, 22, pp. 353–366.

Bischof, J. C., J. Bastacky and B. Rubinsky (1992). An analytical study of cryosurgery in the lung. *Journal of Biomechanical Engineering*, 114, pp. 467–472.

Bischof, J., K. Christov and B. Rubinsky (1993). A morphological study of cooling rate response in normal and neoplastic human liver tissue: cryosurgical implications. *Cryobiology,* 30, pp. 482–492.

Caso, J. R., M. Tsivian, V. Mouraviev, M. Kimura and T. J. Polascik (2012a). Complications and postoperative events after cryosurgery for prostate cancer. *BJU International*, 109, pp. 840–845.

Caso, J. R., M. Tsivian, V. Mouraviev and T. J. Polascik (2012b). Predicting biopsy-proven prostate cancer recurrence following cryosurgery. *Urologic Oncology*, 30, pp. 391–395.

Chang, Z., J. J. Finkelstein, H. Ma and J. G. Baust (1994). Development of a high-performance multiprobe cryosurgical device. *Biomedical Instrumentation and Technology,* 289(5), pp. 383–390.

Chu, K. F. and D. E. Dupuy (2014). Thermal ablation of tumors: biological mechanisms and advances in therapy. *Nature Reviews Cancer*, 4(3), pp. 199–208.

Clarke, D. M., J. M. Baust, R. G. Van Buskirk and J. G. Baust (2001). Chemo-cryo combination therapy: an adjunctive model for the treatment of prostate cancer. *Cryobiology*, 42, pp. 274–285.

Clarke, D. M., J. M. Baust, R. G. Van Buskirk and J. G. Baust (2004). Addition of anticancer agents enhances freezing-induced prostate cancer cell death: implications of mitochondrial involvement. *Cryobiology*, 49, pp. 45–61.

Clarke, D. M., A. T. Robilotto, E. Rhee, R. G. Van Buskirk, J. G. Baust, A. A. Gage *et al.* (2007a). Cryoablation of renal cancer: variables involved in freezing-induced cell death. *Technology in Cancer Research and Treatment,* 6, pp. 69–79.

Clarke, D. M., A. T. Robilotto, R. G. Van Buskirk, J. G. Baust, A. A. Gage and J .M. Baust (2007b). Targeted induction of apoptosis via TRAIL and cryoablation: a novel strategy for the treatment of prostate cancer. *Prostate Cancer and Prostatic Diseases,* 10, pp. 175–184.

Cohen, J. K., R. J. Miller Jr., S. Ahmed, M. J. Lotz and J. Baust (2008). Ten-year biochemical disease control for patients with prostate cancer treated with cryosurgery as primary therapy. *Urology*, 71, pp. 515–518.

Cooper, I. S. (1965). Cryogenic surgery for cancer. *Federation Proceedings,* 24, pp. S237–240.

Cooper, I. S. and A. S. Lee (1961). Cryostatic congelation: a system for producing a limited, controlled region of cooling or freezing of biologic tissues. *Journal of Nervous and Mental Disorder,* 133, pp. 259–263.

Corn, P. G. (2012). The tumor microenvironment in prostate cancer: elucidating molecular pathways for therapy development. *Cancer Management and Research,* 4, pp. 183–193.

den Brok, M. H., R. P. Sutmuller, S. Nierkens, E. J. Bennink, L. W. Toonen, C. G. Figdor *et al.* (2006). Synergy between in situ cryoablation and TLR9 stimulation results in a highly effective in vivo dendritic cell vaccine. *Cancer Research,* 2006, 66, pp. 7285–7292.

Dimond, P. F. (2013). Acquired drug resistance in cancer. *GEN Eng News*, 33, pp. 6–7.

Forest, V., M. Peoc'h, C. Ardiet, L. Campos, D. Guyotat and J. M. Vergnon (2005). In vivo cryochemotherapy of a human lung cancer model. *Cryobiology*, 51, pp. 92–101.

Forest, V., M. Peoc'h, L. Campos, D. Guyotat and J. M. Vergnon (2006). Benefit of a combined treatment of cryotherapy and chemotherapy on tumour growth and late cryo-induced angiogenesis in a non-small-cell lung cancer model. *Lung Cancer,* 54, pp. 79–86.

Gage, A. A. (1978). Current status of cryosurgery for cancer. *Comprehensive Therapy*, 4, pp. 24–30.

Gage, A. A. and J. Baust (1998). Mechanisms of tissue injury in cryosurgery. *Cryobiology*, 37, pp. 171–186.

Gage, A. A. and J. G. Baust (2002). Cryosurgery — a review of recent advances and current issues. *CryoLetters*, 23, pp. 69–78.

Gage, A. A. and J. G. Baust (2007). Cryosurgery for tumors. *Journal of American College of Surgeons*, 205, pp. 342–356.

Gage, A. A., J. M. Baust and J. G. Baust (2009). Experimental cryosurgery investigations in vivo. *Cryobiology*, 59, pp. 229–243.

Gazzaniga, S., A. Bravo, S. R. Goldszmid, F. Maschi, J. Martinelli, J. Mordoh *et al.* (2001). Inflammatory changes after cryosurgery-induced necrosis in human melanoma xenografted in nude mice. *Journal of Investigative Dermatology*, 116, pp. 664–671.

Ghisolfi, L., A. C. Keates, X. Hu, D. K. Lee and C. J. Li (2012). Ionizing radiation induces stemness in cancer cells. *PLoS One*, 7, e43628.

Gilbert, J. C., B. Rubinsky, M. S. Roos, S. T. Wong and K. M. Brennan (1993). MRI-monitored cryosurgery in the rabbit brain. *Magnetic Resonance Imaging*, 11, pp. 1155–1164.

Goel, R., K. Anderson, J. Slaton, F. Schmidlin, G. Vercellotti, J. Belcher *et al.* (2009). Adjuvant approaches to enhance cryosurgery. *Journal of Biomechanical Engineering*, 131, 074003.

Goel, R., D. Swanlund, J. Coad *et al.* (2007). TNF-alpha-based accentuation in cryoinjury — dose, delivery, and response. *Molecular Cancer Therapy*, 6, pp. 2039–2047.

Gu, X.Y., Z. Jiang and W. Fang (2011). Cryoablation combined with molecular target therapy improves the curative effect in patients with advanced non-small cell lung cancer. *Journal of International Medical Research*, 39(5), pp. 1736–1743.

Han, B. and J. C. Bischof (2004). Direct cell injury associated with eutectic crystallization during freezing. *Cryobiology*, 48, pp. 8–21.

Han, B., A. Iftekhar and J. C. Bischof (2004). Improved cryosurgery by use of thermophysical and inflammatory adjuvants. *Technology in Cancer Research and Treatment*, 3, pp. 103–111.

Han, B., D. J. Swanlund and J. C. Bischof (2007). Cryoinjury of MCF-7 human breast cancer cells and inhibition of post-thaw recovery using TNF-alpha. *Technology in Cancer Research and Treatment*, 6, pp. 625–634.

Hanahan, D. and R. A. Weinberg (2011). Hallmarks of cancer: the next generation. *Cell*, 144, pp. 646–674.

Hanai, A., W. L. Yang and T. S. Ravikumar (2001). Induction of apoptosis in human colon carcinoma cells HT29 by sublethal cryo-injury: mediation by cytochrome c release. *International Journal of Cancer*, 93(4), pp. 526–533.

Hoffmann, N. E. and J. C. Bischof (2002). The cryobiology of cryosurgical injury. *Urology*, 60, pp. 40–49.

Hollister, W. R., A. J. Mathew, J. G. Baust and R. G. Van Buskirk (1998). Effects of freezing on cell viability and mechanisms of cell death in a human prostate cell line, *Molecular Urology*, 2, pp. 3–18.

Ikekawa, S., K. Ishihara, S. Tanaka and S. Ikeda (1985). Basic studies of cryochemotherapy in a murine tumor system. *Cryobiology*, 22, pp. 477–483.

Jiang, J., R. Goel, M. A. Iftekhar, R. Visaria, J. D. Belcher, G. M. Vercellotti *et al.* (2008). Tumor necrosis factor-alpha-induced accentuation in cryoinjury: mechanisms in vitro and in vivo. *Molecular Cancer Therapeutics*, 7, pp. 2547–2555.

Jones, J. S. (2011). Radiorecurrent prostate cancer: an emerging and largely mismanaged epidemic. *European Urology*, 60, pp. 411–412.

Kimura, M., Z. Rabbani, V. Mouraviev, M. Tsivian, J. Caso, T. Satoh *et al.* (2010). Role of vitamin D_3 as a sensitizer to cryoablation in a murine prostate cancer model: preliminary in vivo study. *Urology*, 76, 764.e14–764.e20.

Klossner, D. P., J. M. Baust, R. G. Van Buskirk, A. A. Gage and J. G. Baust (2008). Cryoablative response of prostate cancer cells is influenced by androgen receptor expression. *BJU International,* 101, pp. 1310–1316.

Klossner, D. P., A. T. Robilotto, D. M. Clarke, R. G. Van Buskirk, J. M. Baust, A. A. Gage *et al.* (2007). Cryosurgical technique: assessment of the fundamental variables using human prostate cancer model systems. *Cryobiology,* 55, pp. 189–199.

Koushafar, H., L. Pham, C. Lee and B. Rubinsky (1997). Chemical adjuvant cryosurgery with antifreeze proteins. *Journal of Surgical Oncology*, 66, pp. 114–121.

Koushafar, H. and B. Rubinsky (1997). Effect of antifreeze proteins on frozen primary prostatic adenocarcinoma cells. *Urology*, 49, pp. 421–425.

Kuflik, E. G. (2004). Cryosurgery for skin cancer: 30-year experience and cure rates, *Dermatologic Surgery*, 30, pp. 297–300.

Lee, S. M., J. Y. Won, D. Y. Lee, K. H. Lee, K. S. Lee, Y. H. Paik *et al.* (2011). Percutaneous cryoablation of small hepatocellular carcinomas using a 17-gauge ultrathin probe, *Clinical Radiology*, 66, pp. 752–759.

Le Pivert, P., R. S. Haddad, A. Aller, K. Titus, J. Doulat, M. Renard *et al.* (2004). Ultrasound guided combined cryoablation and microencapsulated 5-Fluorouracil inhibits growth of human prostate tumors in xenogenic mouse model assessed by luminescence imaging. *Technology in Cancer Research and Treatment*, 3, pp. 135–142.

Levy, D. A., L. L. Pisters and J. S. Jones (2009). Primary cryoablation nadir prostate specific antigen and biochemical failure. *Journal of Urology,* 182, pp. 931–937.

Littrup, P. J., B. Jallad, P. Chandiwala-Mody, M. D'Agostini, B. A. Adam, B. Bouwman (2009). Cryotherapy for breast cancer: a feasibility study without excision. *Journal of Vascular Interventional Radiology*, 20, pp. 1329–1341.

Mala, T. (2006). Cryoablation of liver tumours — a review of mechanisms, techniques and clinical outcome. *Minimally Invasive Therapy and Allied Technologies*, 15, pp. 9–17.

Mir, L. M. and B. Rubinsky (2002). Treatment of cancer with cryochemotherapy. *British Journal of Cancer*, 86, pp. 1658–1660.

Moltzahn, F. and G. N. Thalmann (2013). Cancer stem cells in prostate cancer. *Translational Andrology and Urology*, 2, pp. 242–253.

Mouraviev, V. G., J. L. Chin, D. B. Downey and M. Moussa (2001a). Image-guided targeted cryoablation for canine prostate model using new cryopreservation cocktail for neurovascular bundle protection. *Advances in Cryomedicine*, pp. 108–110.

Mouraviev, V. G., D. Prokhorov and G. A. Prokhorov (2001b). Rectal and urethral sphincter protection during aggressive prostate cryosurgery: the possible clinical application of special antifreeze cocktail for an injection into anterior and posterior paraprostatic spaces. *Advances in Cryomedicine*, pp. 111–113.

Muldrew, K., J. Rewcastle, B. J. Donnelly, J. C. Saliken, S. Liang, S. Goldie *et al.* (2001). Flounder antifreeze peptides increase the efficacy of cryosurgery. *Cryobiology*, 42, pp. 182–189.

O'Donoghue, E.P., L. A. Milleman, R. H. Flocks, D. A. Culp and W. W. Bonney (1975). Cryosurgery for carcioma of prostate. *Urology,* 5, pp. 308–316.

Onik, G. M., J. K. Cohen, G. D. Reyes, B. Rubinsky, Z. Chang and J. G. Baust (1993). Transrectal ultrasound-guided percutaneous radical cryosurgical ablation of the prostate. *Cancer*, 72, pp. 1291–1299.

Onik, G., B. Porterfield, B. Rubinsky and J. Cohen (1991a). Percutaneous transperineal prostate cryosurgery using transrectal ultrasound guidance: animal model. *Urology*, 37, pp. 277–281.

Onik, G., B. Rubinsky, R. Zemel, L. Weaver, D. Diamond, C. Cobb *et al.* (1991b). Ultrasound-guided hepatic cryosurgery in the treatment of metastatic colon carcinoma. Preliminary results. *Cancer*, 67, pp. 901–907.

Pham, L., R. Dahiya and B. Rubinsky (1999). An in vivo study of antifreeze protein adjuvant cryosurgery. *Cryobiology*, 38, 169–175.

Pitman, M., E. Y. Shapiro, G. W. Hruby, M. D. Truesdale, P. J. Cheetham, S. Saad *et al.* (2012). Comparison of biochemical failure definitions for predicting local cancer recurrence following cryoablation of the prostate. *Prostate*, 72, pp. 1802–1808.

Redondo, P., J. del Olmo, A. Lopez-Diaz de Cerio, S. Inoges, M. Marquina, I. Melero *et al.* (2007). Imiquimod enhances the systemic immunity attained by local cryosurgery destruction of melanoma lesions. *Journal of Investigative Dermatology*, 127, pp. 1673–1680.

Robilotto, A. T., J. M. Baust, R. G. Van Buskirk, A. A. Gage and J. G. Baust (2013). Temperature-dependent activation of differential apoptotic pathways during cryoablation in a human prostate cancer model. *Prostate Cancer and Prostatic Diseases*, 16, pp. 41–49.

Rodriguez-Bigas, M. A., D. Klippenstein, N. J. Meropol, T. K. Weber and N. J. Petrelli (1996). A pilot study of cryochemotherapy for hepatic metastases from colorectal cancer. *Cryobiology*, 33, pp. 600–606.

Rubinsky, B. (2006) Use of cryoprotective agent compounds during cryosurgery. 6,041,787 ed.2000.

Sabel, M. S. (2009). Cryo-immunology: a review of the literature and proposed mechanisms for stimulatory versus suppressive immune responses. *Cryobiology*, 58, pp. 1–11.

Saliken, J. C., B. J. Donnelly and J. C. Rewcastle (2002). The evolution and state of modern technology for prostate cryosurgery. *Urology*, 60, pp. 26–33.

Santucci, K. L., K. K. Snyder, J. M. Baust, R. G. Van Buskirk, V. Mouraviev, T. J. Polascik *et al.* (2011). Use of 1,25alpha dihydroxyvitamin D_3 as a cryosensitizing agent in a murine prostate cancer model. *Prostate Cancer and Prostatic Diseases*, 14, pp. 97–104.

Shackleton, M., E. Quintana, E. R. Fearon and S. J. Morrison (2009). Heterogeneity in cancer: cancer stem cells versus clonal evolution. *Cell*, 138, pp. 822–829.

Shenoi, M. M., I. Iltis, J. Choi, N. A. Koonce, G. J. Metzger, R. J. Griffin *et al.* (2013). Nanoparticle delivered vascular disrupting agents (VDAs): use of TNF-alpha conjugated gold nanoparticles for multimodal cancer therapy. *Molecular Pharmaceutics*, 10, pp. 1683–1694.

Smith, D. J., W. M. Fahssi, D. J. Swanlund and J. C. Bischof (1999). A parametric study of freezing injury in AT-1 rat prostate tumor cells. *Cryobiology*, 39, pp. 13–28.

Sun, Y., J. Campisi, C. Higano, T. M. Beer, P. Porter, I. Coleman *et al.* (2012). Treatment-induced damage to the tumor microenvironment promotes prostate cancer therapy resistance through WNT16B. *Nature Medicine*, 18, pp. 1359–1368.

Tian, T., S. Olson, J. M. Whitacre and A. Harding (2011). The origins of cancer robustness and evolvability. *Integrative Biology (Camb)*, 3, pp. 17–30.

Udagawa, M., C. Kudo-Saito, G. Hasegawa, K. Yano, A. Yamamoto, M. Yaguchi *et al.* (2006). Enhancement of immunologic tumor regression by intratumoral administration of dendritic cells in combination with cryoablative tumor pretreatment and Bacillus Calmette-Guerin cell wall skeleton stimulation. *Clinical Cancer Research*, 12, pp. 7465–7475.

Waitz, R., S. B. Solomon, E. N. Petre, A. E. Trumble, M. Fasso, L. Norton *et al.* (2012). Potent induction of tumor immunity by combining tumor cryoablation with anti-CTLA-4 therapy. *Cancer Research,* 72, pp. 430–439.

Wang, C. L., K. Y. Teo and B. Han (2008). An amino acidic adjuvant to augment cryoinjury of MCF-7 breast cancer cells. *Cryobiology,* 57, pp. 52–59.

Xu, K. C., L. Z. Niu, Y. Z. Hu, W. B. He, Y. S. He, Y. F. Li and J. S. Zuo (2008a). A pilot study on combination of cryosurgery and (125) iodine seed implantation for treatment of locally advanced pancreatic cancer. *World Journal of Gastroenterology,* 14(10), pp. 1603–1611.

Xu, K. C., L. Z. Niu, Y. Z. Hu, W. B. He, Y. S. He, Y. F. Li and J. S. Zuo (2008b). Cryosurgery with combination of (125) iodine seed implantation for the treatment of locally advanced pancreatic cancer. *Journal of Digestive Diseases,* 9(1), pp. 32–40.

Yang, W. L., T. Addona, D. G. Nair, L. Qi and T. S. Ravikumar (2003). Apoptosis induced by cryo-injury in human colorectal cancer cells is associated with mitochondrial dysfunction. *International Journal of Cancer,* 103, pp. 360–369.

Yuan, F., W. Zhou, J. Zhang, Z. Zhang, C. Zou, L. Huang *et al.* (2008). Anticancer drugs are synergistic with freezing in induction of apoptosis in HCC cells. *Cryobiology,* 57, pp. 60–65.

Yuanying, Y., N. Lizhi, M. Feng, W. Xiaohua, Z. Jianying, Y. Fei, J. Feng, H. Lihua, C. Jibing, L. Jialiang and X. Kecheng (2013). Therapeutic outcomes of combining cryotherapy, chemotherapy and DC-CIK immunotherapy in the treatment of metastatic non-small cell lung cancer. *Cryobiology,* 67(2), pp. 235–240.

Chapter 2

Freezing-Induced Changes of Extracellular Matrix and Functional Properties of Engineered Tissues

Soham Ghosh, Seungman Park and Bumsoo Han

1. Introduction

Cryopreservation, preserving biomaterials in a frozen state, can provide reliable long-term storage of cell/tissue engineering products. Such long-term storage is crucial to scale up tissue engineering and regenerative medicine (TE/RM) from laboratory research to manufacturing [Archer and Williams, 2005; Nerem, 2006; Pancrazio *et al.*, 2007]. Success of preserving a few tissue systems has been reported [Wang *et al.*, 2002, Kubo and Kuroyanagi, 2005]; but it is still extremely challenging to develop successful cryopreservation protocols for any given type of engineered tissues (ETs). The two most critical challenges are difficulties (1) in maintaining the functional properties of ETs throughout the preservation processes and (2) in controling highly tissue-type-specific preservation outcomes [Han and Bischof, 2004; Mikos *et al.*, 2006].

Research efforts to address these challenges have been focused on determining optimal freeze/thaw (F/T) protocols and cryoprotective agent (CPA) mixtures for a given cell/tissue-type by comparing the outcomes of cryopreservation protocols, mainly gauged by cell viability. The parameters studied include cooling rates, freezing temperatures, duration of storage, CPA concentration and mixture composition, and CPA loading/unloading methods [Wang *et al.*, 2007a; Wang *et al.*, 2007b; Gustavo *et al.*, 2008; Morato *et al.*, 2008; Serrato *et al.*, 2009; Brockbank *et al.*, 2010, Szarko *et al.*, 2010]. In addition, efforts have been made to determine these protocol parameters for vitrification (i.e. glassifying water into amorphous solid with a high concentration of CPAs) or rapid cooling [Gandolfi *et al.*, 2006; Pegg *et al.*, 2006a, 2006b, 2006c; Schenke-Layland *et al.*, 2007; Brockbank *et al.*, 2008; Isachenko *et al.*, 2008; Keros *et al.*, 2009; Brockbank *et al.*, 2010; Kim *et al.*, 2011]. These approaches have greatly improved the outcome of tissue preservation (i.e. cell viability) but it is still extremely challenging to design and optimize the protocol for various types of ETs [Han and Bischof, 2004; Mikos *et al.*, 2006; Bischof, 2000]. For example, significant post-thaw microstructural changes were observed in case of cryopreserved [Laouar *et al.*, 2007] and vitrified cartilages [Brockbank *et al.*, 2008]. On the contrary, microstructural integrity was retained for human heart valve tissue [Gerson *et al.*, 2009] and arteries [Venkatasubramanian *et al.*, 2010]. Since advances in TE/RM have enabled a wide variety of ETs with diverse and unique functionalities, a more general mechanistic approach is needed for cryopreservation. Although one of the most important outcomes of cryopreservation has been post-thaw cellular viability, it is realized that other features beyond viability may be required for the functionality of advanced ETs. Specifically, it is recently reported that extracellular matrix (ECM) structures are crucial to the tissue's functionality as well as cell viability [Dainese *et al.*, 2006; Laouar *et al.*, 2007; Schenke-Layland *et al.*, 2007; Brockbank *et al.*, 2008; Gustavo *et al.*, 2008; Gerson *et al.*, 2009; Han *et al.*, 2009; Szarko *et al.*, 2010; Venkatasubramanian *et al.*, 2010; Teo *et al.*, 2011; Bischof *et al.*, 1990; Han *et al.*, 2005]. Preserving these structures as well as cell viability should lead to successful cryopreservation of functional ETs.

These recent studies are based on the observation that, when tissue is being frozen, complex spatiotemporal biophysical phenomena occur near the phase change interface as illustrated in Figure 1. These are (1) volume expansion during water-ice phase change; (2) interstitial fluid transport from the freezing interface; (3) swelling of the unfrozen ECM to accommodate the interstitial fluid and; (4) reciprocal action of the ECM through the interstitial fluid pressure (IFP)-stress-strain balance and hydraulic permeability, briefly explained in Figure 1. Though not shown in Figure 1, the cells will also actively participate in the fluid-matrix interaction through the osmotic pressure-driven water transport and exert mechanical force on the matrix. Therefore, proper control of this interaction will enable to control the freezing-induced microstructural changes of ECM and ultimately preserve functional properties of ETs. In this chapter, thus, we

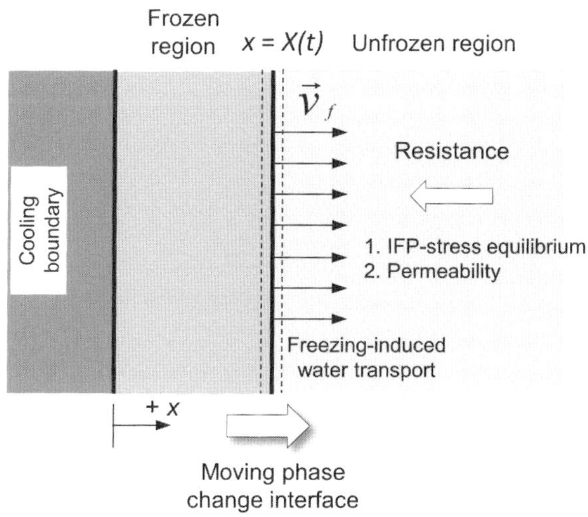

Figure 1. Schematic of freezing-induced cell-fluid-matrix interaction. Due to the volume expansion during ice formation, interstitial fluid transport is induced at the phase change interface. This fluid transport can swell the ECM microstructure, induce mechanical stress and alter hydraulic conductivity. [Adapted from Han *et al.*, 2009].

summarize recent studies on the freezing-induced changes of the ECM microstructure and subsequent functional properties of ETs. Studies on freezing-induced interactions among cells, ECM and interstitial fluid are also reviewed. The results are discussed to establish quantitative understanding, which helps to achieve successful preservation of functional ETs. Although there are already many excellent reviews and recent articles on cell-level biophysics [Bischof, 2000; Karlsson, 2015; Gao and Critser, 2000; Fuller and Paynter, 2004; Abazari *et al.*, 2013; Hunt, 2011; Hagiwara *et al.*, 2009; Balasubramanian *et al.*, 2009] in cell suspension, only a few studies can be found on the biophysics of cells embedded in extracellular matrix. This context-dependent cellular-level biophysics is begining to be recognized [Seawright *et al.*, 2013, Ghosh *et al.*, 2014], but still further research is warranted.

2. Structure-functionality Relationships in ECM

Many functional properties of both native and engineered tissues are associated with the microstructure of the ECM as summarized in following sections. Mechanical and transport properties have been correlated to the structural parameters of ECM. Since the ECM is also connected to the cells through complex adhesion molecules and regulates the transport of various growth factors and cytokines [Frantz *et al.*, 2010], it not only provides a structural scaffold to cells but also plays a major role in regulating cell physiology. These include morphogenesis, growth [Berthiaume *et al.*, 1996; Borene *et al.*, 2004], intercellular signaling [Meredith *et al.*, 1993], tissue remodeling, and wound healing [Grinnell, 1994; Tranquillo, 1998; Petroll and Ma, 2003]. Therefore, for successful cryopreservation of functional tissues, ECM microstructure and cell viability should be preserved [Narine *et al.*, 2006].

2.1. *Structure-permeability Relationship*

Permeability is the ability of a porous material to allow fluids to pass through it. It has been reported that permeability is strongly related to microstructural parameters such as porosity, matrix orientation, composition and deformation as summarized in Table 1. In general, the majority of experimental and computational correlations have shown a similar trend with respect to these microstructural parameters. For example, these studies have shown that larger porosity or pore size, smaller fibril diameter, and lower polymer concentration lead to higher permeability (k). Moreover, fiber alignment parallel to the fluid flow tends to induce increased permeability, compared to an alignment perpendicular to the flow. Recently, it was reported that the interfibril branching is another significant parameter affecting permeability [Whittington *et al.*, 2013a; 2013b; Park *et al.*, 2015a].

2.2. *Structure-diffusivity Relationship*

According to the Stokes-Einstein equation, the free diffusivity (i.e. diffusivity of molecules in liquid) can be expressed as a function of temperature, dynamic viscosity, and hydrodynamic radius:

$$D_{ij} = \frac{k_B T}{6\pi \mu_j r_i}$$

where k_B is Boltzmann's constant, T is the absolute temperature, μ_j is the dynamic viscosity of solvent j, and r_i is the particle radius. In addition, similar to the permeability, the diffusivity is dependent on the microstructural parameters such as pore size, matrix concentration, and orientation. Many

Table 1. Porous microstructure-permeability relationship.

Author	Correlation	Comment
Carmen-Kozeny [Scheidegger, 1974; Levick, 1987; Swartz and Fleury, 2007; Swartz *et al.*, 1999]	$k = \frac{\phi^3}{GS^2} = \frac{\phi r_h^2}{G}$	ϕ: Porosity G: Kozeny factor (related to geometry) S: Wetted surface area per unit volume r_h: Mean hydraulic radius of the pores
Davies [Davies; 1952, Hosseini and Tafreshi, 2010]	$\frac{k}{r_f^2} = \frac{1}{16(1-\phi)^{1.5}\left(1+56(1-\phi)^3\right)}$	r_f: Fiber radius ϕ Porosity
Spielman and Goren [Spielman and Goren; 1968; Pedersen *et al.*, 2010]	$\frac{1}{3} + \frac{5}{6}\left[\frac{\sqrt{k}K_1\left(r_f / \sqrt{k}\right)}{r_f K_0\left(r_f / \sqrt{k}\right)}\right] = \frac{1}{4(1-\phi)}$	r_f: Fiber radius ϕ: Porosity K_p: Modified Bessel function of the second kind with order p
Jackson [Jackson and James, 1986; Stylianopoulos *et al.*, 2008b]	$\frac{k}{r_f^2} = \frac{3}{20(1-\phi)}(-\ln(1-\phi) - 0.931)$	r_f: Fiber radius ϕ Porosity
Lee and Yang [Lee and Yang, 1997]	$\frac{k}{r_f^2} = \frac{4\phi^3(\phi - 0.2146)}{31(1-\phi)^{1.3}}$	r_f: Fiber radius ϕ: Porosity Valid for $0.435 < \phi < 0.937$
Sahraoui and Kaviany [Sahraoui and Kaviany, 1992]	$\frac{k}{r_f^2} = \frac{0.189\phi^{5.1}}{1-\phi}$	r_f: Fiber radius ϕ: Porosity Valid for $0.4 < \phi < 0.8$
Tomadakis [Tomadakis and Robertson, 2005]	$\frac{k}{r_f^2} = \frac{\phi}{8\ln^2\phi}\frac{(\phi-\phi_p)^{\alpha+2}}{(1-\phi_p)^{\alpha}\left[(\alpha+1)\phi-\phi_p\right]^2}$	r_f: Fiber radius ϕ: Porosity ϕ_p: Percolation threshold (0.037 for 3D) α: Constant (0.661 for 3D)

experimental studies have been performed to measure the diffusivity of biomaterials using time-lapse fluorescence microscopy techniques such as an integrative optical imaging (IOI) method [Nicholson and Tao, 1993] and fluorescence recovery after photobleaching (FRAP) technique [Bu and Russo, 1994]. Generally, it has been known that as the polymer concentration of the biomaterials increases, the diffusivity decreases. Effects of other structural parameters on the diffusivity are summarized in Table 2.

Diffusivity depends on many structural parameters of the extracellular matrix. Among these parameters, the porosity is affected mostly by freezing-induced cell-fluid-matrix interactions. Thus, it may be hypothesized that the post-thaw functional properties of tissues can be predicted by knowing the freezing-induced changes of porosity. The relation between the porosity of microstructure and the diffusivity can be described using tortuosity. Tortuosity can be defined as a function of the free and effective diffusivity with the following form [Thorne and Nicholson, 2006]:

$$\lambda = \sqrt{\frac{D_0}{D}}$$

where D_0 is the free diffusivity and D is the effective diffusivity. In addition, it has been known that tortuosity has an inverse relation with the porosity according to Archie [1942].

$$\lambda = \left(\frac{1}{\phi}\right)^{\alpha}$$

where α: 0.5 in 1D or 2D and 0.3 in 3D, and ϕ is the porosity [Tamayol and Bahrami, 2010]. By combining the previous two equations above, the relationship between porosity and diffusivity can be derived as the following equation:

$$\frac{D}{D_0} = \phi^{2\alpha}.$$

This equation indicates that the higher the porosity, the higher the effective diffusivity will be.

Table 2. Predictive models for diffusivity estimation.

Author	Correlation	Comment
Ogston [Ogston *et al.*, 1973]	$\frac{D}{D_0} = A\exp(-brc^{0.5})$	D: Effective diffusivity D_0: Free diffusivity A: Dimensionless constant (between 1 and 2) b: Dimensionless constant (often π) c: Length of matrix fiber per volume r: Solute radius
Kosto [Kosto and Deen, 2004]	$\frac{D}{D_0} = \exp\left(-(1-\phi)^{0.5}\left(1+\frac{R_H}{r_f}\right)\right)$	D: Effective diffusivity D_0: Free diffusivity ϕ: Porosity R_H: Hydrodynamic radius of the pores r_f: Fiber radius
Clague and Phillips [Clague and Phillips, 1996; Phillips, 2000]	$D = SFD_0\begin{cases} F = 1+\frac{r_s}{\sqrt{k}}+\frac{1}{9}\left(\frac{r_s}{\sqrt{k}}\right)^2 \\ S(f) = e^{-0.84f^{1.09}} \\ f = \left(1+\frac{r_s}{r_f}\right)^2(1-\phi) \end{cases}$	D: Effective diffusivity D_0: Free diffusivity S: Steric effect coefficient F: Hydrodynamic effect coefficient k: Permeability r_s: Solute radius r_f: Fiber radius ϕ: Porosity

2.3. *Structure-mechanical Property Relationship*

The mechanical properties have also been known to be strongly affected by various microstructural features (e.g. the fiber volume fraction, orientation, distribution, pore size and pore interconnectivity) and matrix components (e.g. cell type, fibril type and various molecules). These mechanical properties of biological materials include elastic modulus in tension or tensile modulus, shear modulus, elastic modulus in compression or compressive modulus, and tensile strength. It has been determined that the tensile modulus of materials increases with higher polymer concentration and cell density which are correlated with an increase in fibril density (or decrease in porosity) [Zhu *et al.*, 2001; Roeder *et al.*, 2002; Wakatsuki and Elson, 2003; Moroni *et al.*, 2006; Evans and Barocas, 2009]. In addition, it has been known that the fiber arrangement or orientation significantly affects the mechanical property such as the tensile strength [Stylianopoulos *et al.*, 2008a] or tensile modulus [Martin and Boardman, 1993].

Several correlations have been proposed between structure and mechanical property of materials [Mackenzie, 1950; Budiansky and Oconnell, 1976; Hashin, 1988]. Many of these correlations have shown that high porosity or low fiber density decreases the mechanical properties such as compressive stiffness (E_c) and yield strength (σ) in compression [Hashin and Shtrikman, 1963; Zein *et al.*, 2002] based on a power-law empirical relationship:

$$E_c = C_1(100 - P)^n$$

$$\sigma = C_2(100 - P)^n$$

where C_1, and C_2, are constants, the exponent values are $n = 3$ for honey comb and $n = 2$ for open-pore foam scaffolds, and P is porosity (%). Furthermore, effective tensile modulus (E) and shear modulus (G) of porous materials decreases with the increasing porosity as follows [Kovacik, 1999; 2001; 2008]:

$$E = E_0\left(1 - \frac{\phi}{\phi_c}\right)^f$$

$$G = G_0\left(1 - \frac{\phi}{\phi_c}\right)^f$$

where E_0, and G_0 are elastic and shear modulus of solid materials, ϕ is the porosity, ϕ_c is the porosity at which the effective tensile modulus or shear modulus becomes zero, and f is a geometry-dependent parameter.

3. Fluid Structure Interaction during F/T

3.1. *F/T Induced Changes in ECM Microstructure*

Freezing-induced structural changes of ECM were reported by assessing mean void area (MVA) and fibril diameter of collagen matrices after F/T [Han *et al.*, 2009; 2013]. Unfrozen collagen matrix exhibits dense and organized collagen network and uniform fibril diameters. With lower freezing temperatures, they show sparse and disorganized network morphology, and thicker fibril diameters (Figure 2). Both the MVA ratio and the fibril diameter increase with a decreasing freezing temperature. The increased void area with decreasing freezing temperature is thought to be caused by rapid localized freezing without water transport, which induce enlarged pore structure. The increased fibril diameter, which is not fully understood, is thought to be associated with freezing-induced denaturation or unfolding of

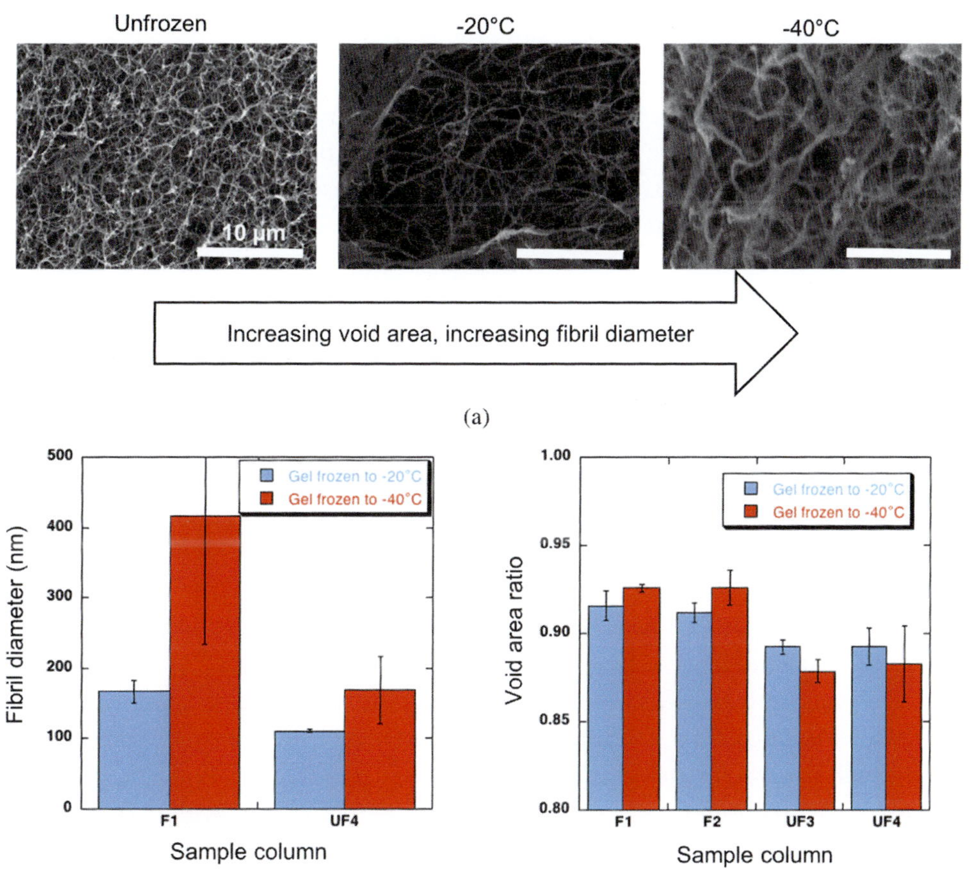

Figure 2. (a) Increase in void area and fibril diameter with lowered freezing temperature. (b) and (c) Quantification of increase in the parameters. [Adapted from Han *et al.*, 2009].

collagen fibrils [Han *et al.*, 2009; Ozcelikkale *et al.*, 2013]. In this context, it can be stated that higher initial porosity will lesser impede the water transport and consequently less structural damage is anticipated.

3.2. *Cell-fluid-matrix Interaction*

In order to explain these structural changes, freezing of the collagen gel has been analyzed by the freezing of poroelastic material saturated with water [Han *et al.*, 2009]. In this study, the post-thaw structural changes were hypothesized to be caused by the volumetric expansion of the interstitial water during freezing. Considering the mass conservation at the freezing interface, the velocity of the interstitial fluid induced at the interface can be formulated as:

$$v_f(X(t)) = \phi\left(1 - \frac{\rho_s}{\rho_l}\right)\frac{dX(t)}{dt}$$

where $X(t)$ is the location of freezing interface, v_f is fluid velocity, ϕ is the porosity, ρ_s is solid phase density and ρ_l is liquid phase density. Subsequent deformation of the unfrozen region can be described by the simplified consolidation equation as follows:

$$\frac{\partial e}{\partial t} = \frac{KE(1-v)}{(1+v)(1-2v)}\frac{\partial^2 e}{\partial x^2} = K\frac{\partial^2 p}{\partial x^2}$$

where the dilatation, e, is given by the divergence of displacement ($e = \nabla \cdot \vec{u}$), K is the hydraulic conductivity, ν is Poisson's ratio and E is the tensile modulus. Based on this model, a theoretical study [Han *et al.*, 2009] suggested that freezing of the tissue induces spatial and temporal redistribution of interstitial fluid and subsequent ECM swelling. The relative role of initial ECM porosity, tensile modulus, and hydraulic conductivity were elucidated through that study.

This theoretical prediction was validated using a real-time noninvasive, dynamic measurement technique to quantify spatiotemporal deformation of dermal equivalent during freezing [Teo *et al.*, 2010]. For this method, named cell image deformetry (CID), engineered tissues were constructed by seeding cells, fluorescently labeled with quantum dots, within collagen matrices. Then, while being exposed to directional freezing, the ETs were continuously imaged on a fluorescence microscope. As illustrated in Figure 3, the time-lapse images were cross-correlated to determine microscale tissue deformation, including deformation, rates, strain rates and dilatation rates as follows:

$$e = \frac{\partial u}{\partial x} + \frac{\partial v}{\partial y} \approx \frac{\Delta u}{\Delta x} + \frac{\Delta v}{\Delta y}$$

The CID technique was further used to quantify the relative significance of cell, ECM and fluid in the freezing-induced cell-fluid-matrix interactions, which leads to post-thaw ECM structural changes [Teo *et al.*, 2011]. To investigate this, the ETs were constructed with various concentrations of cell and collagen. The resulting deformation, F/T induced microstructural changes and cell viability were studied (Figure 3). It was found that ETs with increased collagen content and cell seeding concentration are subject to higher extent of post-thaw structural changes. Although high collagen content makes the structure stiff, it also lowers the hydraulic conductivity due to its dense structure. This low hydraulic conductivity may impede the extent of freezing-induced water transport compared to the low collagen groups and result in more local deformation. Thus, the deformation and post-thaw ECM microstructure are thought to be more altered. However, at the cell level, some interesting observations were made. The ETs with high collagen content showed higher post-thaw viability and better preserved cytoskeletal structure. This improved preservation outcome is thought to result from the stronger mechanical support of the ECM by a denser collagen matrix via an increased number of specific cell-ECM adhesion binding sites as well as cytoskeletal structure (i.e. stress fibers) than other experimental groups. Although the ETs with high cell seeding density had similar mean void area prior to F/T, the cell viability was notably lower because of lack of stress fiber development. This result implies that complex cell-fluid-matrix interaction is as significant as thermal insults.

To understand the effect of thermal insult, the same ETs (same by composition) were submitted to the F/T process by lowering the freezing rate. As expected, lowering the freezing rate has a more detrimental effect on ECM damage, but the extent of the damage was profound (Figure 4). For example, the lower freezing temperature increased the ECM deformation by several times. Cell viability drastically dropped and cytoskeletal structures were completely destroyed. The cells in the unfrozen ET have typical

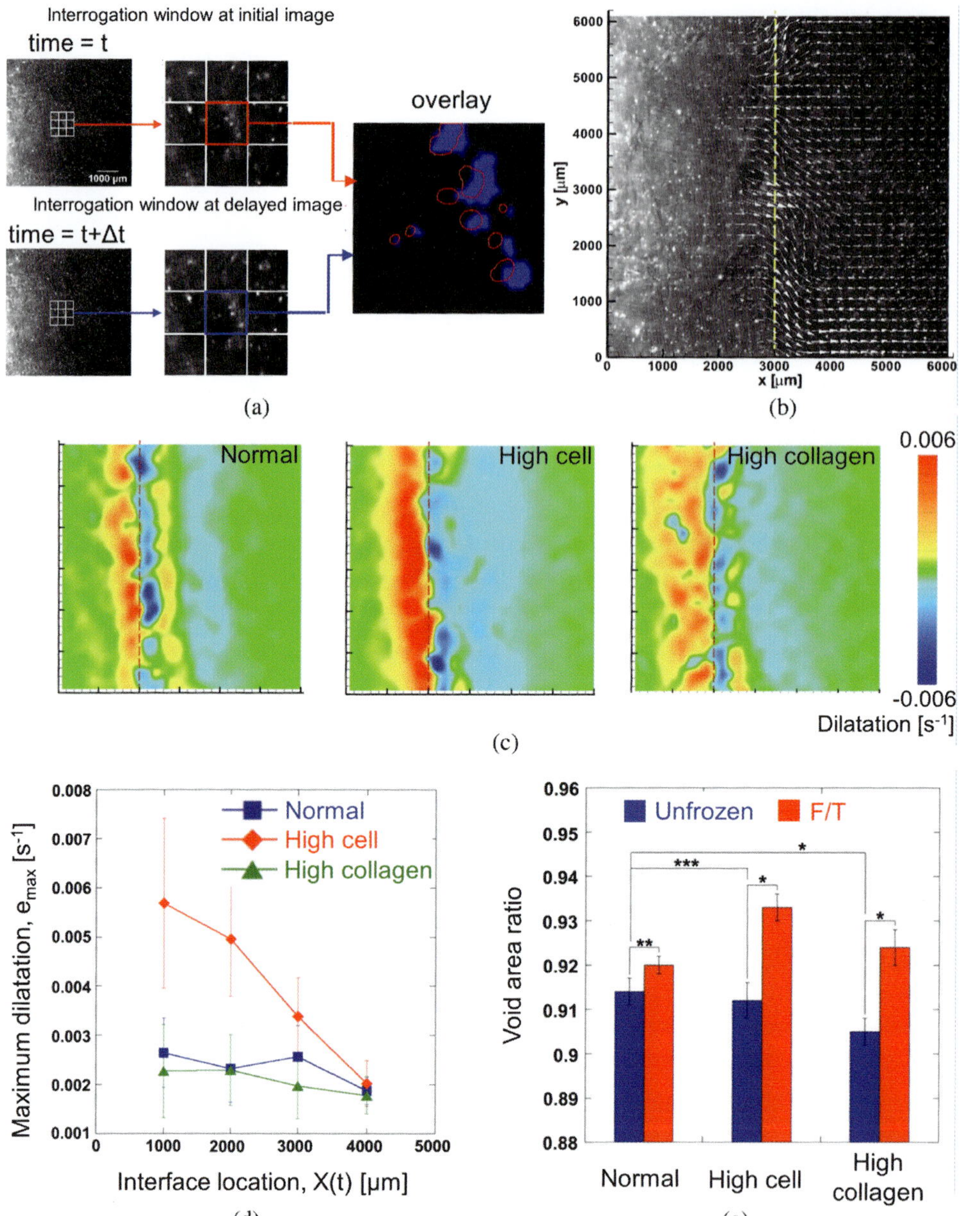

(a) (b)

(c)

(d) (e)

Figure 3. (a) Cell Image Deformetry (CID) method to measure spatiotemporal ECM deformation. (b) Deformation vectors during freezing. Effect of changing ET components on (c) spatial dilatation rate (d) maximum dilatation rate (e) void area ratio change. [Adapted from Teo *et al.*, 2010; 2011].

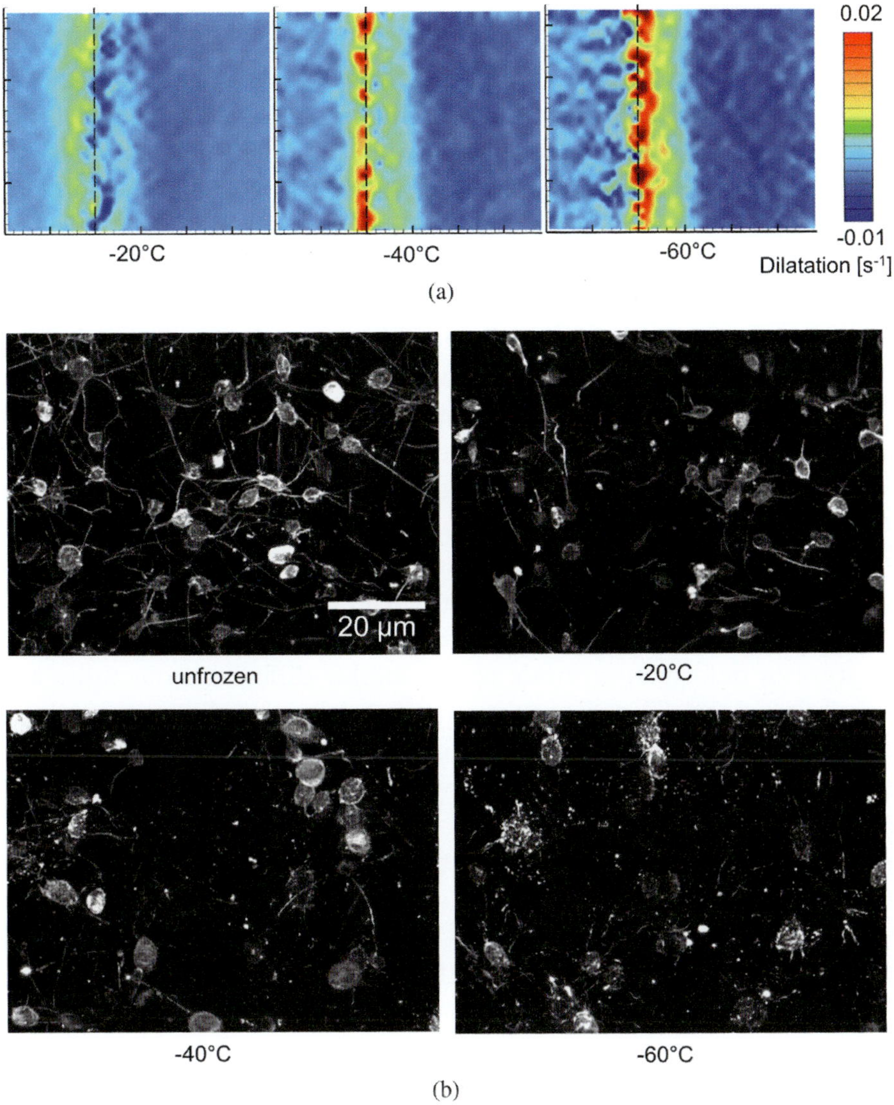

Figure 4. Effects of lowering the freezing temperature on (a) deformation (dilatation rate) in ECM, (b) cell morphology and actin structure. [Adapted from Teo *et al.*, 2011; Han *et al.*, 2013].

dendritic morphology with long extensions, but the cells in the frozen/thawed ETs lose the extensions and have circular morphology regardless of the freezing temperatures, though the effect is more prominent at lower freezing temperatures. Cryoprotectants can be used to preserve ECM and cell level structures. This issue is discussed later.

3.3. *Interaction of Cells with Fluid and ECM*

Although the significance of osmotic pressure driven cell dehydration has been extensively studied in cryobiology [Mazur, 1984], the mechical role of cells in this freezing-induced interaction in the context of tissue has only been recently acknowledged. Seawright *et al.* [2013] reported that the osmotic pressure driven water transport might be significant in native tissue with extremely high cell density, but that was insignificant at the cell density relevant to engineered tissues (i.e. low cell density). However, increased mechanical resistance via cell-matrix adhesion became signicant to determine post-thaw ECM microstructure. This warranted the measurement of cell level deformation in a spatiotemporal fashion to understand the detailed structural damage in and near individual cells within the matrix.

A new method to measure spatiotemporal deformation inside cells during the freezing process was developed by Ghosh *et al.* [2014], based on particle tracking named as particle tracking deformetry (PTD) as explained in Figure 5. The intracellular deformation can be presented in multiple ways including stretch ratio, strain, and stretch invariants (I_1). It was found that the nature of deformation inside cells is starkly different from those at the extracellular space. The deformation is spatially very heterogeneous and varies significantly over only a few micron length. In some instances, cell detachment from the ECM was observed, characterized by rapidly increased deformation over a very short period of time. The deformation in ECM is mostly aligned with freeze front direction but intracellular deformation was significantly multidirectional. Further studies are required to understand the role of cell level poroelastic parameters (cell stiffness and porosity) along with status of cell ECM adhesion complex that may decide F/T induced cell-level deformation, cytoskeletal structure, and cell viability.

4. Post-thaw Functionality of ETs

4.1. *F/T Induced Changes in Functional Properties*

The freezing-induced changes of mechanical, thermal, chemical, optical and transport properties of various tissues have been studied. It was reported that the lower freezing temperature had a negative impact on mechanical properties and cell viability of tissue equivalents [Devireddy *et al.*, 2003]. In contrast, freezing resulted in a significant increase in the average tensile modulus of the porcine femoral artery [Venkatasubramanian *et al.*, 2006]. To date, various types of tissues such as skin, cartilage, ligament, tendon, or bone tissue have been studied to characterize the effect of F/T on their

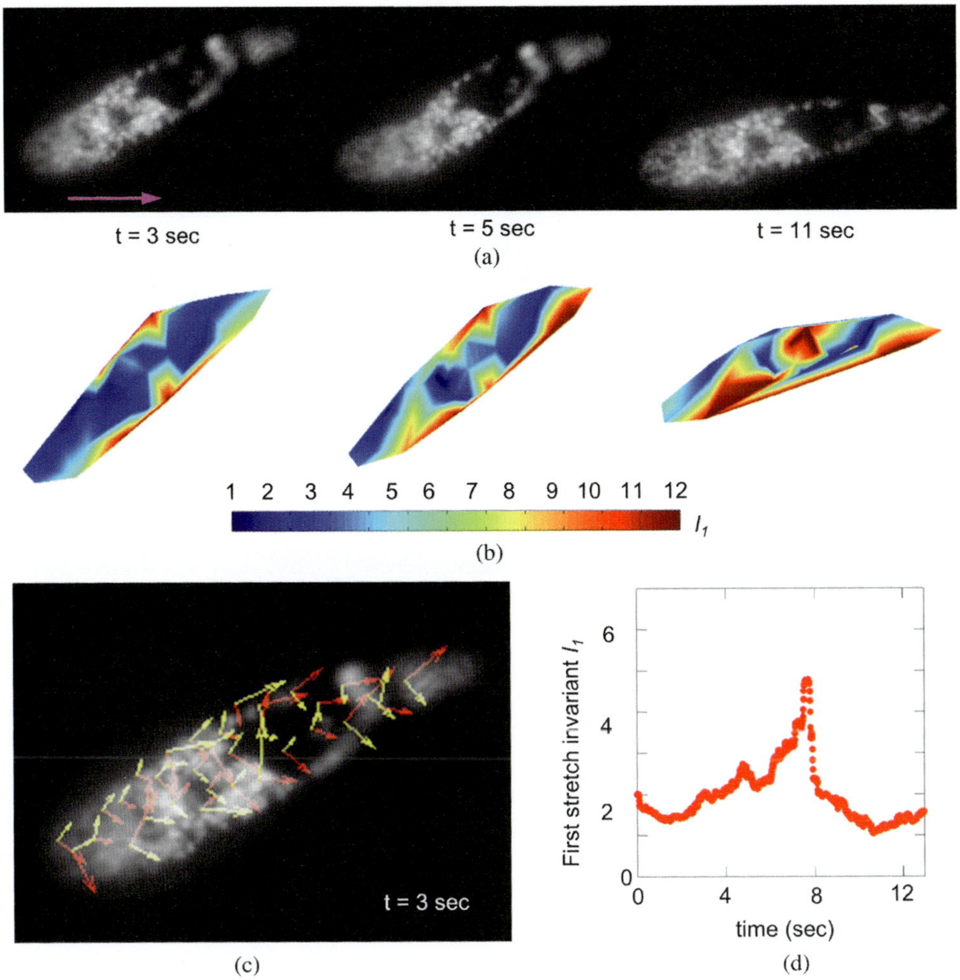

Figure 5. Particle Tracking Deformetry (PTD) method to measure spatiotemporal deformation in cells. (a) Fluorescence micrographs of a cell, arrow shows freeze front propagation direction. (b) Deformation surface plots. (c) Inhomogeneity of deformation directions. (d) Detachment of cell from ECM, quantified by sudden deformation change. [Adapted from Ghosh *et al.*, 2014].

functional properties. Foutz *et al.* [1992] found that the rat skin tissue after freezing had increased fracture strength, but not ultimate strength, or loading response. Szarko *et al.* [2010] investigated the effect of F/T on the mechanical properties (i.e. stiffness) of articular cartilage. They showed that the stiffness did not differ between unfrozen controls and samples frozen to –20°C and –80°C. Some of the studies used ligaments [Viidik and Lewin, 1966; Noyes and Grood, 1976; Dorlot *et al.*, 1980] and tendons [Matthews and Ellis, 1968; Leitschuh *et al.*, 1996; Clavert *et al.*, 2001] to discover that just a single cycle of F/T does not influence the mechanical properties. Subsequently, Moon *et al.* [2006] investigated the effects of refreezing on the viscoelastic and tensile properties of ligaments. However, the results also indicated that even refreezing of the specimens had little effect on the mechanical (i.e. tangent modulus and ultimate tensile strength) and structural properties (i.e. stiffness). As for bone tissues, several studies have shown no difference in mechanical properties after freezing [Sedlin, 1965], while some studies have found significant differences after freezing [Flynn *et al.*, 1990; Rozen *et al.*, 2009].

The effects of F/T on the ECM microstructure and consequent viscoelastic properties of dermal equivalent are systematically investigated in a recent study as illustrated in Figure 6 [Park *et al.*, 2015a]. In this study, creep and relaxation responses of ETs after F/T were measured and analyzed to determine the lumped viscoelastic properties of the ETs, based on the

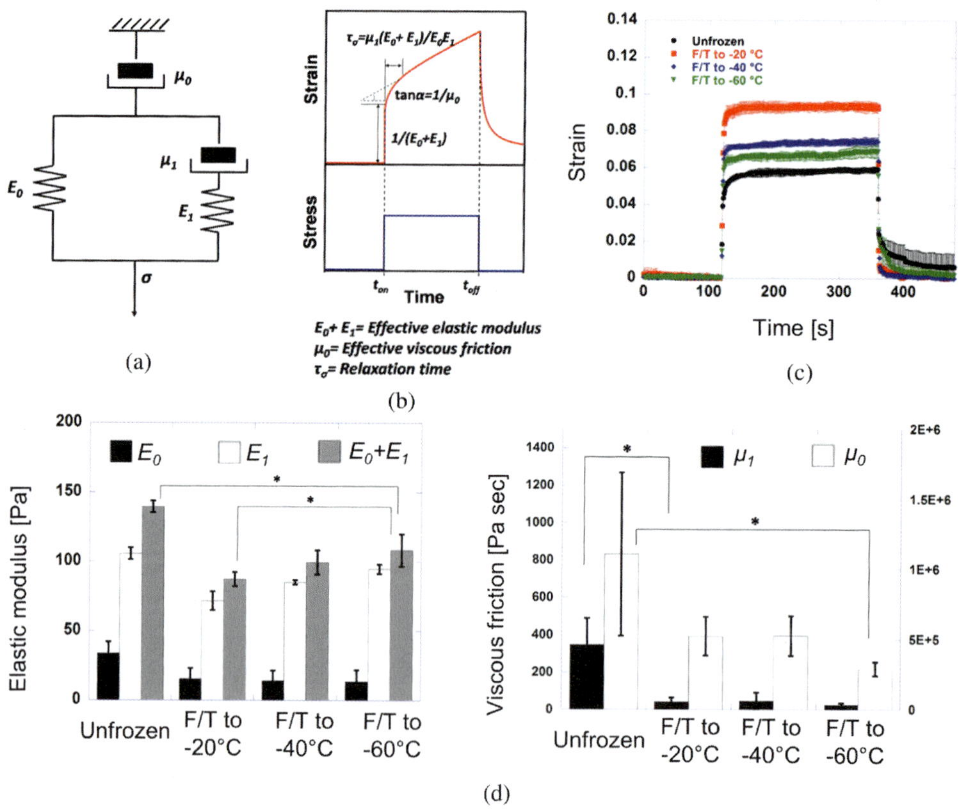

Figure 6. (a) Kelvin–Voigt four-element model consisting of springs (E_0, E_1) and dashpots (μ_0, μ_1). (b) Creep-relaxation responses of a typical viscoelastic material. (c) Creep-relaxation responses of the unfrozen and frozen ETs. (d) F/T induced changes of functional properties: elastic modulus in compression and viscous friction. [Adapted from Park *et al.*, 2015a].

Kelvin–Voigt four-element model as shown in Figures 6(a) and (b) [Bausch *et al.*, 1998]. Figure 6(c) shows representative creep-relaxation responses of unfrozen and frozen ETs. All groups show a typical viscoelastic behavior, but the frozen groups undergo a higher deformation. This is thought to be due to the fact that freezing-induced ice formation damages the integrity of ECM including cell–matrix adhesion. Noticeably, the lower the freezing temperature, the smaller the deformation corresponding to the applied force and the lower the loss in compressive modulus. This is because the faster freezing with the lower freezing temperature (e.g. –60°C) can mitigate tissue dehydration. In other words, water might not have enough time to squeeze out of the tissue at the fast freezing rate. Therefore, the tissue might have more fluid post thaw, which prevents the loss of compressive modulus. In contrast, the viscous properties significantly decreased after F/T but no notable differences were observed due to the freezing temperatures. It should be noted that the initial mechanical properties of the ECM is most likely to determine the extent of freezing-induced deformation by regulating interstitial water transport through hydraulic conductivity and transmitting mechanical stress via elastic modulus. The interplay of these eventually determines the post-thaw porosity of the ECM microstructure.

4.2. *Minimizing Structural/Functional Changes Using CPAs*

Cryopreservation protocols typically employ chemical additives (CPAs) to minimize the freezing-induced changes of cells and tissues [Poirot *et al.*, 2002]. It has been known that CPAs decrease the water/ice phase change temperature, and subsequently reduce the probability of intracellular ice formation. In addition, they stabilize cellular proteins and lipids and minimize the effects of elevated electrolyte concentrations [Fuller, 2004]. Currently many studies have sought to find an appropriate CPA type [Shah, 1983] and concentration [Liseth *et al.*, 2005] for preservation of tissue and cell functionality. Besides CPA itself, various parameters such as a freezing temperature [Borel Rinkes *et al.*, 1992], cooling/thawing rate [Henry *et al.*, 1993; Gook *et al.*, 2008], cryopreservation method (e.g., vitrification [Baicu *et al.*, 2008] and super cooling [Berendsen *et al.*, 2014]) have been considered to reduce the freezing-induced property or structure change. Despite these efforts, it is still very time-consuming to optimize the effective CPA concentration, composition, and corresponding F/T protocols.

However, the effects on CPAs on the freezing-induced cell-fluid-matrix interaction were studied recently using DMSO as a model CPA [Park *et al.*, 2015a]. It was found that there was a threshold DMSO concentration to achieve cryoprotection during the cell-fluid-matrix interaction. The threshold values are dependent on the freezing temperature (e.g. 2M for –20°C and 3M for –40°C for a given engineered tissue) (Figure 7). Cryoprotection is thought to be caused by the reduction of freezing-induced deformation as illustrated in Figure 7(c), and subsequent preservation of cytoskeletal and ECM structures as shown in Figure 7(a) and (b). These structural

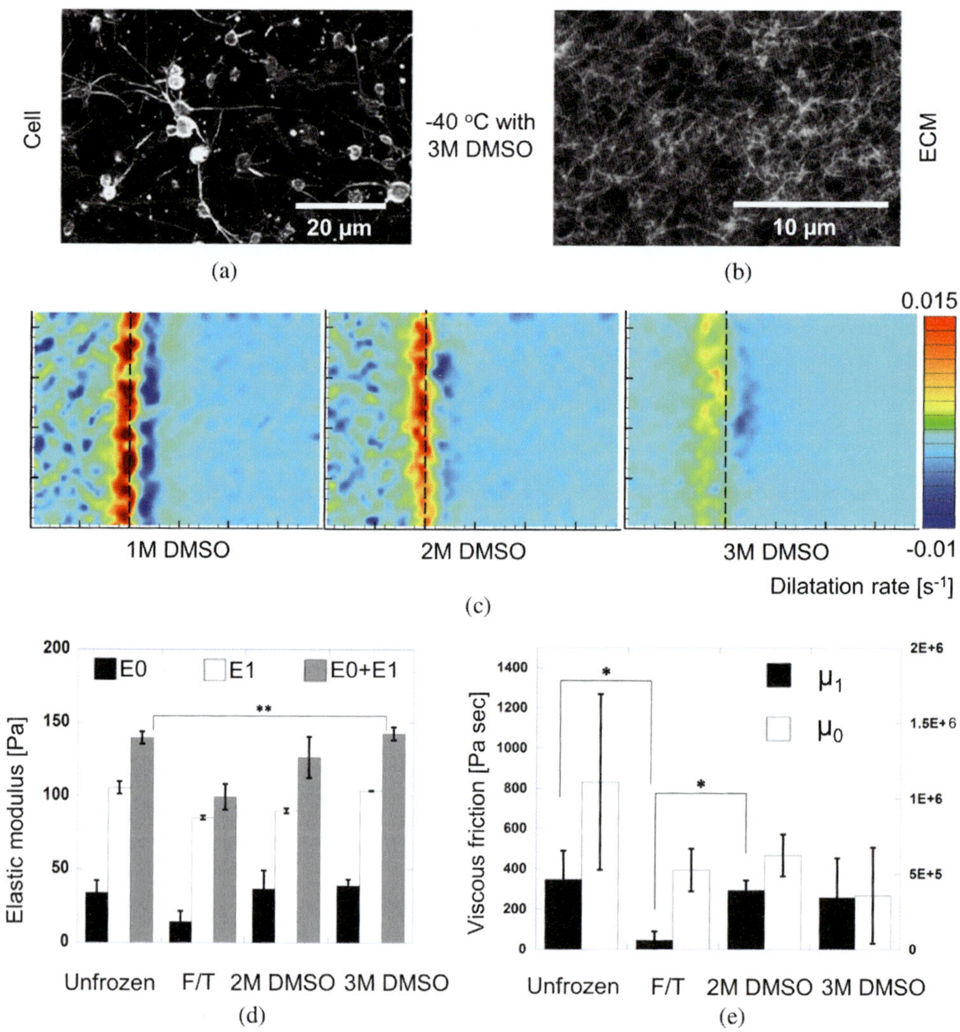

Figure 7. Effects of CPA on (a) preservation of cell and cytoskeletal structure (b) preservation of ECM structure. (c) Threshold effect of CPA concentration on ECM deformation (all cases freezing at –40°C). (d) and (e) Retention of functional properties, elastic modulus in compression and viscous friction by using CPA (all cases freezing at –40°C). [Adapted from Han *et al.*, 2013, Park *et al.*, 2015a].

preservations lead to improved preservation of compressive moduli. These results suggest that for a given freezing temperature, CPA only requires the corresponding threshold concentration to preserve tissue functionality. Additional CPA may not significantly contribute to the preservation of tissue functionality.

4.3. *Effect of Actin Cytoskeletal Structure on the Post-thaw Property of ETs*

Since cells play active roles in the freezing-induced cell-fluid-matrix interaction by providing mechanical support to the matrix via cell-matrix binding, it is plausible to hypothesize that strengthening cytoskeleton may further improve the cryopreservation outcome by providing additional mechanical support. This may provide a new alterative way to minimize the use of CPA for reduction of toxicity. It has been known that actin cytoskeletal structures of fibroblasts vary based on the physical state of the matrix as shown in Figure 8. Fibroblasts attached on glass slides or collagen matrices develop

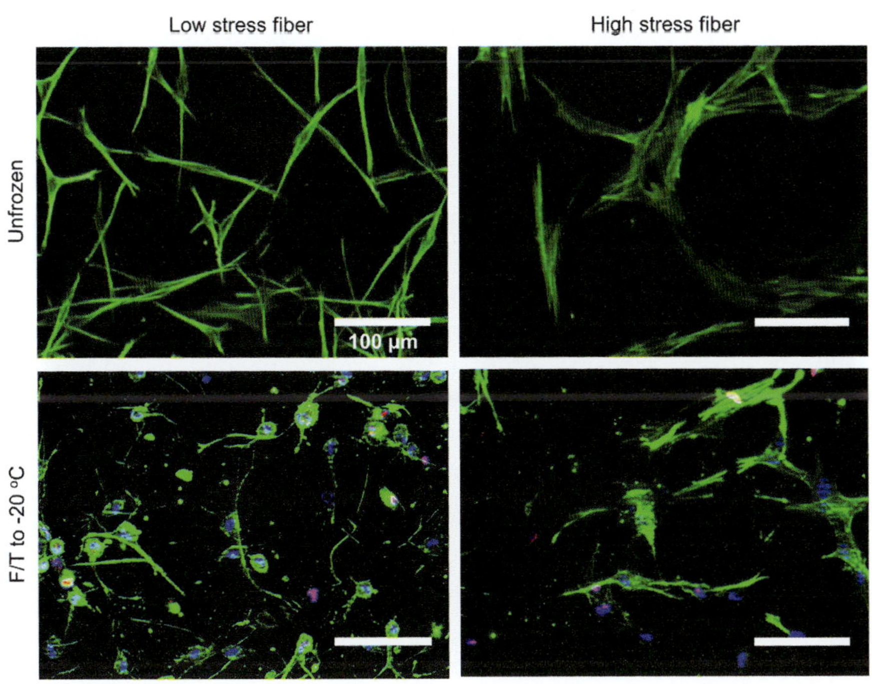

Figure 8. Effects of cytoskeletal structure on ECM preservation. Low stress fiber and high stress fiber formation in cells and resulting F/T induced changes in actin structure. [Adapted from Park *et al.*, 2015].

strong cytoskeleton, so called stress fibers, with thick and condensed actins [Fringer and Grinnell, 2001; Rhee and Grinnell, 2007; Grinnell and Petroll, 2010; Park *et al.*, 2015a]. Based on this knowledge, the effects of the cytoskeleton structures on the freezing-induced interactions were recently studied [Park *et al.*, 2015a]. It was found that the group of ETs with high stress fiber showed less cytoskeleton damage, compared to that of low stress fiber when both groups were exposed to the same freezing condition to –20°C. However, as the freezing temperature was lowered to –40°C and –60°C, actin structure of both groups were severely damaged. This suggests that the protective effect of the stress fiber can mitigate the freezing-induced deformation below certain thresholds. If the deformation is larger than the thresholds, the cells eventually undergo unrecoverable damage [Teo *et al.*, 2011]. Therefore, even though apparent actin structure preservation in the high stress fiber at –20°C was observed, this preservation may no longer be effective at lower temperatures (–40°C and –60°C).

5. Conclusion

Long-term preservation of functional tissues is one of the most actively pursued goals of TE/RM. Success of non-freezing methods by vitrification [Rahimi *et al.*, 2004; Isachenko *et al.*, 2009], desiccation [Acker *et al.*, 2002; Bhowmick *et al.*, 2003] and hypothermic preservation [Berendsen *et al.*, 2014] has been reported, but cryopreservation is still being considered as a leading candidate for tissue preservation due to its advantages in storage duration and temperature, convenience of protocols, and wide availability of freezing equipment. However, in order to further the application of the method, its tissue-specific outcome and limitation preserving in the functional properties of tissues should be addressed. Understanding of the biophysical interactions during freezing of tissue is greatly needed to address these challenges. In addition, understanding the effects of long-term storage at freezing temperature, which is currently lacking, is also crucial. Thus, we reported that freezing-induced cell-fluid-matrix interaction can explain and may predict post-thaw ECM structural changes as well as functional properties. Proper control and possible modulation of this interaction may provide a new avenue to exploit for cryopreservation of functional engineered tissues.

Acknowledgements

This work was supported by NIH R01 EB008388, NSF CBET-1009465 and an OVPR Incentive Grant from Purdue University. Bumsoo Han was also supported by the B.S.F. Schaefer Award and Discovery Park Fellowship from Purdue University.

References

Abazari, A., N. M. Jomha, J. A. W. Elliot and L. E. McGann (2013). Cryopreservation of articular cartilage. *Cryobiogy*, 66(3), pp. 201–209.

Acker, J. P., A. Fowler, B. Lauman, S. Cheley and M. Toner (2002). Survival of desiccated mammalian cells: beneficial effects of isotonic media. *Cell Preservation Technology*, 1(2), pp. 129–140.

Archer, R. and D. J. Williams (2005). Why tissue engineering needs process engineering? *Nature Biotechnology*, 23, pp. 1353–1355.

Archie, G. E. (1942). The electrical resistivity log as an aid in determining some reservoir characteristics. *Transactions of the American Institute of Mining and Metallurgical Engineers*, 146, pp. 54–61.

Baicu, S., M. J. Taylor, Z. Chen and Y. Rabin (2008). Cryopreservation of carotid artery segments via vitrification subject to marginal thermal conditions: Correlation of freezing visualization with functional recovery. *Cryobiology*, 57(1), pp. 1–8.

Balasubramanian, S. K., W. F. Wolkers and J. C. Bischof (2009), Membrane hydration correlates to cellular biophysics during freezing in mammalian cells. *Biochimica et Biophysica Acta*, 1788(5), pp. 945–953.

Bausch, A.R., F. Ziemann, A. A. Boulbitch, K. Jacobson and F. Sackmann (1998). Local measurements of viscoelastic parameters of adherent cell surfaces by magnetic bead microrheometry. *Biophysical Journal*, 75, 2038–2049.

Berendsen, T. A., B. G. Bruinsma, C. F. Puts, N. Saeidi, O. B. Usta, B. E. Uygun, M. L. Izamis, M. Toner, M. L. Yarmush and K. Uygun (2014). Supercooling enables long-term transplantation survival following 4 days of liver preservation. *Nature Medicine*, 20(7), pp. 790–793.

Berthiaume, F., P. V. Moghe, M. Toner and M. L. Yarmush (1996). Effect of extracellular matrix topology on cell structure, function, and physiological responsiveness: hepatocytes cultured in a sandwich configuration. *The FASEB Journal*, 10, pp. 1471–1484.

Bhowmick, S., L. Zhu, L. McGinnis, J. Lawitts, B. D. Nath, M. Toner and J. Biggers (2003). Desiccation tolerance of spermatozoa dried at ambient temperature: production of fetal mice. *Biology of Reproduction*, 68(5), pp. 1779–1786.

Bischof, J. C. (2000). Quantitative measurement and prediction of biophysical response during freezing in tissues. *Annual Review of Biomedical Engineering*, 2, pp. 257–288.

Bischof, J. C., C. J. Hunt, B. Rubinsky, A. Burgess and D. E. Pegg (1990). Effects of cooling rate and glycerol concentration on the structure of the frozen kidney: assessment by cryoscanning electron microscopy. *Cryobiology*, 27(3), pp. 301–310.

Borel Rinkes, I. H., M. Toner, S. J. Sheeha, R. G. Tompkins and M. L. Yarmush (1992). Long-term functional recovery of hepatocytes after cryopreservation in a three-dimensional culture configuration. *Cell Transplant*, 1(4), pp. 281–292.

Borene, M. L., V. H. Barocas and A. Hubel (2004). Mechanical and cellular changes during compaction of a collagen-sponge-based corneal stromal equivalent. *Annals of Biomedical Engineering*, 32(2), pp. 274–283.

Brockbank, K. G., Z. Z. Chen and Y. C. Song (2010). Vitrification of porcine articular cartilage. *Cryobiology*, 60(2), pp. 217–221.

Brockbank, K. G., W. R. MacLellan, J. Xie, S. F. Hamm-Alvarez, Z. Z. Chen and K. Schenke-Layland (2008). Quantitative second harmonic generation imaging of cartilage damage. *Cell and Tissue Banking*, 9(4), pp. 299–307.

Bu, Z. and P. S. Russo (1994). Diffusion of dextran in aqueous (hydroxypropyl) cellulose. *Macromolecules*, 27(5), pp. 1187–1194.

Budiansky, B. and R. J. Oconnell (1976). Elastic moduli of a cracked solid. *International Journal of Solids and Structures*, 12(2), pp. 81–97.

Clague, D. S. and R. J. Phillips (1996). Hindered diffusion of spherical macromolecules through dilute fibrous media. *Physics of Fluids*, 8(7), pp. 1720–1731.

Clavert, P., J. F. Kempf, F. Bonnomet, P. Boutemy, L. Marcelin and J. L. Kahn (2001). Effects of freezing/thawing on the biomechanical properties of human tendons. *Surgical and Radiologic Anatomy*, 23(4), pp. 259–262.

Dainese, L., F. Barili, T. K. Topkara, F. H. Cheema, M. Formato, E. Aljaber, M. Fusari, B. Micheli, A. Guarino, P. Biglioli and G. Polvani (2006). Effect of cryopreservation techniques on aortic valve glycosaminoglycans. *Artificial Organs*, 30(4), pp. 259–264.

Davies, C. N. (1952). The separation of airborne dust and particle. *Proceedings of Institute of Mechanical Engineers*, pp. 185–123.

Devireddy, R. V., M. R. Neidert, J. C. Bischof and R. T. Tranquillo (2003). Cryopreservation of collagen-based tissue equivalents. I. Effect of freezing in the absence of cryoprotective agents. *Tissue Engineering*, 9(6), pp. 1089–1100.

Dorlot, J. M., M. Aitbasidi, G. M. Tremblay and G. Drouin (1980). Load elongation behavior of the canine anterior cruciate ligament. *Journal of Biomechanical Engineering-Transactions of the ASME*, 102(3), pp. 190–193.

Evans, M. C. and V. H. Barocas (2009). The modulus of fibroblast-populated collagen gels is not determined by final collagen and cell concentration: experiments and an inclusion-based model. *Journal of Biomechanical Engineering-Transactions of the ASME*, 131(10), pp. 101014.

Flynn, J., M. J. Rudert, E. Olson, M. Baratz and E. Hanley (1990). The effects of freezing or freeze-drying on the biomechanical properties of the canine intervertebral-disk. *Spine*, 15(6), pp. 567–570.

Foutz, T. L., E. A. Stone and C. F. Abrams (1992). Effects of freezing on mechanical properties of rat skin. *American Journal of Veterinary Research*, 53(5), pp. 788–792.

Frantz, C., K. M. Stewart and V. M. Weaver (2010). The extracellular matrix at a glance. *Journal of Cell Science*, 123(24), pp. 4195–4200.

Fringer, J. and F. Grinnell (2001). Fibroblast quiescence in floating or released collagen matrices. *Journal of Biological Chemistry*, 276(33), pp. 31047–31052.

Fuller, B. J. (2004). Cryoprotectants: the essential antifreezes to protect life in the frozen state. *CryoLetters*, 25(6), pp. 375–388.

Fuller, B. and S. Paynter (2004). Fundamentals of cryobiology in reproductive medicine. *Reproductive BioMedicine Online*, 9(6), pp. 680–691.

Gao, D. and J. K. Critser (2000). Mechanisms of cryoinjury in living cells. *Institute for Laboratory Animal Research Journal*, 41(4), pp. 187–196.

Gandolfi, F., A. Paffoni, E. P. Brambilla, S. Bonetti, T. A. Brevini and G. Ragni (2006). Efficiency of equilibrium cooling and vitrification procedures for the cryopreservation of ovarian tissue: comparative analysis between human and animal models. *Fertility and Sterility*, 85, pp. 1150–1156.

Gerson, C. J., S. Goldstein and A. E. Heacox (2009). Retained structural integrity of collagen and elastin within cryopreserved human heart valve tissue as detected by two-photon laser scanning confocal microscopy. *Cryobiology*, 59(2), pp. 171–179.

Ghosh, S., B. Han and J. C. Dutton (2014). Measurement of spatiotemporal intracellular deformation of cells adhered to collagen matrix during freezing of biomaterials. *Journal of Biomechanical Engineering-Transactions of the ASME*, 136(2), pp. 021025.

Gook, D. A., D. H. Edgar, J. Borg and M. Martic (2008). Detection of zona pellucida proteins during human folliculogenesis. *Human Reproduction*, 23(2), pp. 394–402.

Grinnell, F. (1994). Fibroblasts, myofibroblasts and wound contraction. *Journal of Cell Biology*, 124, pp. 401–404.

Grinnell, F. and W. M. Petroll (2010). Cell motility and mechanics in three-dimensional collagen matrices. *Annual Review of Cell and Developmental Biology*, 26, pp. 335–361.

Gustavo, M., M. G. Andrade, C. N. Sa, A. M. Marchionni, T. C. dos Santos Calmon de Bittencourt and M. Sadigursky (2008). Effects of freezing on bone histological morphology. *Cell and Tissue Banking*, 9(4), pp. 279–287.

Hagiwara, M., J. H. Choi, R. V. Devireddy, K. P. Roberts, W. F. Wolkers, A. Makhlouf and J. C. Bischof (2009). Cell biophysics during freezing of rat and mouse sperm predicts post-thaw motility. *Biology of Reproduction*, 81(4), pp. 700–706.

Han, B. and J. C. Bischof (2004). Engineering challenges in tissue preservation. *Cell Preservation Technology*, 2, pp. 91–112.

Han, B., E. D. Grassl, V. H. Barocas, J. Coad and J. C. Bischof (2005). A cryoinjury model using engineered tissue equivalents for cryosurgical applications. *Annals of Biomedical Engineering*, 33(7), pp. 972–982.

Han, B., J. D. Miller and J. K. Jung (2009). Freezing-induced fluid-matrix interaction in poroelastic material. *Journal of Biomechanical Engineering-Transactions of the ASME*, 131(2), pp. 021002.

Han, B., K. Y. Teo, S. Ghosh, J. C. Dutton and F. Grinnell (2013). Thermomechanical analysis of freezing-induced cell-fluid-matrix interactions in engineered tissues. *Journal of Mechanical Behavior of Biomedical Materials*, 18, pp. 67–80.

Hashin, Z. (1988). Citation classic — the elastic moduli of heterogeneous materials. *Current Contents/Engineering Technology and Applied Sciences*, (39), pp. 14–14.

Hashin, Z. and S. Shtrikman (1963). A variational approach to the theory of the elastic behaviour of multiphase materials. *Journal of the Mechanics and Physics of Solids*, 11(2), pp. 127–140.

Henry, M. A., E. E. Noiles, D. Gao, P. Mazur and J. K. Critser (1993). Cryopreservation of human spermatozoa. IV. The effects of cooling rate and warming rate on the maintenance of motility, plasma membrane integrity, and mitochondrial function. *Fertility and Sterility*, 60(5), pp. 911–918.

Hosseini, S. A. and H. V. Tafreshi (2010). Modeling permeability of 3-D nanofiber media in slip flow regime. *Chemical Engineering Science*, 65(6), pp. 2249–2254.

Hunt, C. J. (2011). Cryopreservation of human stem cells for clinical application: a review. *Transfusion Medicine and Homotherapy*, 38(2), pp. 107–123.

Isachenko, V., E. Isachenko, J. Reinsberg, M. Montag, F. Braun and H. van der Ven (2008). Cryopreservation of human ovarian tissue: effect of spontaneous and initiated ice formation. *Reproductive Biomedicine Online*, 16(3), pp. 336–345.

Isachenko, V., I. Lapidus, E. Isachenko, A. Krivokharchenko, R. Kreienberg, M. Woriedh, M. Bader and J. Weiss (2009). Human ovarian tissue vitrification versus conventional freezing: morphological, endocrinological, and molecular biological evaluation. *Reproduction*, 138(2), pp. 319–327.

Jackson, G. W. and D. F. James (1986). The Permeability of fibrous porous media. *Canadian Journal of Chemical Engineering*, 64(3), pp. 364–374.

Karlsson, J. O. M. (2015). Measurement of intracellular ice formation by high-speed video cryomicroscopy. *Cryopreservation and Freeze-Drying Protocols: Methods in Molecular Biology*, 1257, pp. 181–227.

Keros, V., S. Xella, K. Hultenby, K. Pettersson, M. Sheikhi, A. Volpe, J. Hreinsson and O. Hovatta (2009). Vitrification versus controlled-rate freezing in cryopreservation of human ovarian tissue. *Human Reproduction*, 24(7), pp. 1670–1683.

Kim, G., H. Kim, J. Kim, G. Lee, E. Lee, J. Ahn, J. Park and J. Lim (2011). Effectiveness of slow freezing and vitrification for long-term preservation of mouse ovarian tissue. *Theriogenology*, 75(6), pp. 1045–1051.

Kosto, K. B. and W. M. Deen (2004). Diffusivities of macromolecules in composite hydrogels. *AIChE Journal*, 50(11), pp. 2648–2658.

Kovacik, J. (1999). Correlation between Young's modulus and porosity in porous materials. *Journal of Materials Science Letters*, 18(13), pp. 1007–1010.

Kovacik, J. (2001). Correlation between shear modulus and porosity in porous materials. *Journal of Materials Science Letters*, 20(21), pp. 1953–1955.

Kovacik, J. (2008). Correlation between elastic modulus, shear modulus, Poisson's ratio and porosity in porous materials. *Advanced Engineering Materials*, 10(3), pp. 250–252.

Kubo, K. and Y. Kuroyanagi (2005). The possibility of long-term cryopreservation of cultured dermal substitute. *Artificial Organs*, 29(10), pp. 800–805.

Laouar, L., K. Fishbein, L. E. McGann, W. E. Horton, R. G. Spencer and N. M. Jomha (2007). Cryopreservation of porcine articular cartilage: MRI and biochemical results after different freezing protocols. *Cryobiology*, 54(1), pp. 36–43.

Lee, S. L. and J. H. Yang (1997). Modeling of Darcy-Forchheimer drag for fluid flow across a bank of circular cylinders. *International Journal of Heat and Mass Transfer*, 40(13), pp. 3149–3155.

Leitschuh, P. H., T. J. Doherty, D. C. Taylor, D. E. Brooks and J. B. Ryan (1996). Effects of postmortem freezing on tensile failure properties of rabbit extensor digitorum longus muscle tendon complex. *Journal of Orthopaedic Research*, 14(5), pp. 830–833.

Levick, J. R. (1987). Flow through interstitium and other fibrous matrices. *Quarterly Journal of Experimental Physiology and Cognate Medical Sciences*, 72(4), pp. 409–438.

Liseth, K., J. F. Abrahamsen, S. Bjorsvik, K. Grottebo and O. Bruserud (2005). The viability of cryopreserved PBPC depends on the DMSO concentration and the concentration of nucleated cells in the graft. *Cytotherapy*, 7(4), pp. 328–333.

Mackenzie, J. K. (1950). The elastic constants of a solid containing spherical holes. *Proceedings of the Physical Society of London Section B*, 63(361), pp. 2–11.

Martin, R. B. and D. L. Boardman (1993). The effects of collagen fiber orientation, porosity, density and mineralization on bovine cortical bone bending properties. *Journal of Biomechanics*, 26(9), pp. 1047–1054.

Matthews, L. S. and D. Ellis (1968). Viscoelastic properties of cat tendon: effects of time after death and preservation by freezing. *Journal of Biomechanics*, 1(2), pp. 65–71.

Mazur, P. (1984). Freezing of living cells: mechanisms and implications. *American Journal of Physiology*, 247(3 Part 1), pp. C125–142.

Meredith, J. E., B. Fazeli and M. A. Schwartz (1993). The extracellular matrix as a cell survival factor. *Molecular Biology of the Cell*, 4(September), pp. 953–961.

Mikos, A. G., S. W. Herring, P. Ochareon, J. Ellisseeff, H. H. Lu, R. Kandel, F. J. Schoen, M. Toner, D. Mooney, A. Atala, M. E. Van Dyke, D. Kaplan and G. Vunjak-Novakovic (2006). Engineering complex tissues. *Tissue Engineering*, 12, pp. 3307–3339.

Moon, D. K., S. L. Y. Woo, Y. Takakura, M. T. Gabriel and S. D. Abramowitch (2006). The effects of refreezing on the viscoelastic and tensile properties of ligaments. *Journal of Biomechanics*, 39(6), pp. 1153–1157.

Morato, R., D. Izquierdo, J. L. Albarracin, B. Anguita, M. J. Palomo, A. R. Jimenez-Macedo, M. T. Paramio and T. Mogas (2008). Effects of pre-treating in vitro-matured bovine oocytes with the cytoskeleton stabilizing agent taxol prior to vitrification. *Molecular Reproduction and Development*, 75(1), pp. 191–201.

Moroni, L., J. R. de Wijn and C. A. van Blitterswijk (2006). 3D fiber-deposited scaffolds for tissue engineering: Influence of pores geometry and architecture on dynamic mechanical properties. *Biomaterials*, 27(7), pp. 974–985.

Narine, K., E. Claeys, M. Cornelissen, F. Desomer, H. Beele, L. Vanlangenhove, S. D. Smet and G. V. Nooten (2006). Readily available porcine aortic valve matrices for use in tissue valve engineering. Is cryopreservation an option? *Cryobiology*, 53(2), pp. 169–181.

Nerem, R. M. (2006). Tissue engineering: the hope, the hype, and the future. *Tissue Engineering*, 12, pp. 1143–1150.

Nicholson, C. and L. Tao (1993). Hindered diffusion of high molecular weight compounds in brain extracellular microenvironment measured with integrative optical imaging. *Biophysical Journal*, 65(6), pp. 2277–2290.

Noyes, F. R. and E. S. Grood (1976). The strength of the anterior cruciate ligament in humans and rhesus monkeys. *Journal of Bone and Joint Surgery*, 58(8), pp. 1074–1082.

O'Brien, F. J., B. A. Harley, M. A. Waller, I. V. Yannas, L. J. Gibson and P. J. Prendergast (2007). The effect of pore size on permeability and cell attachment in collagen scaffolds for tissue engineering. *Technology and Health Care*, 15(1), pp. 3–17.

Ogston, A. G., B. N. Preston, J. D. Wells, A. G. Ogston, B. N. Preston, J. M. Snowden and J. D. Wells (1973). Transport of compact particles through solutions of chain polymers. *Proceedings of the Royal Society of London Series A — Mathematical Physical and Engineering Sciences*, 333(1594), pp. 297–316.

Ozcelikkale, A., Y. Li, X. Xu and B. Han (2013). Effects of freezing on collagen nanoscale structure in engineered tissues. *ASME 2013 2nd Global Congress on NanoEngineering for Medicine and Biology*, American Society of Mechanical Engineers, Boston.

Pancrazio, J. J., F. Wang and C. A. Kelley (2007). Enabling tools for tissue engineering. *Biosensors and Bioelectronics*, 22, pp. 2803–2811.

Park, S., A. Seawright, S. Park, J. C. Dutton, F. Grinnell and B. Han (2015a). Preservation of tissue microstructure and functionality during freezing by modulation of cytoskeletal structure. *Journal of the Mechanical Behavior of Biomedical Materials*, 45C, pp. 32–44.

Park, S., C. Whittington, S. L. Voytik-Harbin and B. Han (2015b). Microstructural parameter-based modeling for transport properties of collagen matrices. *Journal of Biomechanical Engineering — Transactions of the ASME*, 137(6), pp. 061003 (1–9).

Pedersen, J. A., S. Lichter and M. A. Swartz (2010). Cells in 3D matrices under interstitial flow: Effects of extracellular matrix alignment on cell shear stress and drag forces. *Journal of Biomechanics*, 43(5), pp. 900–905.

Pegg, D. E., L. Wang and D. Vaughan (2006a). Cryopreservation of articular cartilage. Part 3: the liquidus-tracking method. *Cryobiology*, 52(3), pp. 360–368.

Pegg, D. E., L. Wang, D. Vaughan and C. J. Hunt (2006b). Cryopreservation of articular cartilage. Part 2: mechanisms of cryoinjury. *Cryobiology*, 52(3), pp. 347–359.

Pegg, D. E., M. C. Wusteman and L. Wang (2006c). Cryopreservation of articular cartilage. Part 1: conventional cryopreservation methods. *Cryobiology*, 52(3), pp. 335–346.

Petroll, W. M. and L. Ma (2003). Direct, dynamic assessment of cell-matrix interactions inside fibrillar collagen lattices. *Cell Motility and the Cytoskeleton*, 55, pp. 254–264.

Phillips, R. J. (2000). A hydrodynamic model for hindered diffusion of proteins and micelles in hydrogels. *Biophysical Journal*, 79(6), pp. 3350–3353.

Poirot, C., M. C. Vacher-Lavenu, P. Helardot, J. Guibert, L. Brugieres and P. Jouannet (2002). Human ovarian tissue cryopreservation: indications and feasibility. *Human Reproduction*, 17(6), pp. 1447–1452.

Rahimi, G., E. Isachenko, V. Isachenko, H. Sauer, M. Wartenberg, S. Tawadros, J. Hescheler, P. Mallmann and F. Nawroth (2004). Comparison of necrosis in human ovarian tissue after conventional slow freezing or vitrification and transplantation in ovariectomized SCID mice. *Reproductive Biomedicine Online*, 9(2), pp. 187–193.

Rhee, S. and F. Grinnell (2007). Fibroblast mechanics in 3D collagen matrices. *Advanced Drug Delivery Reviews*, 59(13), pp. 1299–1305.

Roeder, B. A., K. Kokini, J. E. Sturgis, J. P. Robinson and S. L. Voytik-Harbin (2002). Tensile mechanical properties of three-dimensional type I collagen extracellular matrices with varied microstructure. *Journal of Biomechanical Engineering — Transactions of the ASME*, 124(2), pp. 214–222.

Rozen, B., T. Brosh, M. Salai, A. Herman and I. Dudkiewicz (2009). The effects of prolonged deep freezing on the biomechanical properties of osteochondral allografts. *Cell and Tissue Banking*, 10(1), pp. 27–31.

Sahraoui, M. and M. Kaviany (1992). Slip and no-slip velocity boundary conditions at interface of porous plain media. *International Journal of Heat and Mass Transfer*, 35(4), pp. 927–943.

Scheidegger, A. E. (1974). *The Physics of Flow through Porous Media*. (University of Toronto Press, Toronto, Buffalo, N.Y.).

Schenke-Layland, K., J. Xie, S. Heydarkhan-Hagvall, S. F. Hamm-Alvarez, U. A. Stock, K. G. Brockbank and W. R. MacLellan (2007). Optimized preservation of extracellular matrix in cardiac tissue: implications for long-term graft durability. *Annals of Thoracic Surgery*, 83, pp. 1641–1650.

Seawright, A., A. Ozcelikkale, J. C. Dutton and B. Han (2013). Role of cells in freezing-induced cell-fluid-matrix interactions within engineered tissues. *Journal of Biomechanical Engineering — Transactions of the ASME*, 135(9), pp. 91001–91012.

Sedlin, E. D. (1965). A rheologic model for cortical bone — a study of physical properties of human femoral samples. *Acta Orthopaedica Scandinavica*, S, pp. 5–12.

Serrato, D., R. Nieto-Aguilar, I. Garzon, O. Roda, A. Campos and M. Alaminos (2009). Comparison of the effect of cryopreservation protocols on the histology of bioengineered tissues. *Histology and Histopathology*, 24(12), pp. 1531–1540.

Shah, S. A. (1983). Comparison of DMSO and glycerol as cryoprotectants for ascites tumor-cells. *Experientia*, 39(3), pp. 327–328.

Spielman, L. and S. L. Goren (1968). Model for predicting pressure drop and filtration efficiency in fibrous media. *Environmental Science and Technology*, 2(4), pp. 279–287.

Stylianopoulos, T., C. A. Bashur, A. S. Goldstein, S. A. Guelcher and V. H. Barocas (2008a). Computational predictions of the tensile properties of electrospun fibre meshes: Effect of fibre diameter and fibre orientation. *Journal of the Mechanical Behavior of Biomedical Materials*, 1(4), pp. 326–335.

Stylianopoulos, T., A. Yeckel, J. J. Derby, X. J. Luo, M. S. Shephard, E. A. Sander and V. H. Barocas (2008b). Permeability calculations in three-dimensional isotropic and oriented fiber networks. *Physics of Fluids*, 20(12).

Swartz, M. A. and M. E. Fleury (2007). Interstitial flow and its effects in soft tissues. *Annual Review of Biomedical Engineering*, 9, pp. 229–256.

Swartz, M. A., A. Kaipainen, P. A. Netti, C. Brekken, Y. Boucher, A. J. Grodzinsky and R. K. Jain (1999). Mechanics of interstitial-lymphatic fluid transport: theoretical foundation and experimental validation. *Journal of Biomechanics*, 32(12), pp. 1297–1307.

Szarko, M., K. Muldrew and J. E. Bertram (2010). Freeze-thaw treatment effects on the dynamic mechanical properties of articular cartilage. *BMC Musculoskeletal Disorders*, 11(1), p. 231.

Tamayol, A and M. Bahrami (2010). Transverse permeabilty of fibrous porous media. *Proceedings (CD) of the 3rd International Conference on Porous Media and its Applications in Science and Engineering ICPM3*, Montecatini, Italy.

Teo, K. Y., T. O. DeHoyos, J. C. Dutton, F. Grinnell and B. Han (2011). Effects of freezing-induced cell-fluid-matrix interactions on the cells and extracellular matrix of engineered tissues. *Biomaterials*, 32(23), pp. 5380–5390.

Teo, K. Y., J. C. Dutton and B. Han (2010). Spatiotemporal measurement of freezing-induced deformation of engineered tissues. *Journal of Biomechanical Engineering — Transactions of the ASME*, 132(3), pp. 031003.

Thorne, R. G. and C. Nicholson (2006). *In vivo* diffusion analysis with quantum dots and dextrans predicts the width of brain extracellular space. *Proceedings of the National Academy of Sciences of the United States of America*, 103(14), pp. 5567–5572.

Tomadakis, M. M. and T. J. Robertson (2005). Viscous permeability of random fiber structures: Comparison of electrical and diffusional estimates with experimental and analytical results. *Journal of Composite Materials*, 39(2), pp. 163–188.

Tranquillo, R. (1998). Self-organization of tissue-equivalents: the nature and role of contact guidance. *Biochemical Society Symposium*.

Venkatasubramanian, R. T., E. D. Grassl, V. H. Barocas, D. Lafontaine and J. C. Bischof (2006). Effects of freezing and cryopreservation on the mechanical properties of arteries. *Annals of Biomedical Engineering*, 34(5), pp. 823–832.

Venkatasubramanian, R. T., W. F. Wolkers, M. M. Shenoi, V. H. Barocas, D. Lafontaine, C. L. Soule, P. A. Iaizzo and J. C. Bischof (2010). Freeze-thaw induced biomechanical changes in arteries: role of collagen matrix and smooth muscle cells. *Annals of Biomedical Engineering*, 38(3), pp. 694–706.

Viidik, A. and T. Lewin (1966). Changes in tensile strength characteristics and histology of rabbit ligaments induced by different modes of postmortal storage. *Acta Orthopaedica Scandinavica*, 37(2), pp. 141–148.

Wakatsuki, T. and E. Elson (2003). Reciprocal interactions between cells and extracellular matrix during remodeling of tissue constructs. *Biophysical Chemistry*, 100(1–3), pp. 593–605.

Wang, L., D. E. Pegg, J. Lorrison, D. Vaughan and P. Rooney (2007a). Further work on the cryopreservation of articular cartilage with particular reference to the liquidus tracking (LT) method. *Cryobiology*, 55(2), pp. 138–147.

Wang, X., H. Chen, H. Yin, S. S. Kim, T. S. Lin and R. G. Gosden (2002). Fertility after intact ovary transplantation. *Nature*, 415(6870), pp. 385–385.

Wang, X., T. C. Hua, D. W. Sun, B. Liu, G. Yang and Y. Cao (2007b). Cryopreservation of tissue-engineered dermal replacement in Me2SO: toxicity study and effects of concentration and coolng rates on cell viability. *Cryobiology*, 55, pp. 60–65.

Whittington, C. F., E. Brandner, K. Y. Teo, B. Han, E. Nauman and S. L. Voytik-Harbin (2013a). Oligomers modulate interfibril branching and mass transport properties of collagen matrices. *Microscopy and Microanalysis*, 19(5), pp. 1323–1333.

Whittington, C. F., M. C. Yoder and S. L. Voytik-Harbin (2013b). Collagen-polymer guidance of vessel network formation and stabilization by endothelial colony forming cells in vitro. *Macromolecular Biosciences*, 13(9), pp. 1135–1149.

Zein, I., D. W. Hutmacher, K. C. Tan and S. H. Teoh (2002). Fused deposition modeling of novel scaffold architectures for tissue engineering applications. *Biomaterials*, 23(4), pp. 1169–1185.

Zhu, Y. K., T. Umino, X. D. Liu, H. J. Wang, D. J. Romberger, J. R. Spurzem and S. I. Rennard (2001). Contraction of fibroblast-containing collagen gels: initial collagen concentration regulates the degree of contraction and cell survival. *In Vitro Cellular and Developmental Biology—Animal*, 37(1), pp. 10–16.

Chapter 3

Strategies for Overcoming Transport Limitations of Convective Desiccation of Trehalose Solutions for Ambient Temperature Preservation of Biologics

Alexander N. Sinkevich, Sankha Bhowmick and Mehdi Raessi

Nomenclature

A	surface area
a_w	water activity
c	molar concentration
$C.A.$	contact angle
C_p	specific heat
d	cell diameter
D	diffusion coefficient
h	height
h_{fg}	latent heat of evaporation
h_m	mass transfer coefficient
h_t	heat transfer coefficient
k	thermal conductivity
L	characteristic length scale
Le	Lewis number
L_p	permeability coefficient
M	molecular weight
p	partial pressure

P	pressure
Pr	Prandtl number
R	universal gas constant
Ra	Rayleigh number
Re	Reynolds number
T	temperature
T_g	glass transition temperature
t	time
V	volume
V_b	dead volume
x	mole fraction
z	local height coordinate

Subscripts

0	initial condition or reference point
A	solvent (water)
B	solute (trehalose)
b	substrate base
g	glass
i	intracellular
j	dummy indice
k	dummy indice
∞	ambient condition
L	length-averaged
o	extracellular
sub	substrate
$surf$	film surface
w, int	solution-vapor interface

Greek Letters and Symbols

δ	thickness
η	dynamic viscosity
ω	mass fraction
Φ	relative humidity
ρ	density

1. Introduction

Biopreservation is the use of microorganisms as a way to preserve complex biologics such as proteins, lipids, liposomes, and mammalian cells. Since the field of reproductive technology is rapidly growing, there is a growing need for continual improvements in biopreservation techniques [Hubel and Waxberg, 2006; Finger, 2008; Copelan, 2006]. Probably the most well-known preservation technique is cryopreservation, in which the biologic in question is stored at very low temperatures and "preserved" until it is needed for a specific use at a later time. A large drawback to this traditionally used method is that cryogenic media such as liquid nitrogen is expensive and difficult to transport [He *et al.*, 2008]. This caveat could be avoided if there was no need for this media and the biologics could be preserved at room temperature. An alternative is lyopreservation, which attempts to achieve what cryopreservation has by mimicking *anhydrobiosis,* a phenomenon that occurs in nature. In nature there are certain organisms that can survive extreme dehydration (up to 99% water loss) for lengthy periods of time (days, weeks, months, even years) and regain metabolic activity within minutes upon rehydration [Crowe *et al.*, 1998; Crowe and Crowe 2000]. Examples of these organisms include yeast, plant seeds, and nematodes. These organisms achieve this extremely desiccated state by accumulating large quantities of non-reducing disaccharides, such as sucrose and trehalose, within their systems. There have been many studies on the properties of these individual disaccharides, and it is hypothesized that they offer protection by forming hydrogen bonds with proteins and lipids during the dehydration process [Crowe *et al.*, 1993; Leslie *et al.*, 1995; Wolkers *et al.*, 2001]. Trehalose has been shown to have superior capabilities regarding biostabilization [Crowe *et al.*, 1996; Puhler *et al.*, 2001; de Castro and Tunnacliffe, 2000; Chen *et al.*, 2001; Guo *et al.*, 2000; Wolkers *et al.*, 2002; Gordon *et al.*, 2001]. As an aqueous trehalose solution dries, an amorphous skin forms at the solution-vapor interface around room temperature (~22°C). This amorphous skin is so viscous that it is essentially glass, and therefore a biologic such as a mammalian cell could be stored underneath. In order for this glassy skin to be formed at or near room temperature, the trehalose solution must be dried to below 10 wt.% water. However, there are two potential problems: One is too much water becoming trapped underneath the glassy layer. This exposes the biologic to high molecular mobilities and concentration gradients, resulting in death. Therefore it is of paramount importance to be able to accurately predict the biothermodynamics and desiccation kinetics of aqueous trehalose solutions. The second potential problem is a "protrusion effect" where the glass will not form in time and the stored biologics will become exposed to the external air and die. This has been observed in a recent unpublished work [Oliveira, 2012]. The majority of mammalian cell desiccation studies revolve around studying post-desiccation viability of a population of cells in a micro-liter sessile droplet and subjected to a controlled desiccation protocol followed by immediate or post-storage rehydration [Chen *et al.*, 2006; Bhowmick *et al.*, 2003; Norris *et al.*, 2006; Elliot *et al.*, 2006; Sitaula *et al.*, 2009]. Due to the complex nature of the drying kinetics and biothermodynamics, it has been inferred that for successful long-term storage, biologics would have to be stored in a cooler environment postdesiccation. This defeats the goal of successful lyopreservation, which is to store cells above 0°C (ideally between 4–22°C). Therefore, a critical goal is achieving the formation of a stable glass for long-term storage stability.

Much information about the drying process can be ascertained by modeling the aqueous trehalose solution as a simple one-dimensional "thin film". Several studies have adopted this approach [Chen *et al.*, 2006; He *et al.*, 2008; Chakraborty *et al.*, 2011] and have stated the implications for lyopreservation techniques. Chen *et al.* [2006] studied the desiccation kinetics in 400 μm thin films both experimentally and numerically. However, their numerical model could only predict the desiccation kinetics accurately in a dilute solution. Deviations were observed in their model as the solution neared the glassy state since an appropriate mutual diffusion coefficient could not be implemented. He *et al.* [2008] numerically investigated the desiccation kinetics in thin films ranging from 1–1000 μm thickness. Their model incorporated the free volume model for vapor pressure and mutual diffusion coefficient. This much more accurately predicted the effects of glass transition on heat and mass transport processes, and the study serves as a strong springboard for further investigation into optimizing desiccation protocol. Chakraborty *et al.* [2011] propose a spin-drying technique for preservation of mammalian cells inside aqueous trehalose solutions. They argue that the spin-drying technique offers greater advantage by circumventing the problem of trapping excess water within the film. However, their model did not consider the final stages of drying, which of course is the vital stage where the diffusion coefficient within the film would be extremely high and the stored biologic would potentially be exposed to high osmotic gradients. Indeed, the cells in their experiments did not survive rehydration, which the authors attributed to incomplete glass formation within the cells. However, they could not directly measure this due to equipment limitations. While the aforementioned studies provide great insight to the desiccation physics of aqueous trehalose solutions which contributes to the optimization of desiccation protocol, the ultimate question still remains: can a biologic ultimately survive the procedure? Biologics, such as mammalian cells are complex in nature. Cells experience volume changes based on intra- and extracellular concentration gradients. As the volume of the aqueous trehalose solution decreases, the cells inside the solution will also experience a volume change due to the reduction of water in the film. A recent study has begun to take this effect into consideration [Vogl *et al.*, 2014] where the shrinkage of a vesicle was modeled using the level-set method. The authors discovered that membrane deformations are greatly affected by the viscosity and diffusivity of the solution.

Although numerous studies have been performed on convective desiccation, recovery of viable biologics has been limited. This has led to a repeated conjecture that a water transport limitation at the bottom of the film where cells are located prevents the formation of a uniform glassy layer [Sitaula *et al.*, 2009; Gordon *et al.*, 2001; Chakraborty *et al.*, 2011; Chen *et al.*, 2001]. It is critical to evaluate the microscale properties adjacent to a cell during the convective desiccation process. It is also important to realize that as the desiccation process is taking place near ambient temperatures, leaving a cell under a high concentration of solute before reaching a glassy state for too long leads to damaging chemical potential and irreversible cellular damage. Therefore in order to allow rapid formation of a uniform glassy layer around the cell, we shall explore alternative boundary conditions by which glass formation at the cellular level can be accelerated for both natural and forced convective cases. Finally, we shall look into alternative droplet profiles that may minimize water transport limitation and lead to a more uniform glass formation. It is important to note that for the current study it has been assumed that the preservation of biologics will be at room temperature corresponding to 298–300K. Therefore, the focus has been on reducing the moisture content in a water-trehalose solution to levels where stable glass can form at these temperatures. This corresponds to water mass fraction of approximately 0.09. This rationale has been used as the paradigm in this work.

2. Theory and Method

2.1. *Model Description, Governing Equations, and Boundary Conditions*

Two models were used in the current study. Both build upon the thin film model presented by He *et al.* [2008] which simulates an evaporating aqueous trehalose solution on a polystyrene support. Figure 1 shows the setup of the current study's numerical models.

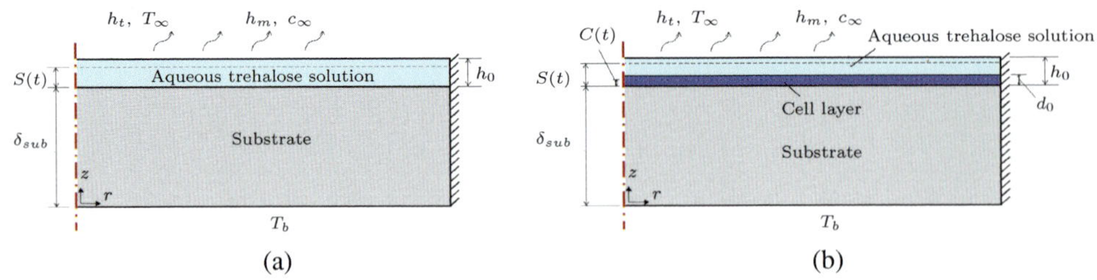

Figure 1. 2D-axisymmetric models used in the current study. The setup shown in (a) was used for the majority of experiments. The setup shown in (b) introduces the biological physics (mammalian cells being represented as a layer of cells). In both models, the right side wall is insulated and the substrate is subject to a temperature boundary condition (either constant or variable). The surface is exposed to convective heat and mass transfer conditions. The substrate thickness, δ_{sub}, was 1 mm.

Commercial finite element solver COMSOL Multiphysics (version 4.4) was used to model the coupled heat transfer physics as well as the thin film evaporation using a deformed geometry technique. Inside the film, the governing equations for heat and multicomponent mass transfer are solved:

$$\rho C_p \frac{\partial T}{\partial t} = k\nabla^2 T \tag{1}$$

$$\frac{\partial}{\partial t}(\rho\omega_j) = -\nabla \cdot \mathbf{j}_\mathrm{j}\,. \tag{2}$$

In Equation (2), ω_j is the mass fraction of the components (water and trehalose) and $\mathbf{j}_\mathrm{j}$ is the relative mass flux vector.

$$\mathbf{j}_\mathrm{j} = -\left(\rho D_j^m \nabla\omega_j + \rho\omega_j D_j^m \frac{\nabla M_n}{M_n}\right) \tag{3}$$

where D_j^m is the mixture-averaged diffusion coefficient, defined as such:

$$D_j^m \frac{1-\omega_j}{\sum_{k\neq j}^N \frac{x_k}{D_{jk}}} \tag{4}$$

where M_n is the mean molar mass of the components, defined in Equation (5), and x_k is the mole fraction of the components. D_{jk} is the multicomponent

Maxwell-Stefan diffusivities which ultimately make up the mutual diffusion coefficient described in Section 2.3.2.

$$M_n = \left(\sum_j \frac{\omega_j}{M_j} \right)^{-1}. \tag{5}$$

At the film surface, $z = S(t)$, the convective heat and mass conditions is applied as well as the evaporation-driven normal mesh velocity.

$$-k\frac{\partial T}{\partial z} = h_t(T - T_\infty) + \rho_A \frac{dS}{dt} h_{fg} \tag{6}$$

$$\rho D\left(\frac{\partial \omega_A}{\partial z} \omega_A \frac{\frac{\partial M_n}{\partial z}}{M_n} \right) = h_m M_A \left(c_{surf} - c_\infty \right) \tag{7}$$

$$\frac{dS}{dt} = -\frac{h_m M_A \left(c_{surf} - c_\infty \right)}{\rho_A}. \tag{8}$$

The conduction within the film is balanced with convective heat transfer and cooling effect due to evaporation, as seen in Equation (6). $\frac{dS}{dt}$ is the change in film height over time and h_{fg} is the latent heat of evaporation, represented by a first-order polynomial [Adams, 2007].

$$h_{fg} = h_1 T + h_2 \tag{9}$$

In Equation (9), h_1 is -2.45×10^3 J kg^{-1} K^{-1} and h_2 is 3.17×10^6 J kg^{-1} [Adams, 2007].

COMSOL calculates the heat transfer coefficient internally for both forced and natural convective scenarios using a characteristic length, L (diameter of the aqueous trehalose solution), and a specified external temperature. Heat transfer coefficients were calculated as follows [Incropera and DeWitt, 2002]:

For forced convection:

$$h_t = \begin{cases} 2\frac{k}{L}\frac{0.3387 Pr^{1/3} Re_L^{1/2}}{\left(1+(0.0468/Pr)^{2/3}\right)^{1/4}}, & Re_L \le 5 \times 10^5 \\ 2\frac{k}{L} Pr^{1/3}\left(0.037 Re_L^{4/5} - 871\right), & Re_L > 5 \times 10^5. \end{cases} \tag{10}$$

For natural convection:

If $T > T_\infty$:

$$h_t = \begin{cases} 2\frac{k}{L}0.54Ra_L^{1/4}, & Ra_L \le 10^7 \\ 2\frac{k}{L}0.15Ra_L^{1/3}, & Ra_L > 10^7 \end{cases}. \tag{11}$$

If $T \le T_\infty$:

$$h_t = \frac{k}{L}0.27Ra_L^{1/4}. \tag{12}$$

The mass transfer coefficient was computed once the heat transfer coefficient was calculated. The heat and mass transfer analogy was used [Bejan, 2004], which is applicable for both laminar and turbulent flows. It applies to natural convection problems as well.

$$h_t = h_m \rho C_p Le^{2/3}. \tag{13}$$

At the film-substrate interface, $z = \delta_{sub}$, there is no penetration of species onto the substrate and there is thermal continuity.

$$\rho D\left(\frac{\partial \omega_A}{\partial z} - \omega_A \frac{\frac{\partial M_n}{\partial z}}{M_n}\right) = 0. \tag{14}$$

At the substrate bottom, $z = 0$, the temperature is specified.

$$T = T_b. \tag{15}$$

The governing equations and boundary conditions from $z = 0$ to $z = C(t)$ (seen in Figure 1b) are described in Section 3.1.

2.2. *Model Parameters*

The density, thermal conductivity and specific heat of the aqueous-trehalose solution were assumed to follow the law of mixtures [He *et al.*, 2008]. Hence, they are linearly dependent on the mass fractions of the mixture components:

$$X = \omega_A X_A + \omega_B X_B. \tag{16}$$

Table 1. Thermophysical properties for materials used in model.

Property	Water	Trehalose	Polystyrene
k (W m^{-1} K^{-1})	0.6	0.29	0.04
ρ (kg m^{-3})	1000	1580	1050
C_p (J kg^{-1} K^{-1})	4200	560	1400

In Equation (16), X represents the solution density, thermal conductivity or specific heat of the solvent (water, denoted by subscript A) and solute (trehalose, denoted by subscript B). Properties are listed in Table 1. ω_A and ω_B are the mass fractions of water and trehalose, respectively, which are solved for by Equation (2). They obey the following relationships:

$$\omega_A = \frac{\text{mass of water}}{\text{mass of solution}} = \frac{m_{water}}{m_{water} + m_{trehalose}} \tag{17}$$

$$\omega_B = 1 - \omega_A. \tag{18}$$

2.3. *Free Volume Model*

In aqueous mixtures, the amount of water that can evaporate at the solution-vapor interface is determined by the water activity, a_w. This parameter is based on atmospheric conditions. Evaporation is driven primarily by the water vapor concentration difference (see Equation (7)). The vapor concentration values at the interface and farfield are determined by the partial pressure at those locations. In aqueous solution problems where complex phenomenon such as glass formation can occur, it is very important to fully capture the physics as accurately as possible. The *free volume model* was used to account for water activity. It was originally developed for polymer-solvent systems and is based on both the free volume theory and Flory-Huggins theory of thermodynamics [He *et al.*, 2008]. The free volume model parameters are listed in Appendix A.

$$a_w = \frac{p_{w,int}}{p_0} = \exp\left[\left(\frac{\Delta\mu_A}{RT}\right)_{T,P} + f\right] \tag{19}$$

where $p_{w,int}$ is the partial vapor pressure of water at the solution-air interface and p_0 is the partial vapor pressure of pure water. A fourth order polynomial was fitted to vapor pressure data from 0 to 100°C listed in Lide [2005].

$\Delta\mu_A$ is the chemical potential of water and f is a function that adjusts for the abrupt change in partial pressure when the solution enters the glassy state around $\omega_A = 0.09$. The chemical potential and function, f, are defined as such [He *et al.*, 2008]:

$$\left(\frac{\Delta\mu_A}{RT}\right)_{T,P} = \ln\phi_A + \left(1 - \frac{\tilde{V}_A}{\tilde{V}_B}\right)(1-\phi_A) + \chi(1-\phi_A)^2 \tag{20}$$

$$f = \begin{cases} M_A\,(1-\omega_A)^2 \dfrac{\left(C_{p,g} - C_p\right)}{RT} \dfrac{dT_g}{d\omega_A}\left(\dfrac{T}{T_g} - 1\right), & T \le T_g \\ 0, \quad T > T_g \end{cases} \tag{21}$$

where T_g is the glass transition temperature (described in Section 2.3.1), ϕ_A is the volume fraction of water, and $\tilde{V}_A$, $\tilde{V}_B$, χ, and $(C_{p\,g} - C_p)$ are free-volume parameters listed in Appendix A. Therefore, water activity is determined from the calculation of the partial pressure (see Equation (19)), which is then used to calculate the molar concentration of water vapor at the solution-air interface.

$$c_{surf} = \frac{p_{w,int}}{RT}. \tag{22}$$

The vapor concentration in the farfield is based off the vapor pressure of pure water calculated at the ambient temperature and relative humidity.

$$c_\infty = \frac{\Phi p_{0,\infty}}{RT_\infty}. \tag{23}$$

2.3.1. *Glass transition temperature*

The glass transition temperature is determined via the Gordon-Taylor equation [Chen *et al.*, 2000]:

$$T_g = \frac{\omega_A T_{g,A} + \kappa\omega_B T_{g,B}}{\omega_A + \kappa\omega_B} \tag{24}$$

where κ is an empirically determined parameter [He *et al.*, 2008] listed in Appendix A and $T_{g,A}$ and $T_{g,B}$ are the glass transition temperatures of pure water and pure trehalose.

2.3.2. *Mutual diffusion coefficient*

The mutual diffusion coefficient in the water-trehalose solution was calculated using the following equations [He *et al.*, 2008]:

$$D_{mut} = 0.5D_A \left[\frac{Q}{\frac{\tilde{V}_A \phi_A^2}{\tilde{V}_B W} + (1-\phi_A)(1+2\phi_A)} + \frac{1+W+\phi_A(W-1)}{1+W-\phi_A(W-1)} \right] \tag{25}$$

where D_A is the self-diffusion coefficient of water and W is defined as the ratio of the self-diffusion coefficients of trehalose and water in an infinitely dilute solution (see Equation (27)).

$$D_A = D_0 \exp\left(\frac{\Delta E}{RT}\right) \times \exp\left(\frac{-\omega_A \hat{V}_A^* + \omega_B \hat{V}_B^*}{\omega_A (K_{11}/\gamma)(K_{21} - T_{g,A} + T) + \omega_B (K_{12}/\gamma)(K_{22} - T_{g,B} + T)}\right) \tag{26}$$

$$W = \left(\frac{D_{0,B}}{D_{0,A}}\right)_{\phi_A \to 1}. \tag{27}$$

Q is defined as:

$$Q = \phi_A \left[\frac{\partial\left(\frac{\mu_A}{RT}\right)}{\partial(\phi_A)} \right]. \tag{28}$$

Note: ΔE, $\hat{V}_A^*$, $\hat{V}_B^*$, ζ, K_{11}/γ, K_{21}, K_{12}/γ, and K_{22} are free volume parameters (listed in Appendix A).

2.3.3. *Viscosity*

As the mutual diffusion coefficient is decreased, the viscosity of the solution will become extremely high, representing glass formation. The viscosity can be predicated using the William-Landel-Ferry model [He *et al.*, 2008; 2006]:

$$\eta = \eta_g \times 10^{\frac{-C_1(T-T_g)}{C_2 - T_g + T}} \tag{29}$$

$$\eta = \eta_{g,0} \exp\left[k_{\eta,g}(x_A \tilde{V}_A + x_B \tilde{V}_B)\right] \tag{30}$$

where η_g is the viscosity of glass, x_A and x_B are the mole fractions of water and trehalose, and $\eta_{g,0}$, $\hat{V}_A$, $\hat{V}_b$, $k_{\eta,g}$, C1, and C_2 are free volume model parameters (listed in Appendix A).

2.4. *Water Transport Kinetics from Cells*

Since the recovery of viable biologics in the literature has been limited, it is important to include the behavior of biologics undergoing desiccation in this study. We introduce the biologics as a "layer" of cells (see Figure 1b and Section 3.1). Recall that there are two primary concerns during the desiccation process: one is avoiding a final desiccated state in which excess water is not trapped underneath the glassy surface. The other is avoiding a protrusion effect during desiccation. This is a challenge since biologics such as cells experience changes in volume during the drying process. Cells will shrink due to an extracellular-intracellular solute concentration difference, but they will reach an equilibrium once enough water is lost (known as the "dead volume"). These physics are described by Starling's Law [Bhowmick *et al.*, 1998].

$$\frac{dV}{dt} = L_p ANRT\left(\frac{1}{V} - \frac{1}{V_b}\right) \tag{31}$$

where L_p is the permeability coefficient, A is the surface area of the cell, N is the number of osmales, R is the universal gas constant, T is the absolute temperature, V is the osmotically active water volume in the cell at time t and V_b is the dead volume of the cell (once osmotic equilibrium is reached) [Noiles *et al.*, 1993]. The incorporation of cell shrinkage in the model is described in Section 3.1.

3. Numerical Model

COMSOL 4.4 was used for the simulations with two models, shown in Figure 1, developed for specific purposes: The first model (seen in Figure 1a) simulates the desiccation kinetics of an aqueous trehalose solution. The second (seen in Figure 1b) is a compartment model which includes the dynamics of the biologic to-be-preserved (model described in Section 3.1). The model seen in Figure 1a is used for the majority of simulations. The water mass fraction data recorded at $z = C(t)$ in the compartment model is comparable to water mass fraction data recorded at a similar location in the model without the cell layer (between 1–3% difference). The inclusion of the cell layer does not significantly change the desiccation kinetics of the aqueous trehalose solution. Due to these observations, and to avoid additional complexities to the numerical model, the model seen in Figure 1a is primarily used. However, the compartment model seen in Figure 1b does provide important insights to the optimization of desiccation methods, as detailed in Section 4.10.

3.1. *Inclusion of Cell Dynamics*

In our numerical compartment model, the mammalian cell is introduced as a "layer" of cells (the height being the diameter) from $z = 0$ to $z = C(t)$ (see Figure 1b). For this 1-D problem, a human lymphocyte was chosen for the cell type and loaded with a certain amount of trehalose. Transport equation (2) was solved using a high diffusion coefficient ($D = 10^{-5}$ m^2 s^{-1} >> D_{mut}) to ensure uniform concentration profiles in the cell layer. Therefore, COMSOL solves the mass transport problem (and calculates water and trehalose mass fractions) in both the aqueous trehalose solution film ($z = C(t)$ to $z = S(t)$) and the cell layer ($z = 0$ to $z = C(t)$). The solution film thickness decreases in the z-direction due to evaporation (natural or forced convection) and the cell layer shrinks due to an extracellular-intracellular solute concentration difference (see Section 2.4 and Equation (32)) [Chen *et al.*, 2006].

$$\frac{dC(t)}{dt} = \begin{cases} -L_p RT\left(c_{i,B} - c_{0,B}\right), & z > d_f \\ 0, & z \leq d_f \end{cases} \tag{32}$$

where d_f is the final diameter of the cell (thickness of the cell layer), corresponding to the cell's respective dead volume.

3.2. *Mesh and Solver Settings*

Triangular mesh elements were used for the substrate, with the maximum element size being 2.92×10^{-5} m. For the solution film, a structured quadrilateral mesh (or *mapped* mesh) was used. The aspect ratio is large, but acceptable, since there are no significant changes in concentration or temperature in the r-direction. A total of 70 elements were used in the z-direction for the aqueous solution film. For the compartment model seen in Figure 1b, these 70 elements were split into a 50 element section for the

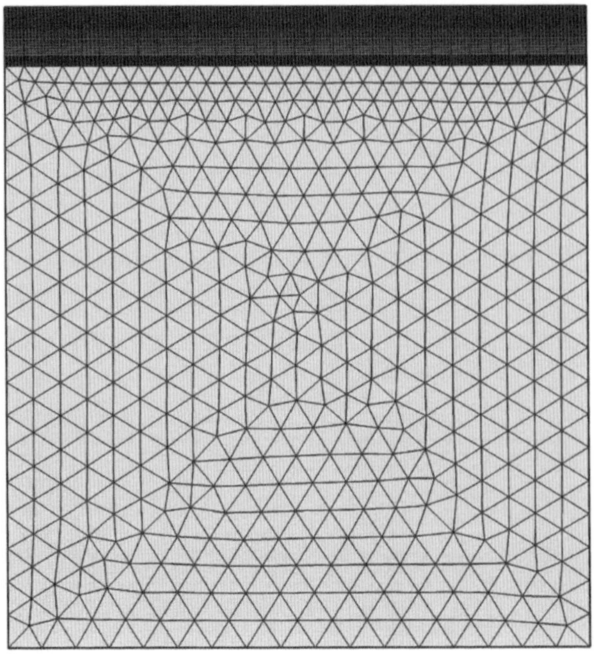

Figure 2. Mesh size used in current study (compartment model shown).

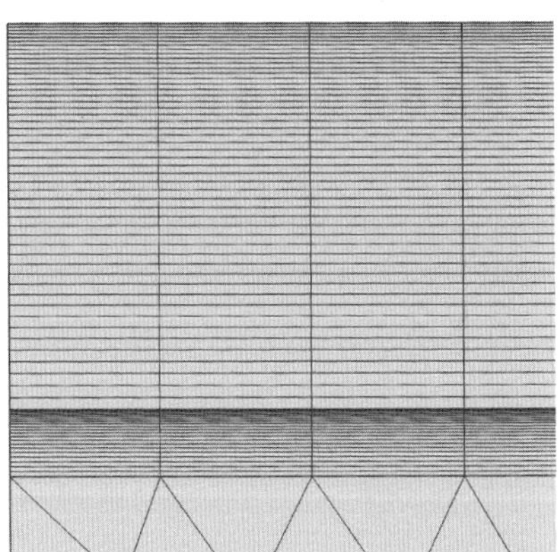

Figure 3. Example of the mesh distribution in the aqueous trehalose film.

solution film and 20 element section for the cell layer (see Figures 2 and 3 and Section 3.1 for additional information). Higher mesh density was used at the top layers to accurately capture the concentration gradients in those regions. A grid refinement study was performed to find the optimal mesh setup for acceptable numerical accuracy and computational run time (results not shown). Winslow smoothing was applied to the moving mesh with first-order elements. For the solution, the PARDISO Direct Solver was chosen and the absolute and relative tolerances were 5×10^{-4} and 0.01 respectively. For the compartment model, the absolute tolerance was lowered to 0.001 to ensure accurate calculation of mass transport across the cell membrane.

4. Results

Sections 4.1 through 4.9 describe the results of tests performed on the numerical model seen in Figure 1a and Section 4.10 describes the results of tests performed on the compartment model seen in Figure 1b. In the literature, there have been many tests similar to the tests performed in Sections 4.2 and 4.3 [He *et al.*, 2008; Chen *et al.*, 2006]. Our results have been compared with these sources (data not shown) before proceeding with the numerical experiments seen in Section 4.4 and beyond.

4.1. *Initial Look at Convective Drying Kinetics*

Figure 4 is a plot of the water mass fraction at the bottom of the thin film under natural and forced convection desiccation. The simulation was for a 10 mm diameter, 50 μm initially thick, 0.2M trehalose solution at room temperature ($T_b = T_\infty = 20°C$) and 0% relative humidity for both forced and natural convection. The forced convection setting used an external flow rate of 6.5 m/s, correlating to a heat transfer coefficient of 100 W/m^2K and mass transfer coefficient of 0.089 m/s. As shown in Figure 4a, over a 48-hour-period the desiccation under natural and forced convective conditions are almost invariant and the solution reaches a final moisture content of 0.0975. As shown in Figure 4b, the majority of the desiccation process takes place within the first hour. Figure 5 is a blow up of the drying kinetics for the forced convective case in the late stages of the simulation (32–48 hr) along with the corresponding glass transition temperature. Here the water mass fraction at the bottom of the film and a location representing the top of the biologic (several micrometers above the film bottom) is plotted along with the corresponding glass transition temperature at these locations.

It can be seen from both Figures 4 and 5 that although the drying kinetics vary over the first hour (with more water being removed by forced convection), the water mass fraction at the bottom of the film reaches a point where the glass transition temperature is between 12–15°C. It can be seen that over a very long period of time (~2 days), whether forced or natural

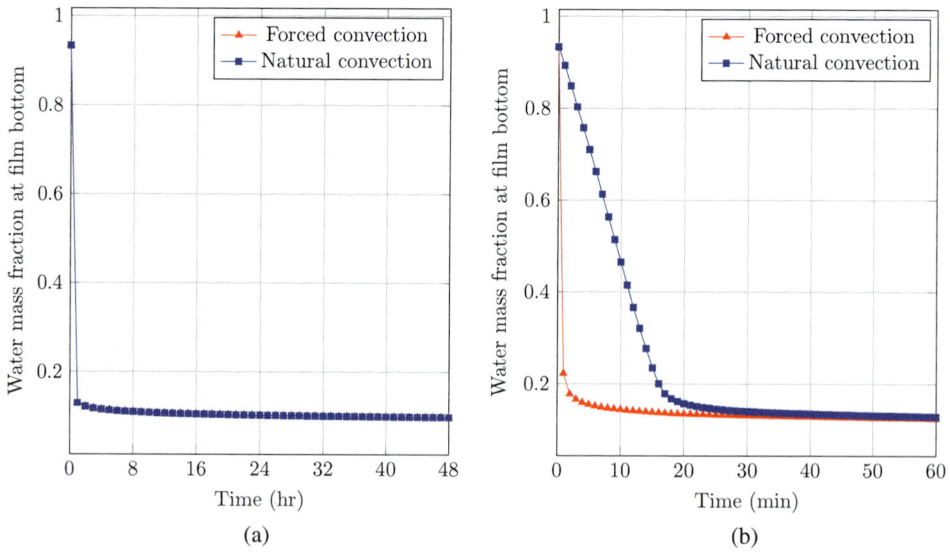

Figure 4. Water mass fraction data for forced and natural convective cases for a simulation time of 2 days. (a) shows that after a long while, the final mass fraction is the same regardless of whether forced or natural convection is used. (b) shows the first hour of desiccation. Forced convection results in faster drying but the final mass fraction does not depend on convective condition, as seen in (a).

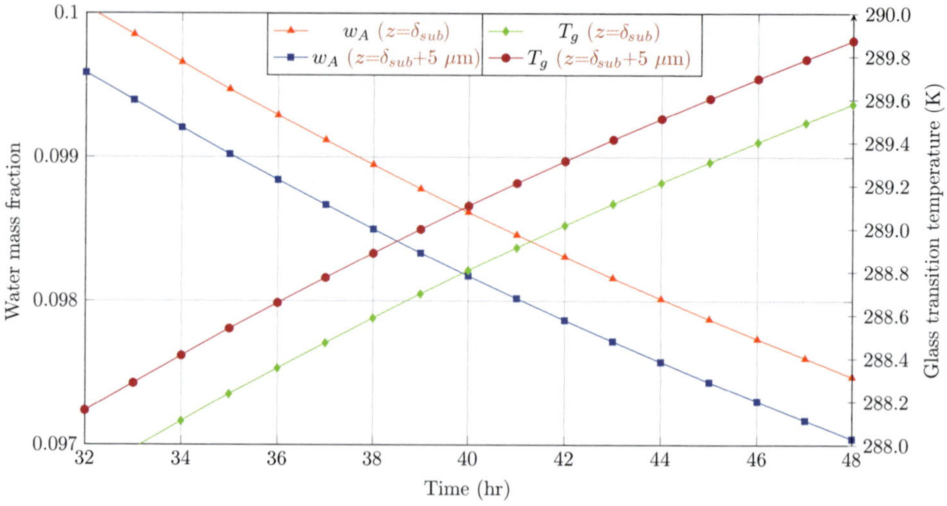

Figure 5. Water mass fraction and glass transition temperature data at the bottom of the solution film and top of the biologic (several μm above solution bottom) for forced convective case seen in Figure 4.

convection is used, the water mass fraction necessary for glass formation at room temperature ($\omega_A \sim 0.09$ corresponding to a $T_g \sim 298$K) is not reached. Therefore, storage is required at or below 10°C, since it is reasonable to store at temperatures 5°C below glass transition temperature. These initial tests show that this is a transport-limited issue and the time scale for glass formation is too long. The next step was to perform a parameter sweep of the initial geometry and boundary conditions to explore the possibility of minimization of mass transport limitations.

4.2. *Variation of Initial Film Thickness*

The initial thickness of the film was varied from 30 μm to 500 μm. This accounted for spin drying conditions where centripetal forces allow the drop to spread out uniformly over a larger area with a small height. The film was unheated at room temperature ($T_b = T_\infty = 20$°C), in an environment with 0% relative humidity, and loaded with 0.2M trehalose. The simulations were run for both forced and natural convection. Simulations were run until the mass fraction of water changed very little (i.e. $\frac{\partial \omega_A}{\partial t} < 1 \times 10^{-6}\,\mathrm{s}^{-1}$) and for an appropriate drying time (up to 1 hour for forced convection cases and up to 5 hours for natural convective cases). As shown in Figure 6, thinner initial film thicknesses will lead to faster desiccation and lower residual water content at

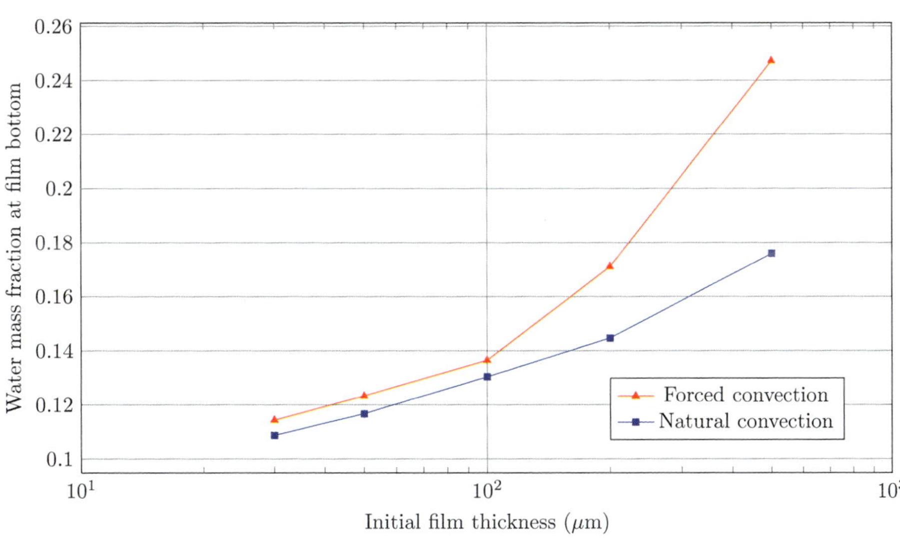

Figure 6. Final water mass fraction at the bottom of the film based on initial film thickness for forced and natural convection scenarios.

the bottom of the film. There is a distinct change in behavior for films less than 100 μm. Compared to natural convection, forced convection trapped a lot more moisture at the film bottom for starting heights above 100 μm. For thinner films, this effect although existent is not as pronounced. Similar results were obtained by He *et al.* [2008].

The trade-off between faster kinetics and trapping more water can be seen by comparing Figures 6 and 7. While forced convection results in faster drying times (see Figure 7 for a comparison), natural convection appears to remove slightly more water (Figure 6). A possible explanation for this is that glass forms more quickly at the film surface of the aqueous trehalose solution in forced convective cases. Since the glassy film starts to form at the top, thicker initial layers lead to a larger volume of water trapped in the film. This shows the merits of spin drying or atomization studies for trapping less water. However, when a biologic-to-be-preserved is taken into consideration, our results show that the biologic ultimately will not be able to survive. It is clear that greater than 10% water is trapped and prevents the formation of uniform glass. Two issues to overcome are removing enough water and reducing the total drying time.

Figure 8 clearly shows that the sudden increase in glass transition temperature at the film surface coincides with the drastic reduction in water loss kinetics from the bottom of the film. Glass is forming on the top of the film

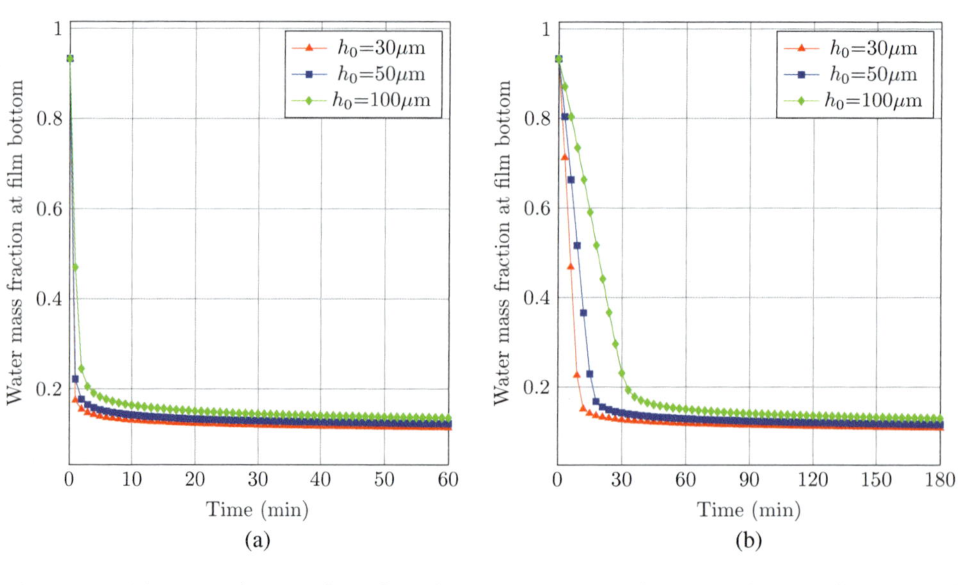

Figure 7. (a) Drying kinetics for 3 forced convection cases. (b) Drying kinetics for 3 natural convection cases.

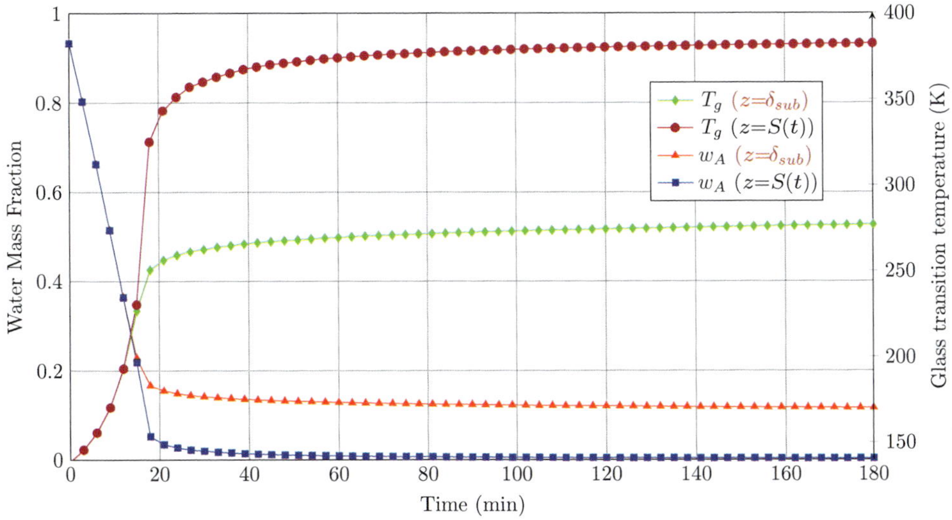

Figure 8. Water mass fraction and glass transition temperature data at top and bottom of a 50 μm initially high solution film with 0.2M trehalose under natural convection.

and arresting further decrease in film height due to evaporation. It is beneficial for trapping a biologic (most mammalian cells being roughly 20 μm in diameter) but only if enough water has been removed from the film. Initially the water mass fraction is decreasing, indicating evaporation is taking place. As the water mass fraction is lowered, the mass fraction of trehalose is increasing, therefore raising the glass transition temperature. Once the glass transition temperature has reached a value close to that of pure trehalose, the resistance is high (a "glass" has formed). Figure 8 illustrates that this occurs at the film surface fairly early on during drying (notice that ω_A goes to zero quickly at the surface). But an appropriate glass transition temperature is not reached at the film bottom, and therefore too much water is trapped. Smaller initial film thicknesses obviously reduce the drying time as well as result in lower residual mass fractions but too much water ultimately remains trapped within the solutions. Further parameters must be adjusted to allow for greater water removal and promote glass formation.

4.3. *Variation of the Molarity of Trehalose*

The amount of trehalose initially loded into the solution affects desiccation kinetics and can lead to different drying profiels which may impact preservation of biologics [Oliveira, 2012]. The amount of trehalose initially loaded into the solution was varied from 0.05M to 1M in films initially 30, 50, and 100 μm

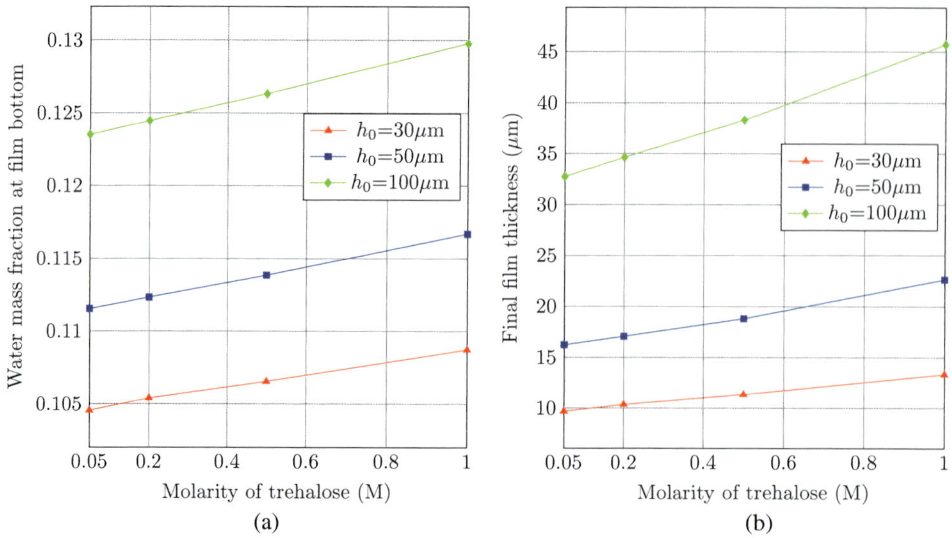

Figure 9. Final mass fraction and film thickness for different initial thicknesses and molarity of trehalose for natural convection conditions.

thick. These thicknesses were chosen as realistic starting points for convective desiccation protocols based on simulations in the previous section. It is seen in Figures 9a and 9b that a lower amount of trehalose will lead to lower final water mass fractions at the film bottom. The higher molarity trehalose solution leads to higher viscosity and slower diffusivity of water. Therefore, as the solution dries, glass starts forming at the film surface early on and traps higher water within the film. This result is similar to that found by He *et al.* [2008] and Chen *et al.* [2006].

It is clearly seen in Figure 9 that there is little difference in the final residual water mass fraction at the film bottom and final film thickness for solutions with initial trehalose molarities between 0.05M and 0.2M. The target water mass fraction values (~ 0.1) can be achieved using solutions 30 μm thick and 0.05M or 0.2M trehalose. However, an initial thickness of 30 μm leads to a final film thickness of around 10 μm. This is too thin for most mammalian cells. In general, a mammalian cell is around 20 μm in diameter and may shrink to a diameter of approximately 15 to 17 μm based on the cell type. This illustrates the difficult design problem present in the optimization of desiccation protocol. A final film thickness of approximately 15–20 μm is important because avoiding a protrusion of the biologic to the ambient air is just as important as removing enough water to form uniform glass. Since solutions initially 50 μm thick shrink to this target thickness and lower amounts of trehalose result in lower residual water mass fractions, the remainder of tests are performed on solutions initially 50 μm thick loaded with and 0.05M trehalose.

4.4. *Variation of Base Temperature*

Further improvement in the desiccation protocol is sought by altering the boundary conditions. Heating the substrate may assist in removing water in excess of that removed by pure convection. It is seen in Figure 10a that the base temperature of the substrate has a stronger effect on the natural convective cases compared to forced convection over the temperature ranges explored.

Based on Figure 10a, the target final water mass fraction at the film bottom appears to be reached at higher base temperatures with natural convective conditions. Looking at the final thickness of the solution (see Figure 10b) it can be seen that a protrusion effect of a mammalian cell may be avoided as well (the final film thicknesses are in the 15–17 μm range, as explained in Section 4.3). In Figure 10a it is seen that for higher base temperatures, lower mass fractions are achievable for natural convection. However, while the mass fractions at the bottom of the film are lower than that of forced convective cases at these temperatures, the final film thickness

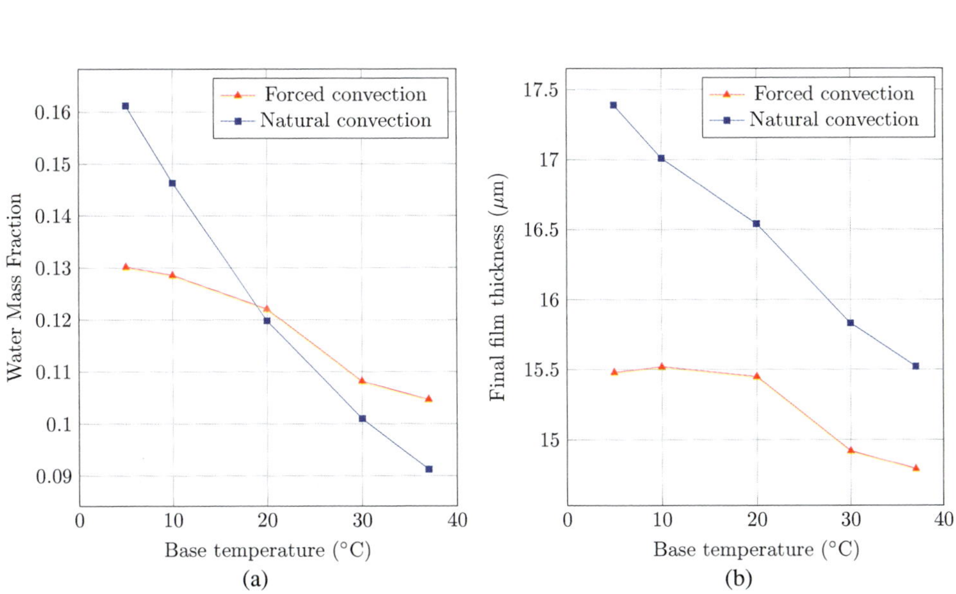

Figure 10. (a) Final water mass fraction at the bottom of the solution film for different substrate base temperatures and convective conditions. (b) Final film thickness for different substrate base temperatures and convective conditions.

Table 2. Water mass fraction at the bottom of the solution film and film thickness recordings at $t = 18$ min for forced and natural convective cases where the base temperature, $T_b = 37°C$.

	Forced convection	Natural convection
Water mass fraction ($z = \delta_{sub}$)	0.1246	0.1122
Film thickness (μm)	15.4	16.1

is still higher than that of forced convective cases (see Figure 10b). This is confusing since logically a greater film thickness indicates that there is more water remaining inside the film. But referring to Figure 10a, this is not the case. We attribute this observation to the difference in desiccation kinetics of forced and natural convection. During forced convection, the film thickness initially decreases very rapidly due to the faster desiccation kinetics. After this initial decrease in thickness, a glass starts forming at the surface of the film, increasing the resistance to evaporation. During natural convection, the desiccation kinetics are slower, allowing water to be removed continuously before glass starts to form at the film surface. Table 2 shows the corresponding water mass fractions and film thickness mid-simulation for the case where the base temperature is 37°C.

It is evident that while the water mass fraction at the bottom of the film is lower during natural convection (as also seen in Figure 10a), the film thickness is still higher. Once a glassy state has been reached at the surface, the solution is extremely viscous, making further evaporation very difficult. During forced convection, there is a greater cooling effect due to evaporation. This reduces the impact of base temperature heating and could prevents further removal of water. Since this is not present during natural convection, more water can be removed from the solution, but the film thickness still decreases more slowly.

4.5. *Variation of Ambient Temperature*

The final water mass fraction at the bottom of the film was observed to be lower when the base temperature of the substrate was higher. The ambient temperature, T_∞, was next varied to see if a similar effect was observed. The fixed parameters were the same as in Section 4.4 (initial film thickness of 50 μm, 0.05M trehalose, 0% relative humidity). Two base temperatures are chosen, $T_b = 10°C$ and 37°C.

Table 3 shows that lower mass fractions are attained at the film bottom in hotter environments (generally when T_b or T_∞ are greater than 30°C).

Table 3. Effect of environmental temperature on final residual water mass fraction at the bottom of the film and final film thickness.

	Forced convection				Natural convection			
	$T_b = 10°C$		$T_b = 37°C$		$T_b = 10°C$		$T_b = 37°C$	
T_∞ (°C)	Final ω_A	Final z (μm)	Final ω_A	Final z (μm)	Final ω_A	Final z (μm)	Final ω_A	Final z (μm)
5	0.1939	17.34	0.1453	15.90	0.1554	17.32	0.0962	15.49
10	0.1780	16.85	0.1173	15.31	0.1365	16.89	0.0945	15.55
20	0.1188	15.28	0.1047	14.79	0.1469	17.11	0.0912	15.52
30	0.1061	14.97	0.0938	14.75	0.1435	16.90	0.0886	15.65
37	0.0982	14.71	0.0870	14.54	0.1405	17.13	0.0886	15.98

He *et al.* [2008] found similar results where lower water mass fractions are attained at higher environmental conditions. Slightly lower water mass fractions at the bottom of the film are also reached during a natural convective situation, which agrees with the trend seen in Figures 10a and 10b. The lowest mass fraction reached is 0.0886, which correlates to a glass transition temperature of approximately 24.5°C. However, the temperatures at the film bottom during those simulations are between 30–37°C, which are potentially dangerous for the biologic to-be-preserved. Metabolic activity will be increased at these temperatures. Determining whether the base temperature, T_b, or ambient temperature, T_∞, has a greater effect on water evaporation will be investigated at a later time.

4.6. *Influence of Radiation*

To investigate how much influence radiation has on desiccation, a surface-to-ambient radiation boundary condition was applied at the film surface.

$$q = \epsilon\sigma(T_{surr}^4 - T^4) \tag{33}$$

where surrounding temperature, T_{surr}, was equal to ambient temperature, T_∞. Simulations were run for both forced and natural convection scenarios. Substrate base temperature was kept at 20°C while the water-trehalose solution had an initial thickness of 50 μm and was loaded with 0.05M trehalose. Simulations were terminated after 20 minutes for forced convection and 40 minutes for natural convection.

Table 4. Effect of material emissivity on final residual water mass fraction at the bottom of the film.

	Forced convection		Natural convection	
	$T_{surr} = 20°C$	$T_{surr} = 37°C$	$T_{surr} = 20°C$	$T_{surr} = 37°C$
Emissivity value	Final ω_A	Final ω_A	Final ω_A	Final ω_A
0	0.1334	0.1111	0.1344	0.1278
0.2	0.1334	0.1108	0.1331	0.1245
0.9	0.1334	0.1110	0.1346	0.1252

As seen in Table 4, different emissivity values at the film's surface did little to influence the final water mass fraction obtained at the film bottom. This can only indicate that convection is the dominant mode of heat transfer.

4.7. *Effect of Substrate Material*

The material of the substrate was changed to see if higher or lower conductivities could aid in the desiccation process. The reasoning is that at higher temperatures (approaching 37°C), metabolic activity is increased for the biologic, which may be detrimental to the preservation process. Using materials with higher conductivities, such as glass (~27 times more conductive than polystyrene) or aluminum (over 5000 times more conductive than polystyrene), may allow for using lower base temperatures. The water-trehalose solution parameters were kept at an initial film height of 50 μm, 0.05M trehalose, $T_\infty = 20°C$, and 0% relative humidity.

It is seen in Table 5 that at lower temperatures, the lowest water mass fractions were mostly attained using polystyrene substrates. At higher base temperatures, water mass fractions tended to be lower for substrates with higher thermal conductivity (glass and aluminum). It also appeared that when substrate conductivity was higher, water mass fraction values were similar for both forced and natural convection. Overall, lower water mass fractions were attained during natural convective cases. It should be noted that the drying times for solutions on top of aluminum and glass substrates were much faster than polystyrene substrates. The explanation is that higher thermal conductivity allow for faster heat transfer and, thus, drying times. This also helps explain that at cooler temperatures, polystyrene substrates led to lower water mass fractions since it they did not transfer the cooling effect as quickly as glass or aluminum. Table 6 shows the final height of the dried solution film for each scenario.

Table 5. Final water mass fraction at the bottom of the film for scenarios using different substrate materials.

	Final water mass fraction at film bottom					
	Forced convection			Natural convection		
T_b (°C)	Polystyrene	Glass	Aluminum	Polystyrene	Glass	Aluminum
5	0.1302	0.1955	0.1999	0.1612	0.1656	0.1658
10	0.1286	0.1735	0.1780	0.1463	0.1493	0.1496
20	0.1221	0.1136	0.1135	0.1198	0.1174	0.1225
30	0.1082	0.0985	0.0970	0.1010	0.0981	0.0982
37	0.1047	0.0889	0.0870	0.0912	0.0881	0.0878

Table 6. Final thickness of solution for scenarios using different substrate materials.

	Final thickness of solution film (μm)					
	Forced convection			Natural convection		
T_b (°C)	Polystyrene	Glass	Aluminum	Polystyrene	Glass	Aluminum
5	15.48	17.57	17.54	17.39	17.53	17.59
10	15.52	16.87	16.88	17.01	17.47	17.19
20	15.45	15.39	15.17	16.54	16.56	16.83
30	14.92	14.97	14.70	15.83	15.75	15.69
37	14.79	14.71	14.69	15.52	15.42	15.35

While using substrates with higher thermal conductivity may indeed help reduce drying times, the problem of trapping too much water at the solution bottom still exists. A constant base temperature is not a solution because it raises the temperature and therefore reduces the weight fraction of water necessary to form glass. However, what is known is that raising the temperature allows for further evaporation of water in the aqueous trehalose solution. A possible solution may be to raise the base temperature of the substrate only when the water mass fraction at the film bottom reaches a certain limit. This raised temperature would only last temporarily and then the simulation would continue at room temperature. Not only may this allow for glass formation at the film bottom, but it may also help with reducing drying time since water may be able to be removed more quickly in a shorter period of time.

4.8. *Controlled Heating of the Substrate*

Intermittent heating of the substrate was explored to see if it circumvents the problems associated with constant base temperature heating. The substrate would be heated intermittently from below via a controller and return to room temperature. Examples of the controlled temperature increases are shown in Figure 11.

Figure 12 shows the effect of these intermittent temperature increases on water mass fraction. It shows that temporary heating certainly removes more water. However, the substrate temperature cannot surpass 40°C, as that will make temperatures at the film bottom too high for the biologic to survive. When the base temperature of a 1 mm thick polystyrene substrate is raised to 40°C, the film bottom will rise to approximately 26°C, which is still safe for the biologic.

The performance of several different wave patterns were reviewed. For study purposes, a 10 minute-20 minute-10 minute heating time pattern was chosen for forced convective cases and a 30 minute-10 minute-30 minute pattern for natural convective cases (see Figure 13). These times correspond to a 20°C-40°C-20°C heating of the substrate bottom. The reason for longer heating times for forced convection and shorter for natural convection is that

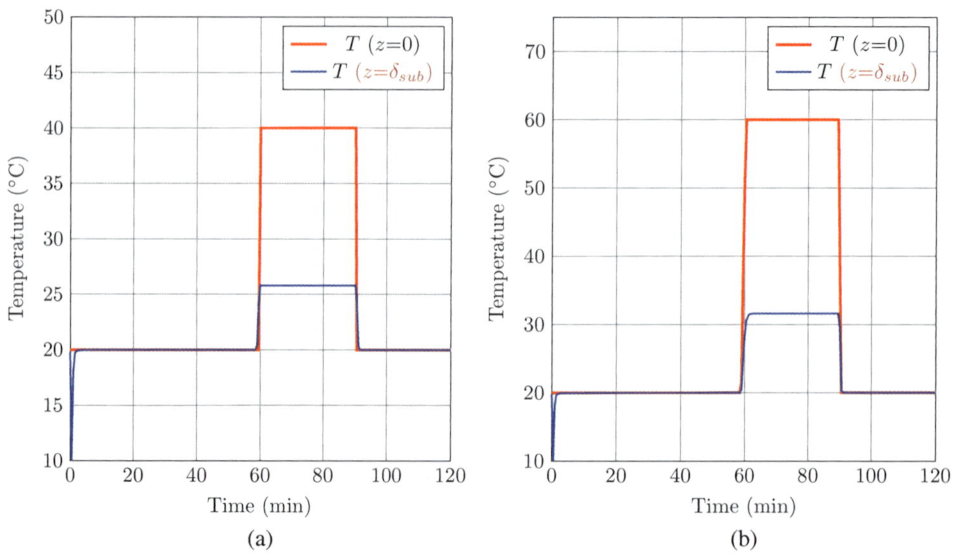

Figure 11. Examples of controlled spikes in substrate heating. (a) Substrate bottom is heated to 40°C, raising the temperature of the film bottom to ~26°C during this time. (b) Substrate bottom is heated to 60°C, raising the temperature of the film bottom to ~32°C during this time.

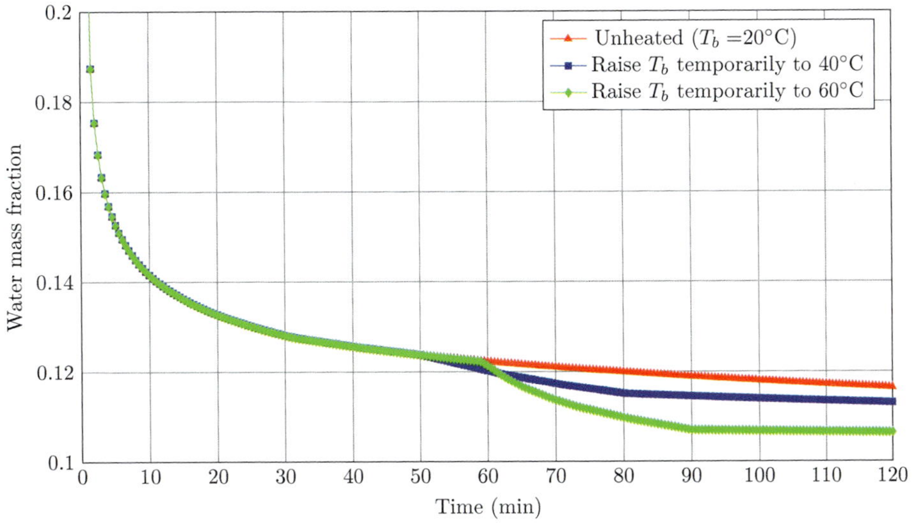

Figure 12. Result of intermittent heating seen in Figure 11 on the water mass fraction at the bottom of the film.

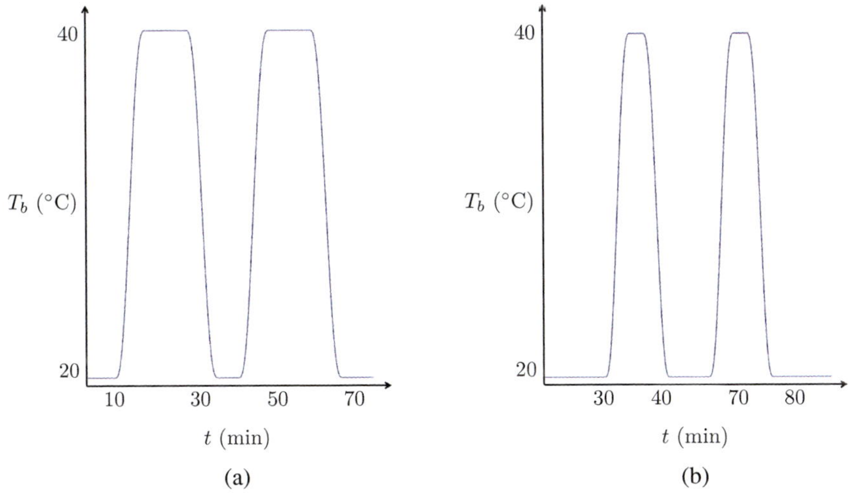

Figure 13. Examples of the heating wave patterns for (a) forced convection and (b) natural convection.

during natural convection scenarios, heat cannot be removed as quickly. Thus, when the temperature is raised, it stays "hotter" for a longer period of time. For example, the temperature at the bottom of the film is 36°C when the substrate is heated to 40°C during natural convection. This is not as safe

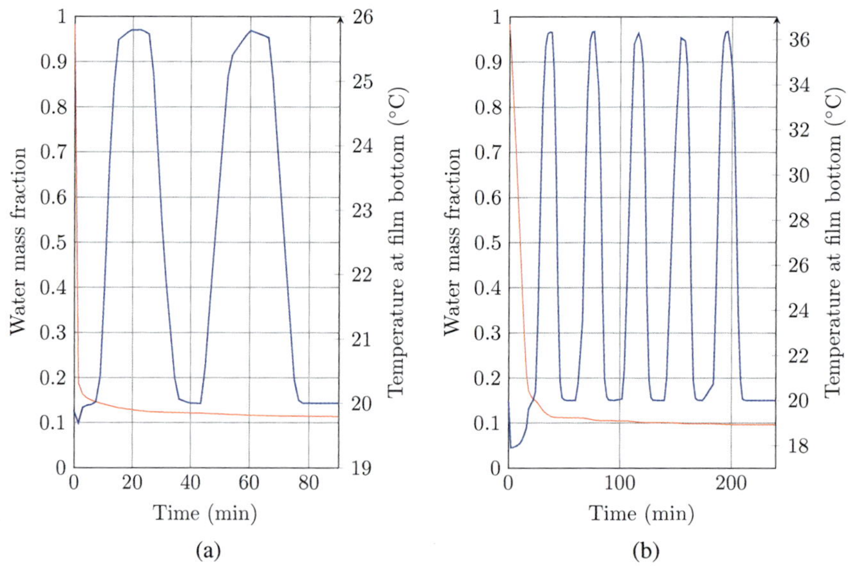

Figure 14. Water mass fraction and temperature at the bottom of the solution over time for typical forced and natural convection scenarios.

Table 7. Summary of results using controlled substrate heating wave patterns (displayed in Figure 13).

	Forced convection		**Natural convection**	
Substrate type	**Polystyrene**	**Aluminum**	**Polystyrene**	**Aluminum**
Final water mass fraction	0.1072	0.0927	0.0976	0.0969
Drying time (min)	180	151	180	169

Note: The maximum temperature in the heating waves was 37°C for cases using aluminum substrates, instead of 40°C, due to the material's higher thermal conductivity.

as the 26°C observed at the film bottom when the substrate is heated to 40°C during forced convective situations. Figure 14 shows the typical behavior for simulations after the base temperature input seen in Figure 13 was applied.

Cases were run for both a polystyrene substrate (insulator) and aluminum substrate (conductor). Table 7 summarizes the results. Each case was run until glass was formed or nearly formed at the film bottom. For both forced and natural convective cases, using an aluminum substrate resulted in shorter drying times. This is attributed to aluminum's higher thermal conductivity. Glass was never reached in a forced convective case using a polystyrene substrate. Temperatures at the bottom of the film should be closely monitored when substrates with higher thermal conductivity are used, for it is easier for the film to reach a temperature which will kill the biologic.

Using controlled heating of the substrate results in the nearing of a glassy state at the bottom of the film in a much more reasonable time (recall the desiccation kinetics for unheated substrates seen in Figure 4). In addition, the substrate temperature is always brought back to room temperature which ensures that the final residual water mass fraction value obtained at the film bottom correlates to a glass transition temperature of roughly 20°C.

4.8.1. *Increasing frequency of base temperature heating pattern*

The drying times can be further reduced by increasing the frequency of the heating wave patterns. By changing the 10 minute-20 minute-10 minute pattern to a 2 minute-10 minute-2 minute pattern for forced convection and the 30 minute-10 minute-30 minute pattern to a 5 minute-5 minute-5 minute pattern for natural convection, the total drying times are shorter (see Table 8). The trends seen in the previous section still exist (i.e. glass was never reached in a forced convection case using a polystyrene substrate). However, comparing Tables 7 and 8, it is clearly seen that the frequency of the base temperature heating patterns certainly has an effect on drying times.

4.9. *Controlled Variation of Ambient Temperature*

Much like the controlled heating of the substrate, we implemented wave patterns for the ambient temperature. This would be equivalent to exposing the film to temporary periods of hot air to drive out additional water during

Table 8. Summary of results using controlled substrate heating wave patterns with increased frequencies (discussed in Section 4.8.1).

	Forced convection		Natural convection	
Substrate type	**Polystyrene**	**Aluminum**	**Polystyrene**	**Aluminum**
Final water mass fraction	0.1086	0.0969	0.0978	0.0991
Drying time (min)	125	72	125	91

Note: The maximum temperature in the heating waves was 37°C for cases using aluminum substrates, instead of 40°C, due to the material's higher thermal conductivity.

Table 9. Summary of results using controlled ambient temperature heating wave patterns.

	Forced convection		Natural convection	
Substrate type	**Polystyrene**	**Aluminum**	**Polystyrene**	**Aluminum**
Final water mass fraction	0.1079	0.1227	0.1085	0.1185

drying. For forced convection, we employed a 5 minute-2 minute-5 minute wave of 20°C-40°C-20°C for the ambient temperature while for natural convection we used a 10 minute-5 minute-10 minute wave of 20°C-100°C-20°C. Using these wave patterns ensured that the temperature at the bottom of the film never exceeded 37°C. Simulations were run for 1 hour for forced convection and 2 hours for natural convection using both polystyrene and aluminum substrates. Table 9 summarizes the final water mass fractions obtained.

As seen in Table 9, it is seen that when the ambient temperature is changed, lower water mass fractions are obtained using polystyrene substrates (the opposite result was obtained in the previous section). This is attributed to aluminum's higher thermal conductivity. It was observed that in the simulations using polystyrene substrates that the bursts of "hot" ambient temperatures increased the temperature at the bottom of the film (maximum temperatures varied between 30–34°C). Conversely, in the simulations using aluminum substrates, the temperature at the bottom of the film remained predominately around 20°C, despite the blasts of "hot" air. This shows that the base temperature of the substrate has a much greater influence on the removal of water than the ambient temperature.

The ambient temperature can affect the final mass fraction obtained, but not as strongly as if the base temperature was increased. However, it is likely that controlled heating of both the ambient and base temperatures may allow for an optimal scenario in which the appropriate water content can be reached in a desirable amount of time. For example, Figure 15 shows a natural convective case (polystyrene substrate) where the ambient temperature is increased to 100°C for a few minutes and then the base temperature is temporarily heated to 40°C for the remainder of time. Table 10 shows a comparison between the final residual water mass fraction at the film bottom for this simulation and similar scenarios. Preliminary results show that using a customized pattern for heating both the environment and substrate can lead to faster drying times.

In summary, more water can be removed from the aqueous trehalose solution by heating the substrate and/or environment. The base temperature has a

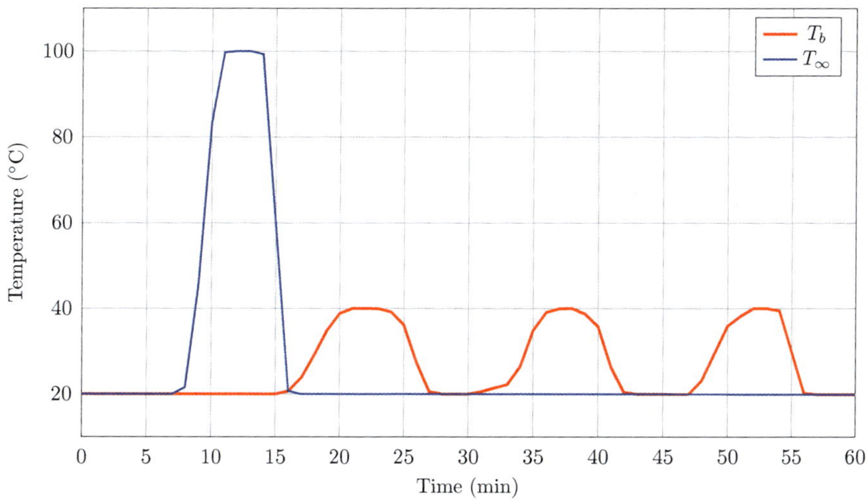

Figure 15. Plot of wave functions for ambient and substrate temperatures.

Table 10. Comparison of final water mass fractions at the bottom of the film for different heating methods for natural convective scenarios on polystyrene substrates.

Case	Final ω_A ($z = \delta_{Sub}$)	Time (hr)
Vary base temperature only (10 min-5 min-10 min wave of 20°C-40°C-20°C	0.1005	2
Vary ambient temperature only (10 min-5 min-10 min wave of 20°C-100°C-20°C	0.1085	2
Vary base and ambient temperature (Figure 15)	0.1067	1

greater effect on removing water than ambient temperature. However, higher temperatures can lead to dangerous metabolic conditions for the biologic-to-be-preserved. Temporary heating of the substrate/ambient temperature is a promising alternative method for removing water because it always periodically returns to room temperature. Not only will this keep the biologic's metabolic activity under control but it ensures that a glass will form when the water content is around 0.09. As seen in Table 10 sinusoidal heating of the substrate base is more effective at reaching lower residual water mass fractions than sinusodial heating of the ambinent air. However, using a combination of heating the environment and substrate may help in reducing the overall desiccation time.

4.10. *Compartment Model*

To simulate the effect of cell shrinkage, the compartment model seen in Figure 1b is utilized. A relatively small mammalian cell (human lymphocyte)

Table 11. Properties of human lymphocyte [Chen *et al.*, 2006]

Isomotic volume, $V_{i,c}$ (μm^3)	210
Isomotic diameter, d_0 (μm)	7.4
Fraction of osmotically inactive volume, $\frac{V_b}{V_{i,c}}$	0.37
Membrane permeability, L_p (m^3/N·s)	1.14×10^{-14}

is selected for the "cell layer" (cell properties listed in Table 11). Due to the size of this cell, a smaller initial thickness for the 10 mm diameter aqueous trehalose solution is selected for these simulations ($h_0 = 30\ \mu$m). Referencing Figure 9b, the final film thickness for solutions initially 30 μm thick is around 10 μm. This should avoid possible "protrusion" effects since lymphocytes are initially around 7 μm in diameter. For the simulations, 0.2M intra- and extracellular trehalose is used. The simulation was run for 2 hours at room temperature ($T_b = T_\infty = 20°$C) and 0% relative humidity under forced and natural convection (the forced convection setting used an external flow rate of 6.5 m/s, correlating to a heat transfer coefficient of 100 W/m^{2}K and mass transfer coefficient of 0.089 m/s) on a 1 mm thick polystyrene substrate. It is important to note that a lymphocyte, like many mammalian cells, undergo a maximum amount of shrinkage caused by the water transport across the membrane. During the simulation, once water transport shrank the lymphocyte cell layer to a height representative of the diameter of the cell's "dead volume" (see Equation (31)), water transport did not continue across the membrane ($z = C(t)$ in Figure 1b).

This characteristic of a semi-permeable membrane presents unique challenges regarding residual water mass fraction. While the final mass fractions at the bottom of the solution and right above the cell layer were low, the water content inside the cell remained high. This is due to water transport being discontinued after the dead volume is reached. While glass could theoretically be close to being reached above and below the cell layer ($\omega_A \sim 10$ wt.%), the water inside the cell is around 60 wt.%. With water content this high, glass will certainly not form inside the cell. Increasing the amount of intracellular trehalose is the best way to achieve intracellular glass (see Table 12). However, if intracellular trehalose is increased, extracellular trehalose must be increased as well to prevent water transport into the cell. This therefore raises the residual water mass fraction above the cell layer and at the bottom of the solution film (reference Section 4.3 and Figure 9).

However, changes can be made to drive further water out the cell. Table 13 shows the results for a 30 μm initially thick solution with a lymphocyte cell layer with different amounts of intra- and extracellular trehalose under forced

Table 12. Final water mass fractions at solution bottom, inside cell, and top of the cell layer after 2 hours for a solution initially 30 μm thick and a lymphocyte cell layer with different amounts of intra- and extracellular trehalose.

	Forced convection			Natural convection		
Initial molarity of trehalose	ω_A $(z = C(t))$	ω_A **(Inside cell layer)**	ω_A $(z = \delta_{sub})$	ω_A $(z = C(t))$	ω_A **(Inside cell layer)**	ω_A $(z = \delta_{sub})$
0.2M	0.1039	0.6012	0.1099	0.1065	0.5999	0.1119
0.5M	0.1051	0.5075	0.1114	0.1077	0.5065	0.1133
1M	0.1072	0.3639	0.1136	0.1098	0.3617	0.1158
2M	0.1139	0.1223	0.1179	0.1173	0.1207	0.1207
2.5M	0.1197	0.1198	0.1198	0.1213	0.1214	0.1217

Table 13. Water mass fractions at the solution bottom, inside cell, and top of the cell layer for a solution initially 30 μm thick and a lymphocyte cell layer with different amounts of intra- and extracellular trehalose. The substrate is made of aluminum and is heated by a 2 min-3 min-2 min (20°C-37°C-20°C) wave pattern during forced convection and a 4 min-2 min-4 min (20°C-37°C-20°C) wave pattern during natural convection.

Forced convection				
Initial molarity of trehalose	ω_A $(z = C(t))$	ω_A **(Inside cell layer)**	ω_A $(z = \delta_{sub})$	t **(min)**
0.5M	0.0954	0.5075	0.1006	25.3
1M	0.0977	0.3650	0.1047	25.4
2M	0.0996	0.1223	0.1067	40.3
2.5M	0.1034	0.1034	0.1060	65.3
Natural convection				
Initial molarity of trehalose	ω_A $(z = C(t))$	ω_A **(Inside cell layer)**	ω_A $(z = \delta_{sub})$	t **(min)**
0.5M	0.0961	0.5058	0.1024	42.8
1M	0.0973	0.3620	0.1034	48.8
2M	0.0999	0.1217	0.1060	66.9
2.5M	0.1037	0.1037	0.1042	96.8

and natural convection. The solution is on a 1 mm thick aluminum substrate being heated with controlled heating wave patterns. The first observation is that the controlled substrate heating certainly removes more water from the solution and in a faster drying time (compare water mass fraction values in

Table 13 and their respective drying times to the water mass fraction values in Table 12, which are after 2 hours of drying). The second observation is that variation of parameters (substrate material, heating wave patterns, etc.) does not appear to have a great effect on lowering the residual water content inside the cell layer for intra- and extracellular trehalose concentrations lower than 2M. This is likely due to the membrane characteristics of the biologic to-be-preserved. Lower concentrations of trehalose correlate to higher concentrations of water and only a certain amount of water is allowed to transport across the semi-permeable membrane. However, increasing trehalose concentrations as well as varying heating conditions does result in greater chances of achieving a glassy state inside the cell layer for both forced and natural convection (compare water mass fraction values in Tables 12 and 13 for 2M and 2.5M trehalose).

To summarize, the model used in Sections 4.1–4.9 was subjected to a variety of boundary conditions in order to find an optimal scenario in which a glassy state could be achieved at the bottom of the solution film. The model used in this section shows that even if the best combination of boundary conditions are made, there is still a vital role that intracellular trehalose plays in achieving a glassy state inside the cell. The limited success of cellular viability seen in the literature substantiates this claim. The most important result found is that even if glass is formed above and below the cell layer, glass may not form inside the biologic due to the membrane properties of mammalian cells. Higher amounts of intracellular trehalose appear to be possible a solution to this issue. However, additional parameters such as variation of base and ambient heating will need to be used in addition to increasing intracellular trehalose for formation of a stable glass within the solution and cell.

5. Discussison

Our work is an extension of the thin film work performed by our group and others [Chen *et al.*, 2006; Oliveira, 2012; He *et al.*, 2008; Chakraborty *et al.*, 2011]. Results show that most of the desiccation work performed experimentally would have moisture content at the film bottom that would require storage closer to refrigeration temperatures. We have performed a variety of theoretical approaches to determine if room temperature glasses could be obtained. Results show that conditions could be optimized to encapsulate the biologics in a glassy film and avoid any protrusion effect. This is predominately achieved through heating the substrate and environment in a controlled, sinusoidal manner. However, while this leads to the rapid formation of a glass at the bottom of aqueous trehalose solutions, it was also observed that water content remained high inside the cell layer. This would prevent formation of a stable glass and poses as a challenge in preserving the biologic. This may be one of the main reasons for the undesirable outcome of many post-desiccation cell viability studies [Bhowmick *et al.*, 2003; Puhler *et al.*, 2001; Gordon *et al.*, 2001; Mittal *et al.*, 2008]. Acker *et al.* [2002] argues that even if glass forms in the solution, glass is perhaps not being completely formed inside the cell. This could be due to the rate of water leaving the cytoplasm being very slow since the solution outside the cell has been almost completely vitrified. Sitaula *et al.* [2009] claimed that membrane integrity is poor due to an osmotic shock, which supports Acker *et al.*'s argument. Many studies have stated that for successful long-term preservation, more attention should be devoted to either post-desiccation protocol or alternative drying methods. The most notable alternative method, spin drying by Charkraborty *et al.* [2011], mainly quickens the desiccation process and smoothens out irregularities on the solution surface. This ensures a more uniform glass coating but poor cell viability remains an issue. Charkraborty *et al.* [2011] could not measure water content of post-desiccated cells due to equipment limitations, but suggested that their cells did not survive rehydration due to incomplete glass formation. Our method of controlled substrate and environment heating achieves a similar goal as Charkraborty *et al.* [2011] by quickly forming a glassy surface and clearly shows that incomplete glass formation inside the cell remains an issue. Based on these studies, a clear recommendation that can emerge for convective desiccation protocols is storage at lower temperatures like 4°C for long-term storage stability.

The issue with the final residual water content inside the cell being too high is challenging to overcome. Our simulations including a human lymphocyte cell layer show that even if there is over 2M of intracellular trehalose, the final mass fraction of water in the cell layer is above 10 wt.%. This is due to transport limitations which vary depending on cell type. Once lymphocytes shrink to roughly 40% of their initial volume, the cell membrane becomes impermeable and further water transport across the membrane discontinues. This would also happen to Pacific oyster spermatozoa cells after they shrink to approximately 60% of their original volume [Chen *et al.*, 2006]. Platelets, on the other hand, have been successfully preserved [Wolkers *et al.*, 2001; Crowe *et al.*, 2005; Zhou *et al.*, 2006; Pietramaggiori *et al.*, 2007]. Their membrane properties make this possible; there is no determinant "dead volume". If our simulations were altered to allow for water transport past the dead volume, the transport will continue until equilibrium is reached (results not shown). Therefore, the final water mass fraction inside the cell would be similar to the water mass fraction at the top of the cell layer and bottom of the solution. This behavior is similar to tardigrades and other low-level organisms in nature who naturally build up large amounts of trehalose to survive in a dehydrated state.

Our simulation results show that for cells with transport-limited membrane permeabilities, such as lymphocytes, more trehalose needs to be added for an increased chance of viability. However, this raises concerns, as increasing the amount of intra- and extracellular trehalose will raise the final residual mass fraction of water above the cell layer. Furthermore, high intracellular concentrations of trehalose can be toxic for the mammalian cell. Cytotoxicity has been reported for intracellular concentrations above 0.15M [Puhler *et al.*, 2001]. However, there is some debate to these values. Matsuo [2001] found that 0.05M, 0.1M, or 0.2M trehalose provided adequate desiccation tolerance to human corneal epithelial cells and reported no obvious toxicity effects at any concentration. McGinnis *et al.* [2005] was able to successfully preserve mouse spermatozoa using 0.5M trehalose. Chen *et al.* [2001] state that intracellular trehalose concentrations greater than 0.2M are necessary for adequate desiccation tolerance. This would agree more with actual anhydrobiotic organisms in nature, who can accumulate roughly 0.5M of trehalose before drying [2001]. It was shown in Section 4.10 that appropriate changes to the desiccation model (i.e. using more conductive substrate materials along with sinusoidal substrate heating) can lead to lower water content within the cell at higher trehalose concentrations. However, the amount of intracellular trehalose required is quite high (< 2M). This may be toxic to the biologic-to-be-preserved. It is also seen Section 4.10 that as the amount of intracellular trehalose increases, so does the resistance to evaporation (more difficult to achieve residual water mass fractions of 0.09–0.10 on top of the cell layer). The final residual water mass fraction within the cell is predominantly dependent on the specific membrane permeability properties of the biologic.

Due to these membrane permeability-related transport limitations, successful desiccation of membranous materials such as mammalian cells is very difficult. In addition, the need for increased trehalose concentrations (to lower the water content in the cell) increases the resistance to evaporation greatly. In order to achieve the goal of preserving a biologic at room temperature, the water mass fraction must be below 10 wt.%. The majority of time during desiccation is going from a water mass fraction of 0.11 to 0.09 (due to extremely high resistance). Instead of storing the desiccated media at room temperature (a glass transition temperature of around 22°C and water mass fraction of approximately 0.09), the biologics could be stored at 4°C (correlating to a water mass fraction of around 0.117). Most of the final residual water mass fractions seen in the simulations were around 0.10, which is close to a glass transition temperature of 14–15°C. These temperatures are still above freezing. While the techniques presented herein (sinusoidal heating) show improvement to the lowering of residual water content values, it looks that storage temperatures lower than room temperature are still necessary post-desiccation.

Another issue that must be taken into account is the liquid water/solute concentration distribution. Ultimately, there is a distribution which effects cellular viability (an issue Charkraborty *et al.* [2011] attempted to remedy). Aksan *et al.* [2005] have stated that the drying rate is affected by glass formation at the early stages of desiccation. This glass formation causes changes in the solution mixture and controls the final droplet shape. This behavior has been seen in a recent study [Oliveira, 2012]. So while the method of using a controlled heating pattern provides promising results, it is very likely that a problem of an inhomogeneous mixture during desiccation still exists.

Thin films are basically droplets with very small contact angles (C.A.). Therefore they are similar to droplets on hydrophilic surfaces. It has been seen in many studies that an internal flow can be induced by surface tension, temperature, and concentration gradients within the solution. This is known as Marangoni flow and has been studied extensively [Hu and Larson, 2002; 2005; Girard *et al.*, 2008; Fanton *et al.*, 1996; Nicolov *et al.*, 2002; Tadmor, 2009; Xu and Luo, 2007]. One explanation for this poor concentration distribution in aqueous trehalose solutions is non-uniform evaporation along the solution-vapor contact line. This non-uniformity causes Marangoni flow to develop which then carries solute particles (trehalose) to the edge of the drop. This solute builds up along the edge due to the small contact angle, causing what is commonly referred to as the "coffee-ring" effect [Deegan *et al.*, 1997]. This also may be responsible for forming depressions in the solution center (which now has more water concentration), causing cellular protrusion and death (as seen in Figures 16 and 17). Adams [2007] studied vesicles being preserved droplets on hydrophilic surfaces (C.A. = 53°) and saw that glass formed from the outer edge and progressed along the contact line. In that study, the Marangoni flow was neglected but the glass formation pattern observed does indicate that solute first builds up near the contact point. This agrees with what has been seen experimentally in [Oliveira, 2012] (see Figure 17).

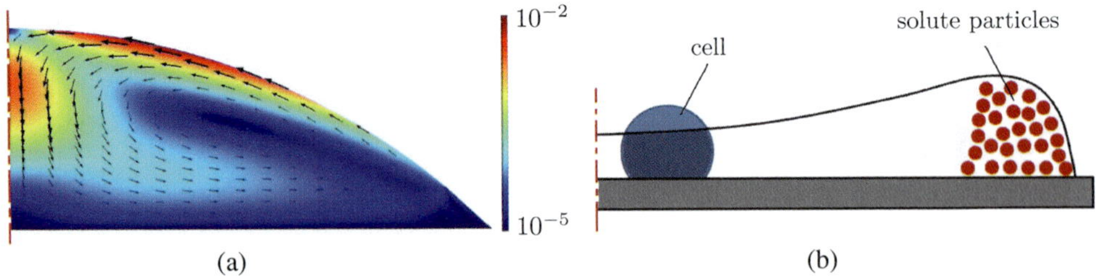

Figure 16. Example of Marangoni flow in a hydrophilic drop (50° C.A.). (a) shows the velocity magnitude (m/s) and associated flow pattern. This flow pattern can cause the solute (trehalose) to build up along the contact point. After a period of time, this causes a depression in the middle of the droplet and expose the biologic to-be-preserved, as shown in (b).

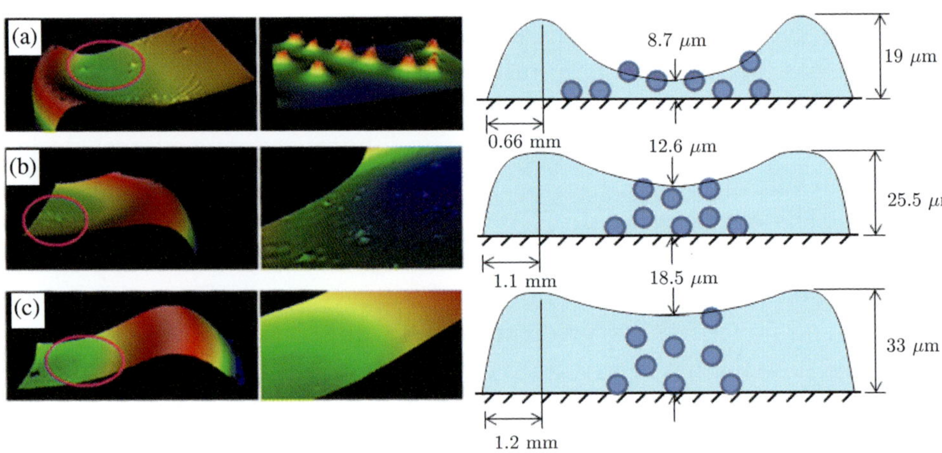

Figure 17. Profilometer scans of dried 10 mm diameter samples on glass with different starting concentrations of trehalose: (a) 0.2M (b) 0.6M (c) 1M (data from [Oliveira, 2012]). The image on the left shows a broad scan of a symmetrical half of the drop; the complete profile with dimensions of the final dried drop are recreated in a cartoon form on the right. A zoomed in view is shown with the pictures in the middle to highlight the protrusion of cells from the samples. Clearly a higher sugar concentration allows uniform droplet desiccation and protection of cells. This leads to higher cellular viability (data not shown) [Oliveira, 2012].

Hydrophobic (C.A. > 90°) and superhydrophobic (C.A. > 150°) surfaces produce droplets with much higher contact angles. If the internal Marangoni flow pattern is similar to that of droplets on hydrophilic surfaces, perhaps the higher contact angles will allow solute to not become trapped at the contact point. The authors are developing a fully coupled two-phase flow aqueous trehalose solution model using COMSOL. The model couples the Two-Phase Flow (Moving Mesh), Heat Transfer in Fluids and Solids, Transport of Diluted Species (water vapor concentration in air), and Transport of Concentrated Species (for liquid water and trehalose concentration in the drop) interfaces. This way a more realistic scenario can be simulated by capturing all the necessary physics that occur in the drying process. Preliminary results show that the Marangoni flow pattern is counterclockwise (see Figure 18). This is supported by experimental findings [Dicuangco *et al.*, 2014] and numerical studies [Tam *et al.*, 2009; Pan *et al.*, 2013]. Since the flow pattern moves upwards along the interface, reaches the top, and has a large magnitude downward, it is likely solute will be deposited over and over again along the droplet bottom, where the cells would be. Figure 19 shows that glass starts to form from the top of the droplet and progress downward, unlike the hydrophilic droplet where glass starts forming from the droplet-surface contact line. The internal flow pattern remains counterclockwise within the drop. This could certainly help avoid cell protrusion and solute buildup.

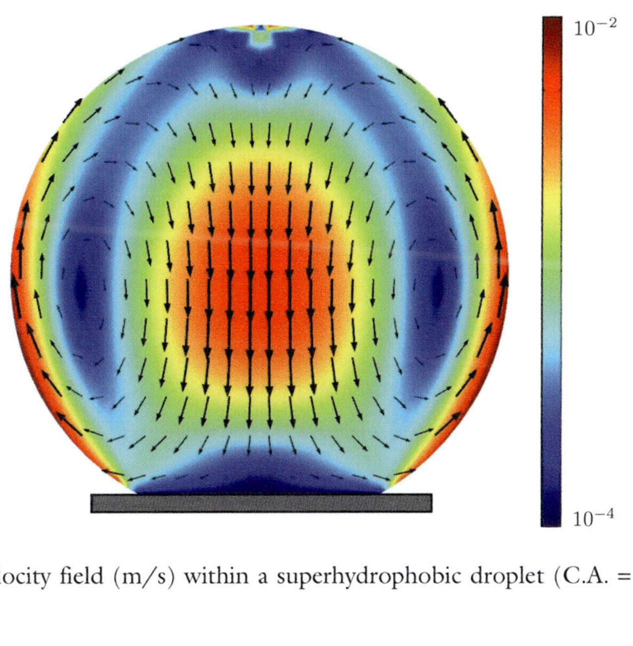

Figure 18. Velocity field (m/s) within a superhydrophobic droplet (C.A. = 150°) at t =10 seconds.

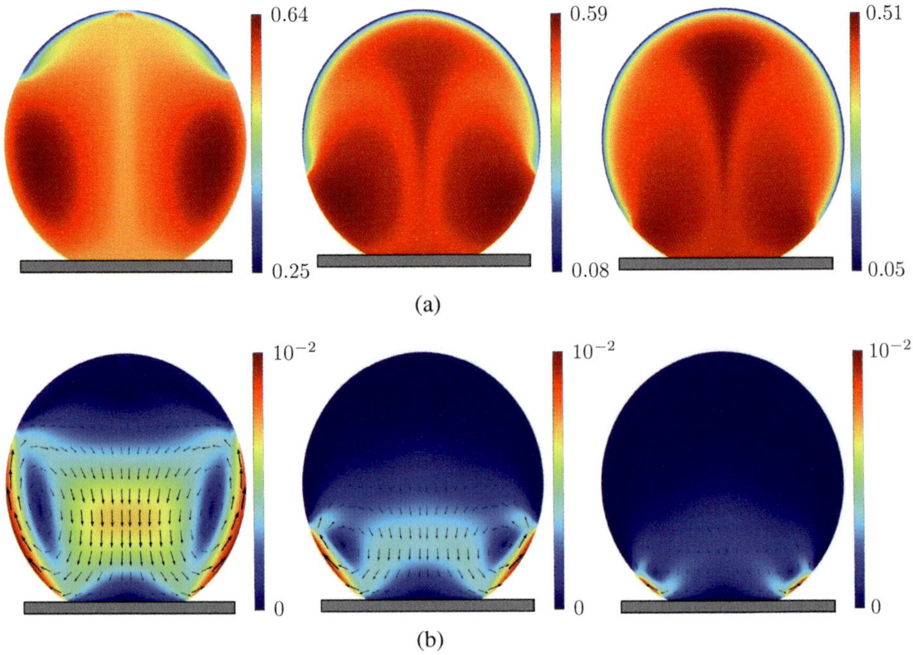

Figure 19. Plots of the water mass fraction and velocity magnitude (m/s) for a superhydrophobic droplet (150° C.A.). (a) shows the progression of glass (very low mass fraction starts at top and moves downward). (b) shows that as time progresses, the glass formation starts to diminish the magnitude of the Marangoni flow near the top of the droplet (due to increased viscosity) but the counterclockwise flow pattern is maintained.

Currently, work is being done to increase the robustness of the model. At this time of writing, the two-phase flow droplet model can only simulate natural convection scenarios. It is known that Marangoni flow can be manipulated to assist in enhancing heat transfer [Thokchom *et al.*, 2014; Mandal and Bakshi, 2012; Lu *et al.*, 2011; Majumder *et al.*, 2012]. For example, Thokchom *et al.* [2014] have modified the Marangoni flow pattern in pinned sessile water drops (and thus, solute particle deposition) by heating the free surface of the droplet using Infrared light. We plan to explore techniques such as this as well as methods described in this paper (controlled heating of the substrate and environment) for our application of preserving biologics. Therefore, it will be important to study the glass formation pattern, evaporation behavior, and solute concentration profiles within the droplet. We plan to ultimately extend the model by simulating a single cell within the droplet to include water transport between the cell and solution. More powerful computational tools may need to be utilized for this step.

6. Conclusion

Results from the current study indicate unequivocally that obtaining room temperature glasses with convective desiccation is significantly limited by transport challenges. The water content that remains at the bottom of the film for realistic desiccation times (~ hours) is fairly significant and would require refrigeration storage for long-term storage stability. Varying boundary conditions, most notably sinusodial heating patterns both from the bottom and the top most notably sped up the kinetics, but lowered the final water content at the bottom of the film (where cells are supposed to reside) marginally. Furthermore, studies using a cell model which is represented by a membrane with Starling's Law transport properties indicate that significant water could be trapped inside unless the intracellular (and therefore extracellular) concentration of trehalose is increased. These numerical findings, however, must be verified experimentally. Nonetheless, these observations are invaluable to the optimization of desiccation protocol. In order to further investigate the possibility of convective desiccation as a tool for obtaining stable glasses near ambient temperature, variation in droplet geometries like using hydrophobic surfaces may be necessary.

References

Acker, J. P., A. Fowler, B. Lauman, S. Cheley and M. Toner (2002). Survival of desiccated mammalian cells: beneficial effects of isotonic media. *Cell Preservation Technology*, 1(2), pp. 129–140. doi:10.1089/153834402320882638.

Adams, D. (2007). *Adsorption, Encapsulated Solute Leakage and Microow of Giant Vesicles during Anhydrobiotic Preservation in Trehalose Solutions.* Massachusetts Institute of Technology, Department of Chemical Engineering. URL http://books.google.com/books?id=5ZONtgAACAAJ

Aksan, M. T. A. and S.C. Morris (2005). Analysis of desiccation and vitrification characteristics of carbohydrate films by shear-wave resonators. *Langmuir*, 21(7), pp. 2847–2854. doi:10.1021/la047760y.

Bejan, A. (2004). *Convection Heat Transfer.* John Wiley and Sons.

Bhowmick, S., C. A. Khamis and J. C. Bischof (1998). Response of a liver tissue slab to an hyperosmotic sucrose boundary condition: microscale cellular and vascular level effects. *Annals of the New York Academy of Sciences*, 858(1), pp. 147–162.

Bhowmick, S., L. Zhu, L. McGinnis, J. Lawitts, B. D. Nath, M. Toner and J. Biggers (2003). Desiccation tolerance of spermatozoa dried at ambient temperature: production of fetal mice. *Biology of Reproduction*, 68(5), pp. 1779–1786.

Chakraborty, N., A. Chang, H. E. M. A. Menze, S. C. Hand and M. Toner (2011). A spin-drying technique for lyopreservation of mammalian cells. *Annals of Biomedical Engineering*, 39(5), pp. 1582–1591. doi:10.1007/s10439-011-0253-1.

Chen, B., A. Fowler and S. Bhowmick (2006). Forced and natural convective drying of trehalose/water thin films: implication in the desiccation preservation of mammalian cells. *Journal of Biomechanical Engineering*, 128(3), pp. 335–346. doi:10.1115/1.2187051.

Chen, T., J. P. Acker, A. Eroglu, S. Cheley, H. Bayley, A. Fowler and M. Toner (2001). Beneficial effect of intracellular trehalose on the membrane integrity of dried mammalian cells. *Cryobiology*, 43(2), pp. 168–181. doi:http://dx.doi.org/10.1006/cryo.2001.2360.

Chen, T., A. Fowler and M. Toner (2000). Literature *Review: Supplemented Phase Diagram of the Trehalose Water Binary Mixture. Cryobiology*, 40(3), pp. 277–282. doi:http://dx.doi.org/10.1006/cryo.2000.2244.

Copelan, E. A. (2006). Hematopoietic stem-cell transplantation. *New England Journal of Medicine*, 354(17), pp. 1813–1826, pMID: 16641398. doi:10.1056/NEJMra052638.

Crowe, J. H., J. F. Carpenter and L. M. Crowe (1998). The role of virtrification in anhydrobiosis. *Annual Review of Physiology*, 60, pp. 73–103. doi:10.1146/annurev.physiol.60.1.73.

Crowe, J. H. and L. M. Crowe (2000). Preservation of mammalian cells—learning Nature's tricks. *Nature Biotechnology*, 18(2), 145–146. doi:10.1038/72580.

Crowe, J. H., L. M. Crowe and J. F. Carpenter (1993). Preserving dry biomaterials: the water replacement hypothesis, part II. *BioPharm International*, 6, pp. 40–44.

Crowe, J. H., L. M. Crowe, W. F. Wolkers, A. E. Oliver, X. Ma, J.-H. Auh, M. Tang, S. Zhu, J. Norris and F. Tablin (2005). Stabilization of dry mammalian cells: lessons from nature. *Integrative and Comparative Biology*, 45(5), pp. 810–820. doi:10.1093/icb/45.5.810.

Crowe, L. M., D. S. Reid and J. H. Crowe (1996). Is trehalose special for preserving dry biomaterials? *Biophysical Journal*, 71(4), pp. 2087–2093. doi:10.1016/S0006-3495(96)79407-9.

de Castro, A. G. and A. Tunnacliffe (2000). Intracellular trehalose improves osmotolerance but not desiccation tolerance in mammalian cells. *FEBS Letters*, 487(2), pp. 199–202. doi:http://dx.doi.org/10.1016/S0014-5793(00)02340-1.

Deegan, R. D., O. Bakajin, T. F. Dupont, G. Huber, S. R. Nagel and T. A. Witten (1997). Capillary flow as the cause of ring stains from dried liquid drops. *Nature*, 389, pp. 827–829. doi:10.1038/39827.

Dicuangco, M., S. Dash, J. A. Weibel and S. V. Garimella (2014). Effect of superhydrophobic surface morphology on evaporative deposition patterns. *Applied Physics Letters*, 104(20). doi:http://dx.doi.org/10.1063/1.4878322.

Elliott, G. D., X.-H. Liu, J. L. Cusick, M. Menze, J. Vincent, T. Witt, S. Hand and M. Toner (2006). Trehalose uptake through P2X7 purinergic channels provides dehydration protection. *Cryobiology*, 52(1), pp. 114–127. doi:http://dx.doi.org/10.1016/j.cryobiol.2005.10.009.

Fanton, X., A. Cazabat and D. Quéré (1996). Thickness and shape of films driven by a Marangoni flow. *Langmuir*, 12(24), pp. 5875–5880.

Finger, E. B. (2008). Organ preservation. URL http://emedicine.medscape.com/article/431140-overview.

Girard, F., M. Antoni and K. Sefiane (2008). On the effect of Marangoni flow on evaporation rates of heated water drops. *Langmuir*, 24(17), pp. 9207–9210.

Gordon, S. L., S. R. Oppenheimer, A. M. Mackay, J. Brunnabend, I. Puhlev and F. Levine (2001). Recovery of human mesenchymal stem cells following dehydration and rehydration. *Cryobiology*, 43(2), pp. 182–187. doi:http://dx.doi.org/10.1006/cryo.2001.2361.

Guo, N., I. Puhlev, D. R. Brown, J. Mansbridge and F. Levine (2000). Trehalose expression confers desiccation tolerance on human cells. *Nature Biotechnology,* 18(2), pp. 168–171.

He, X., A. Fowler, M. Menze, S. Hand and M. Toner (2008). Desiccation kinetics and biothermodynamics of glass forming trehalose solutions in thin films. *Annals of Biomedical Engineering,* 36(8), pp. 1428–1439.

He, X., A. Fowler and M. Toner (2006). Water activity and mobility in solutions of glycerol and small molecular weight sugars: implication for cryo- and lyopreservation. *Journal of Applied Physics,* 100(7), 074702.

Hu, H. and R. G. Larson (2002). Evaporation of a sessile droplet on a substrate. *The Journal of Physical Chemistry B,* 106(6), pp. 1334–1344.

Hu, H. and R. G. Larson (2005). Analysis of the effects of Marangoni stresses on the microow in an evaporating sessile droplet. *Langmuir,* 21(9), pp. 3972–3980.

Hubel, A. and M. Waxberg (2006). Preservation of cell-based products for human therapeutic application. *Regulatory Affairs Focus,* pp. 36–37.

Incropera, F. and D. DeWitt (2002). *Fundamentals of Heat and Mass Transfer.* John Wiley and Sons.

Leslie, S. B., E. Israeli, B. Lighthart, J. H. Crowe and L. M. Crowe (1995). Trehalose and sucrose protect both membranes and proteins in intact bacteria during drying. *Applied Environmental Microbiology,* 61(10), pp. 3592–3597.

Lide, D. R. (2005). *CRC Handbook of Chemistry and Physics,* Internet Version.

Lu, G., Y.-Y. Duan, X.-D. Wang and D.-J. Lee. (2011). Internal flow in evaporating droplet on heated solid surface. *International Journal of Heat and Mass Transfer,* 54(19), pp. 4437–4447.

Majumder, M., C. S. Rendall, J. A. Eukel, J. Y. Wang, N. Behabtu, C. L. Pint, T.-Y. Liu, A. W. Orbaek, F. Mirri, J. Nam, *et al.* (2012). Overcoming the "coffee-stain" effect by compositional Marangoni flow-assisted drop-drying. *The Journal of Physical Chemistry B,* 116(22), pp. 6536–6542.

Mandal, D. K. and S. Bakshi (2012). Internal circulation in a single droplet evaporating in a closed chamber. *International Journal of Multiphase Flow,* 42, pp. 42–51.

Matsuo, T. (2001). Trehalose protects corneal epithelial cells from death by drying. *British Journal of Ophthalmology,* 85(5), pp. 610–612.

McGinnis, L. K., L. Zhu, J. A. Lawitts, S. Bhowmick, M. Toner and J. D. Biggers (2005). Mouse sperm desiccated and stored in trehalose medium without freezing. *Biology of Reproduction,* 73(4), pp. 627–633.

Mittal, S. and R. V. Devireddy (2008). Desiccation tolerance of adult stem cells in the presence of trehalose and glycerol. *The Open Biotechnology Journal,* 2, pp. 211–218.

Nikolov, A. D., D. T. Wasan, A. Chengara, K. Koczo, G. A. Policello and I. Kolossvary (2002). Superspreading driven by Marangoni flow. *Advances in Colloid and Interface Science,* 96(1), pp. 325–338.

Noiles, E., P. Mazur, P. Watson, F. Kleinhans and J. Critser (1993). Determination of water permeability coefficient for human spermatozoa and its activation energy. *Biology of Reproduction,* 48(1), pp. 99–109.

Norris, M., A. Aksan, K. Sugimachi and M. Toner (2006). 3-O-methyl-D-glucose improves desiccation tolerance of keratinocytes. *Tissue Engineering* 12(7), pp. 1873–1879. doi:10.1089/ten.2006.12.1873.

Oliveira, M. (2012). Biophysical impact of trehalose in mammalian cell desiccation preservation: convective desiccation of a sessile droplet, Master's Thesis, University of Massachusetts Dartmouth, North Dartmouth, MA.

Pan, Z., S. Dash, J. A. Weibel and S. V. Garimella (2013). Assessment of water droplet evaporation mechanisms on hydrophobic and superhydrophobic substrates. *Langmuir*, 29(51), pp. 15831–15841.

Pietramaggiori, G., A. Kaipainen, D. Ho, C. Orser, W. Pebley, A. Rudolph and D. P. Orgill (2007). Trehalose lyophilized platelets for wound healing. *Wound Repair and Regeneration*, 15(2), pp. 213–220.

Puhlev, I., N. Guo, D. R. Brown and F. Levine (2001). Desiccation tolerance in human cells, *Cryobiology*, 42(3), pp. 207–217. doi:http://dx.doi.org/10.1006/cryo.2001.2324.

Sitaula, R., H. Elmoazzen, M. Toner and S. Bhowmick (2009). Desiccation tolerance in bovine sperm: A study of the effect of intracellular sugars and the supplemental roles of an antioxidant and a chelator, *Cryobiology*, 58(3), pp. 322–330. doi:http://dx.doi.org/10.1016/j.cryobiol.2009.03.002.

Tadmor, R. (2009). Marangoni flow revisited. *Journal of Colloid and Interface Science*, 332(2), pp. 451–454.

Tam, D., V. von Arnim, G. McKinley and A. Hosoi (2009). Marangoni convection in droplets on superhydrophobic surfaces. *Journal of Fluid Mechanics*, 624, pp. 101–123.

Thokchom, A. K., A. Gupta, P. J. Jaijus and A. Singh (2014). Analysis of fluid flow and particle transport in evaporating droplets exposed to infrared heating. *International Journal of Heat and Mass Transfer*, 68, pp. 67–77.

Vogl, C., M. Miksis, S. Davis and D. Salac (2014). The effect of glass-forming sugars on vesicle morphology and water distribution during drying. *Journal of the Royal Society Interface*, 11(99). doi:10.1098/rsif.2014.0646.

Wolkers, W., S. Looper, A. McKiernan, N. Tsvetkova, F. Tablin and J. Crowe (2002). Membrane and protein properties of freeze-dried mouse platelets. *Molecular Membrane Biology*, 19(3), pp. 201–210.

Wolkers, W. F., N. J. Walker, F. Tablin and J. H. Crowe (2001). Human platelets loaded with trehalose survive freeze-drying. *Cryobiology* 42(2), pp. 79–87. doi:http://dx.doi.org/10.1006/cryo.2001.2306.

Xu, X. and J. Luo (2007). Marangoni flow in an evaporating water droplet. *Applied Physics Letters*, 91(12), pp. 124–102.

Zhou, X.-L., H. Zhu, S.-Z. Zhang, F.-M. Zhu, G.-M. Chen and L.-X. Yan (2006). Freeze-drying of human platelets: influence of intracellular trehalose and extracellular protectants. *CryoLetters*, 27(1), pp. 43–50.

Appendix A: Free Volume Model Parameters

Parameter	Value	Units
$\hat{V}_A$	1	mL/g
$\hat{V}_B$	0.616	mL/g
M_A	18.02	g/mol
M_B	342.3	g/mol
$\tilde{V}_A$	18	mL/mol
$\tilde{V}_B$	211	mL/mol
$\hat{V}_A^*$	0.91	mL/g
$\hat{V}_B^*$	0.587	mL/g
$T_{g,A}$	136	K
$T_{g,B}$	387.15	K
D_0	1.39×10^{-7}	m^2 /s
ΔE	1.98×10^3	J/mol
K_{11}/γ	1.945×10^{-3}	mL/(g·K)
K_{21}	–19.73	K
K_{12}/γ	2.16×10^{-4}	mL/(gvK)
K_{22}	89.9	K
$(C_{p,g} - C_p)$	–0.566	J/(g·K)
κ	0.17	—
χ	0.22	—
ζ	0.75	—
W	0.198	—
$\eta_{g,0}$	2.27×10^4	Pa·s
$k_{n,g}$	0.13	mol/mL
C_1	13.15	—
C_2	K_{22}	K

List adapted from Adams [2007] and He *et al.* [2008].

Section 2

Microscale

Chapter 4

Microscale Materials and Devices for Cell Cryopreservation by Vitrification

Haishui Huang and Xiaoming He

1. Introduction

Cell cryopreservation is an enabling technology for the burgeoning field of cell-based medicine, including assisted reproduction [Boldt *et al.*, 2003; Menezo, 2004; Rizzino, 2010; Hunt, 2011]. This is because continuous expansion of cells by culturing at 37 °C is expensive and time-consuming, and may result in differentiation of stem cells that are especially in high demand due to their capability of differentiating into various types of cells for tissue regeneration [Niwa *et al.*, 2000]. Cell cryopreservation can be utilized to overcome these challenges by maintaining cells in a state of suspended animation at cryogenic temperatures [He *et al.*, 2008; Lee, *et al.*, 2010b; Zhang *et al.*, 2010]. It is achieved by first cooling and storing biological materials (e.g., live cells and tissues) at deep cryogenic temperatures (e.g., −196 °C) where all biophysical and biochemical activities and associated damages are arrested, and then warming them back to desired temperatures (e.g., 37 °C) to resume normal cellular activities at a desired future time [Mazur, 1970]. For assisted reproduction, ovarian tissues, follicles, sperms, oocytes, or embryos can be isolated from animals or humans, cooled to and stored at cryogenic temperatures, and warmed back for maturation, fertilization, and/or development at a desired time or location [Chen, 1986; Isachenko *et al.*, 2009; Riggs *et al.*, 2010; Vutyavanich *et al.*, 2010; Edgar and Gook, 2012; Bastings *et al.*, 2013].

However, a non-optimized cryopreservation procedure could cause damage to the biological materials. Similar to hyperthermic injuries, the cryothermic damages can be investigated at the molecular, cellular, and tissue (both *in vitro* and *in vivo*) levels [He, 2011b]. At molecular level, both lipid phase change and phase separation that can induce damages to the cell membrane have been observed during cell dehydration at the low temperature of cryopreservation [Steponkus and Lynch, 1989; Yamada *et al.*, 2002; Bischof, 2006]. Protein denaturation that is usually demonstrated by the conformational alterations from α-helix to β-sheet has been observed in cells with freezing [Bischof *et al.*, 2002]. The conformational change is a result of the excursions of temperature, pH, and solute concentrations away from the physiologic condition during cell cryopreservation [Bischof and He, 2005]. At cellular level, cell injury is usually attributed to the two classical mechanisms: freeze concentration-induced cell dehydration and intracellular ice formation (IIF) during cryopreservation [Mazur *et al.*, 1972]. Cell injuries due to freeze concentration are usually associated with lipid and protein alterations at the molecular level. IIF is often regarded as a lethal event to a living cell during cryopreservation because it can not only lead to freeze concentration inside cells due to the precipitation of solutes as a result of the formation of ice crystals, but also mechanically destroy the cellular plasma membrane and sub-cellular organelles due to the volumetric expansion associated with phase change from water to ice and sharp dendritic front of ice crystals [Mazur, 1984; Karlsson *et al.*, 1994; Yang *et al.*, 2009]. At the tissue level, besides possible crack formation [Rabin *et al.*, 1996; He and Bischof, 2005], cryopreservation could cause more injuries (in terms of necrosis) to *in vivo* tissues than *in vitro* ones [Hoffmann and Bischof, 2001; Rupp *et al.*, 2002; Chao *et al.*, 2004]. This is because the freezing and thawing processes can cause damage to the vasculature of an *in vivo* tissue, which results in the deprivation of nutrients and oxygen to cells [Hoffmann and Bischof, 2001]. This phenomenon induces a secondary killing mechanism to cells *in vivo*, in addition to the aforementioned direct cell injury. However, in contrast to the *in vivo* hyperthermic lesion, the wound healing has been observed throughout the whole *in vivo* cryolesion. This is probably due to the temporary reperfusion immediately after thawing and no vascular stasis would occur until a few hours later [Rupp *et al.*, 2002; Coad and Bischof, 2003].

Two approaches have been adopted to improve the viability of cryopreserved cells: slow freezing with extracellular ice formation and ice-free vitrification [Fahy *et al.*, 1984; Mazur, 1984; Rall and Fahy, 1985; Karlsson and Toner, 1996; Fahy *et al.*, 2004b]. In addition, a certain amount of cryoprotective agents (CPAs) such as dimethylsulfoxide (DMSO), 1,2-propanediol (PROH), glycerol, and ethylene glycol (EG) is often added in the cryopreservation solution. A typical slow-freezing procedure is illustrated in the phase diagram

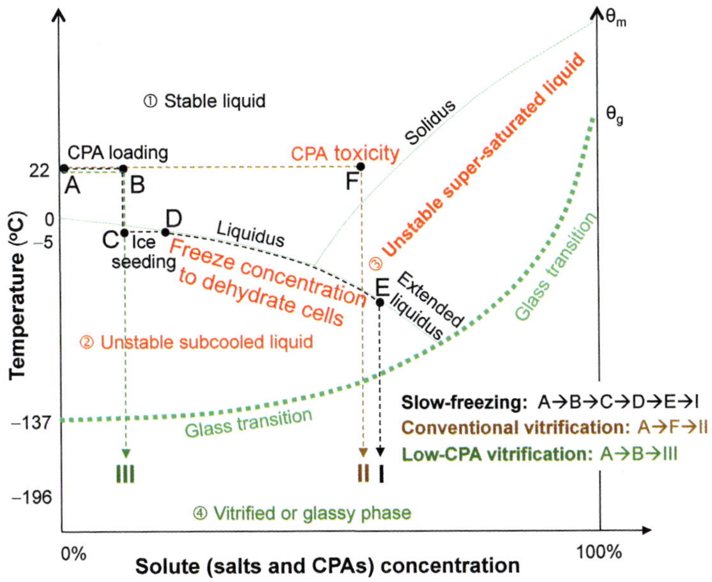

Figure 1. A schematic illustration on the extended phase diagram of the three different approaches to achieve cell cryopreservation from an initial liquid state (A) to a final glassy phase (I, II, or III). The phase diagram is divided into four thermodynamic regimens by the liquidus, extended liquidus, solidus, and the glass transition curve; the four regimens are the liquid, subcooled liquid, super-saturated liquid, and the glassy phase; cells (in a liquid solution initially) must enter the glassy phase for long-term cryopreservation; CPA represents cryoprotective agent, and θg and θm represent the glass transition and melting temperature of solute, respectively. Figure reprinted from [He, 2011b] with permission from Xiaoming He [the author].

(Figure 1) as the procedure A→B→C→D→E→I. After loaded with a low concentration of CPAs (usually < ~2 M, A→B), the biological samples are first cooled to a high subzero temperature (usually > −10 °C, B→C) for seeding ice in the extracellular solution (point C in Figure 1) by locally cooling the sample with a deeply cooled object (e.g., needles or metal blocks pre-cooled in liquid nitrogen). Ice seeding can also be realized by adding a small amount of a natural ice nucleator (e.g., silver iodide and *Pseudomonas syringae* that nucleate ice at ~ −6 and −2 °C, respectively, [Searles *et al.*, 2001; Rambhatla *et al.*, 2004; Liu *et al.*, 2005]) in the cryopreservation solution. After ice seeding, the intracellular space is still free of ice while the extracellular solution is partially transformed into ice crystals. As a result, the extracellular osmolality is higher than that inside cells due to the precipitation of solutes from ice crystals into the unfrozen extracellular solution. Therefore, the intracellular water starts to flow out because of the imbalance of osmolality across plasma membrane (C→D in Figure 1). After ice seeding, the samples are further cooled to between –40 and –100 °C with a slow cooling rate (typically < ~10 °C/min). As the temperature decreases, more extracellular ice would form and the extracellular osmolality continuously increases, and the intracellular water continuously outflows during this period (D→E). As a result of this dehydration, little intracellular water would be available for IIF. Finally, the samples are transferred into liquid nitrogen for long-term storage, where the dehydrated cells will enter a vitreous state (E→I). It is worth noting that there is an optimal cooling rate that is cell-type dependent for the slow-freezing approach. If the cooling rate is too high, the intracellular water does not have enough time for exosmosis and significant IIF ensues. On the other hand, if the cooling rate is too slow, severe shrinkage and freeze concentration can deviate a cell far from its normal physiological state, causing serious cryoinjuries at the molecular level. As a result, there is an inverted U-shaped relationship between cooling rate and cell viability [Mazur, 1984; Gao and Critser, 2000; He, 2011b].

During conventional vitrification (A→F→II, Figure 1), the biological samples are transformed into an amorphous (i.e., vitrified or glassy) phase without intracellular and extracellular ice formation altogether during the whole procedure. This is achieved by adding a high concentration of CPAs (A→F) into the vitrification solution, followed by rapidly cooling the samples to a cryogenic temperature (usually –196 °C of liquid nitrogen, F→II) so that the water molecules do not have time to reconfigure their orientations and positions to form ice crystals [Bhat *et al.*, 2005; Yang *et al.*, 2009]. Although this method can avoid the inevitable ice formation associated with slow freezing, it requires a very high concentration of CPAs (up to 8 M) (point F in Figure 1) to achieve vitrification at a rapid cooling rate (1,000–100,000 °C/min) [Fahy *et al.*, 1984; Rall and Fahy, 1985; Fahy *et al.*, 2004b]. This unusually high concentration of CPA is toxic to most mammalian cells even for a short exposure time (ranging from seconds to minutes dependent on the cell type) [Chen *et al.*, 2000; Fahy *et al.*, 2004a; Fowler and Toner, 2005; Heng *et al.*, 2005; Hunt *et al.*, 2006]. As a result, multiple steps of CPA loading (before cooling) and unloading (after warming) are needed to minimize metabolic and osmotic shocks to cells, which makes this procedure complicated, stressful, time-consuming, and difficult to reproduce [Rall and Fahy, 1985; Karlsson and Toner, 1996; Fowler and Toner, 2005; Heng *et al.*, 2005; Hunt *et al.*, 2006].

More recently, low-CPA vitrification (A→B→III, Figure 1) is attracting more and more attention for cell cryopreservation as it combines all the advantages of slow-freezing (low CPA) and conventional vitrification (no ice formation), while avoiding all of their drawbacks (ice formation and high concentrations of CPAs) [He, 2011a; 2011b]. Over the past three decades, several microscale materials and devices have been explored to reduce the CPA concentration required for achieving cell vitrification by either creating a high cooling/warming rate or modulating the kinetics of ice formation. These microscale materials and devices together with that used for conventional vitrification are discussed in more detail below.

2. Plastic Straw (PS)

Cell cryopreservation by vitrification using plastic straw (PS) with volume varying from 0.25–0.5 ml was first introduced in the middle of the 1980s [Fahy *et al.*, 1984; Rall and Fahy, 1985]. These plastic straws are usually called conventional or French straws. The cooling and warming rate that can be created with these straws is not very fast (up to ~2,000 °C/min albeit rapid compared to that for slow freezing), for which the CPA concentration required to achieve apparent vitrification is usually very high (5–8M) [Rall and Fahy, 1985; Sakai *et al.*, 1990; Carvalho *et al.*, 2011].

Due to the presence of high concentration of CPAs in the solution for PS vitrification, multiple steps of CPA loading before cooling and unloading after warming are crucial to ensure a high cell survival post-vitrification. Multiple-step CPA loading and unloading can reduce the severe osmotic stress to living cells, which can sometimes cause even more cell injuries than the cryopreservation procedure per se [Rall and Fahy, 1985]. Therefore, precise control over the CPA concentration together with the loading/unloading temperature and time at each step is crucial [Rall and Fahy, 1985; Sakai *et al.*, 1990]. In addition, the CPA loading and unloading protocol should be cell type dependent as the permeability of plasma membrane varies vastly among different types of cells [Bischof, 2000]. These variations make the conventional vitrification using PS complicated, time-consuming, stressful, and difficult to predict and/or reproduce.

3. Open Pulled Straw (OPS)

To improve the cooling and warming rates of vitrification, the PS has been pulled after heat-softening to reduce its diameter and wall thickness, and the resultant straw is often called open pulled straw (OPS) [Vajta *et al.*, 1998]. Besides improving the cooling/warming rate, the reduced size of OPS can decrease the heterogeneity of thermal distribution in the sample during cooling and warming. OPS has been successfully used to vitrify various biological materials, such as human embryonic stem cells [Reubinoff *et al.*, 2001], mouse embryos at various stages [Zhou *et al.*, 2005], and bovine ova and embryos [Vajta *et al.*, 1998]. Usually, the PS can be pulled to OPS such that the outer diameter can de decreased from 2 mm to ~0.8 mm and the wall thickness can be reduced from 0.15 mm to 0.07 mm. In addition, a small sample volume (1–2 μl) is used for vitrification (Figure 2). As a result, the cooling and warming rates can be increased by ~10 times (over 20,000 °C/min) [Vajta *et al.*, 1998]. The CPA solution usually contains 2.5–3 M EG and 2–2.5 M DMSO as penetrating CPAs and 0.3–0.5 M sucrose or trehalose as the non-penetrating CPA. Although the total penetrating CPA concentration is reduced to ~5 M, a multi-step protocol is still needed for loading and unloading the penetrating CPA to minimize osmotic shock to cells [Vajta *et al.*, 1998; Reubinoff *et al.*, 2001; Zhou *et al.*, 2005].

As shown in Figure 2 [Vajta *et al.*, 1998], a small volume (1–2 μl) of CPA-laden sample is first loaded into OPS by capillary effect. The OPS is then plunged into liquid nitrogen to vitrify the sample within a fraction of seconds. Afterwards, it is thawed back to ambient temperature in an aqueous solution, which usually contains a certain amount of sucrose as thawing solution [Vajta *et al.*, 1998; Reubinoff *et al.*, 2001; Zhou *et al.*, 2005]. After thawing and multiple steps of dilution to remove the CPAs, the cells are transferred into culture medium for further use.

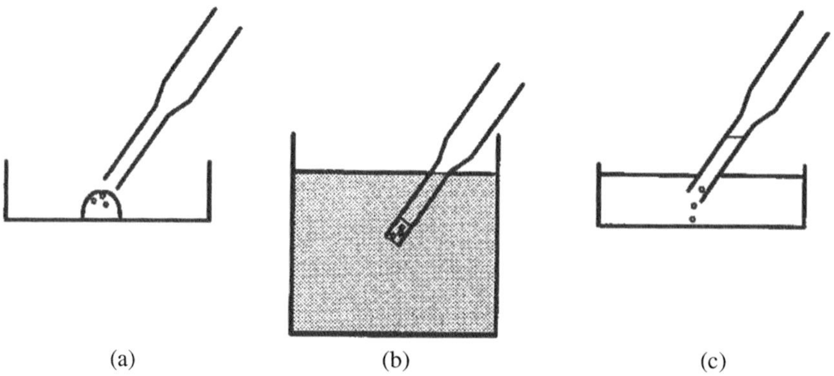

Figure 2. The open pulled straw (OPS) method. (a) Oocytes or embryos are loaded into the narrow end of the straw from a droplet by means of the capillary effect. (b) Cooling is performed by directly plunging the OPS into liquid nitrogen. (c) At warming, the OPS is plunged into cell culture medium, and the vitrification medium becomes liquid and is diluted immediately by the medium entering the straw as a result of the capillary effect. Subsequently, the oocytes or embryos move out of the straw into the culture dish as a result of gravity. Figure reprinted and redrawn from [Vajta *et al.*, 1998] with permission from John Wiley and Sons.

4. Electron Microscopy (EM) Grid

Electron microscopy (EM) grids have been used to vitrify bovine oocytes and blastocytes and human multipronuclear zygotes and blastocytes [Park *et al.*, 1999; Park *et al.*, 2000; Son *et al.*, 2003]. In order to enhance the cooling/warming rate, the EM grids can be made of materials with high thermal conductivity at cryogenic temperature, such as copper. During the experiment, samples can be attached to the EM grids and then plunged into liquid nitrogen to achieve ultra-fast cooling rate. For the cryopreservation of human multipronuclear zygotes on EM grids, the CPA solution includes 5 M EG and 0.5 M sucrose and the cells were immersed in the CPA solution for ~30 s before cooling [Park *et al.*, 2000]. The results show that the viability of multipronuclear zygotes is over 85% post-vitrification, and the cleavage rate and *in vitro* development of vitrified samples are comparable to that of fresh zygotes [Park *et al.*, 2000]. For the vitrification of human blastocysts on EM grids, the volume of blastocoele on expanded blastocysts should be reduced as extensive ice crystals can form in the blastocoele during cooling (Figure 3) [Son *et al.*, 2003]. The CPA solution includes 6.5 M EG and 0.3 M sucrose. The results showed that 90% expanded blastocysts could survive post-vitrification, and 12 out of 25 patients got pregnant (48%) [Son *et al.*, 2003]. Similar to the PS and OPS, the CPA should be loaded and unloaded by multiple steps to minimize the osmotic shock to cells given the high concentration of CPAs used.

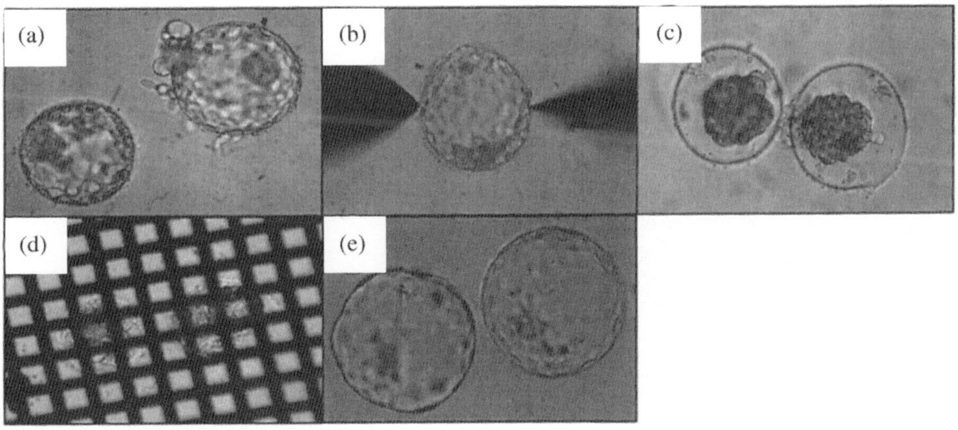

Figure 3. Cooling and warming of human expanded blastocysts on electron microscopy (EM) grids after artificial shrinkage. Human expanded blastocysts (a) before artificial shrinkage, (b) during artificial shrinkage, (c) after artificial shrinkage, (d) loaded onto EM grids, and (e) hatched blastocysts 18 h after warming/vitrification. Original magnification: 100×. Figure reprinted and redrawn from [Son *et al.*, 2003] with permission from Oxford University Press.

5. Cryoloop

In contrast to the conventional slow freezing in cryovial, cryoloop has been used to vitrify sperm at a fast cooling rate with comparable viability and mobility [Schuster *et al.*, 2003; Desai *et al.*, 2004]. As shown in Figure 4 [Schuster *et al.*, 2003], ~5 μl of sperm suspension can be inoculated on a cryoloop to form a thin film of fluid. The sperm loaded cryoloop can be cooled using two methods: (1) plunging into liquid nitrogen directly and (2) first precooling the specimen in liquid nitrogen vapor for 5 min and then plunging it into liquid nitrogen. Although higher cooling rate can be achieved by the first method, it was demonstrated that better cell viability can be obtained using the second approach [Schuster *et al.*, 2003]. For the second method, the optimal CPA is 1.64 M glycerol and ~40% sperms show motility post-warming [Schuster *et al.*, 2003]. This method is particularly useful to cryopreserve a small number of sperm for severe oligozoospermia patients.

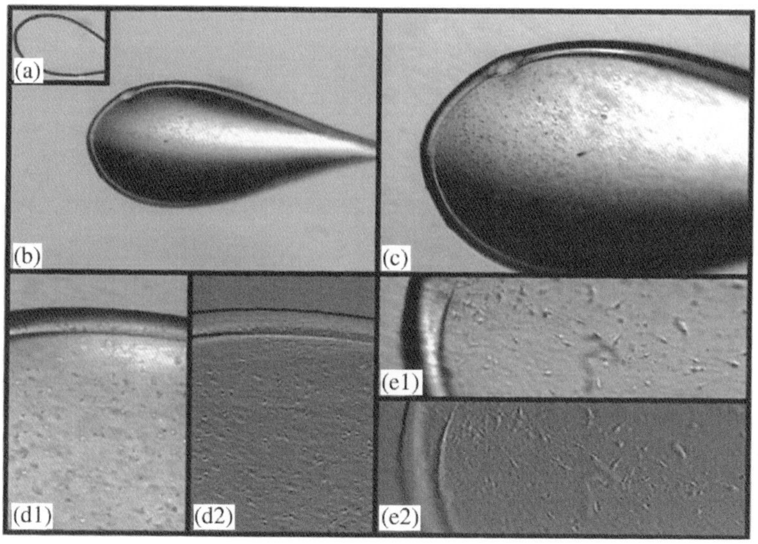

Figure 4. Ultra-rapid cooling with cryoloop. Empty cryoloop prior to sperm inoculation (a) (100×). Inoculated cryoloops at 100× (b), 200x (c), 400× (d1) and 640× (e1). To enhance visualization of individual spermatazoa, embossed photographs of 400× and 640× are shown in d2 and e2, respectively. Figure reprinted and redrawn from [Schuster *et al.*, 2003] with pemission from Oxford University Press.

6. Droplet Vitrification

As discussed before, a smaller sample size can not only enhance the cooling/warming rate but also reduce the thermal heterogeneity and the probability of volume-catalyzed nucleation of ice. Therefore, smaller samples are more likely to vitrify under the same cooling condition. Droplet vitrification is to vitrify the aqueous microdroplets containing live cells [Demirci and Montesano, 2007]. As shown in Figure 5 [Demirci and Montesano, 2007], the CPA-laden cell suspension is ejected through a needle by a pulse generator to produce microdroplets that fall into the liquid nitrogen directly for vitrification. The solidified (vitrified or un-vitrified) microparticles are collected in a cell strainer and transferred to the thawing solution for warming. Finally, the cells are re-suspended in cell culture medium by centrifugation and aspiration, before transferring into a culture flask.

As expected, the smaller the droplet size, the lower the concentration of CPA is required for vitrification (Figure 6) [Demirci and Montesano, 2007]. If the size of microdroplet can be decreased to 100 μm, the CPA (PROH here) concentration can be reduced to ~1.5 M that is similar to the CPA

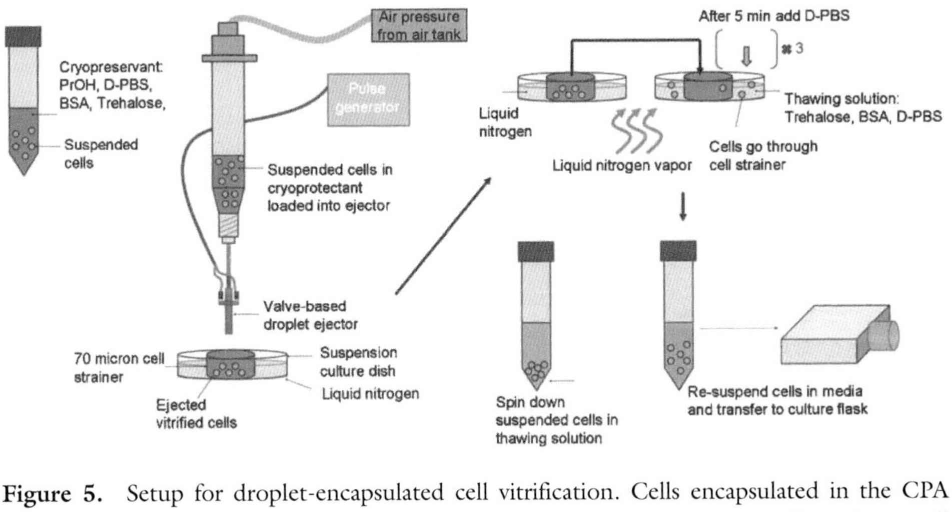

Figure 5. Setup for droplet-encapsulated cell vitrification. Cells encapsulated in the CPA droplets were ejected directly into liquid nitrogen. The ejected cells were collected in a cell strainer and were rapidly transferred (in nitrogen vapor) to the thawing media for warming. The collected cells were unloaded from the CPA following a step-by-step process to avoid osmotic shock. They were finally transferred to a flask and cultured in their media for subsequent analysis. Figure reprinted and redrawn from [Demirci and Montesano, 2007] with permission from Royal Society of Chemistry.

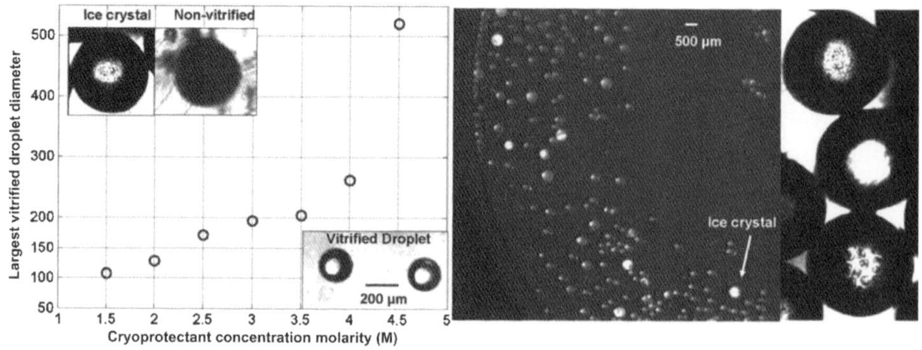

(a) (b) (c)

Figure 6. Droplet size-dependence of apparent vitrification. (a) Small droplets vitrify at low CPA concentration. The grey circles indicate the largest droplets that vitrified, indicating the vitrification size limit. The inserted image (lower right corner) shows vitrified droplets (194 mm in diameter, 3 M PROH), whereas those in the top-left corner are non-vitrified (525 mm in diameter, 4.5 M PROH). (b) Droplet morphology can be used to judge if there was ice crystal formation within the droplet (phase filter imaging). The vitrified droplets appear clear and translucent whereas the frozen non-vitrified droplets appear cloudy and dark. One non-vitrified droplet is marked with an arrow on at the bottom right. (c) A zoom-in view of the droplets reveals droplet structure. Again, vitrified droplets are clear and non-vitrified droplets appear either as dark dots or with internal crystals. The top and bottom droplets are not vitrified, whereas the middle droplet is vitrified. Figure reprinted and redrawn from [Demirci and Montesano, 2007] with permission from Royal Society of Chemistry.

concentration used for slow freezing. Five different types of mammalian cells were tested: AML-12 hepatocytes, NIH 3T3 fibroblasts, HL-1 cardiomyocytes, mouse embryonic stem cells, and RAJI cells. The diameter of the cell containing droplets is ~70 μm and 1.5 M PROH and 0.5 M trehalose were used as the CPAs. The results show that cell viability can decrease by 2–7% (depending on the cell type) after ejection only, and by 5–9% after ejection and vitrification altogether. Furthermore, high throughput is implied in this vitrification technique as the droplet generation rate can go up to 1,000 droplets/s. These results demonstrate the microdroplet vitrification with a low CPA concentration could be used to achieve a high cell viability (>90% on average) post-vitrification. However, the open nature of the approach makes it vulnerable to possible contamination from the environment including liquid nitrogen. This issue needs to be resolved before the approach is acceptable for clinical applications.

7. Cryogenic Oscillating Heat Pipe (COHP)

When a two-phase mixture in the form of liquid slugs and vapor bubbles is enclosed in a serpentine capillary tube, oscillating flow can be generated in the pipe if there are heat receiving and releasing portions within the closed pipe [Akachi, 1990]. The fluid circulation is generated by both the pressure increase associated with the volume expansion during evaporation of the liquid phase at the heat-receiving portion (evaporator) and the pressure decrease associated with the volume contraction during condensation of the vapor phase at the heat-releasing portion (condenser). The enclosed fluid can circulate at high velocity within the loop-type pipe under repeating evaporation and condensation, carrying the heat from evaporator to condenser with an ultra-high heat transfer coefficient (>10^4 W/m^2/C) [Zuo *et al.*, 2001; Natsume *et al.*, 2011]. Thus, it is called oscillating heat pipe (OHP). If cryogenic fluid, such as liquid nitrogen, is used as the working fluid, the cooling rate can be increased to more than 10^5 °C/min, which can be used to vitrify biological samples with a low concentration of CPA (1~2 M) [Boutron, 1986; Jiao *et al.*, 2006]. Given its adjustable size and integrative nature, cryogenic OHP (COHP) bears promising applications for low-CPA vitrification of various biological materials.

As shown in Figure 7 [Han *et al.*, 2008], the biological samples are compressed into a thin layer before vitrification because the thinner the sample is, the higher the heat transfer coefficient would be [Han *et al.*, 2008]. The thickness of the sample layer also depends on the cell size: for stem cells with a diameter of ~10 μm, it can be set as 50 μm; and for mouse oocytes, it can be 100 μm. The material of the container wall can be silicon as it has excellent thermal conductivity (1680 W/m/K) at cryogenic temperature (–196 °C). During the experiment, the thin layer of liquid nitrogen would be evaporated into vapor and both the liquid slug and vapor are under oscillating motion in the channel, which could result in an extremely high cooling rate (up to

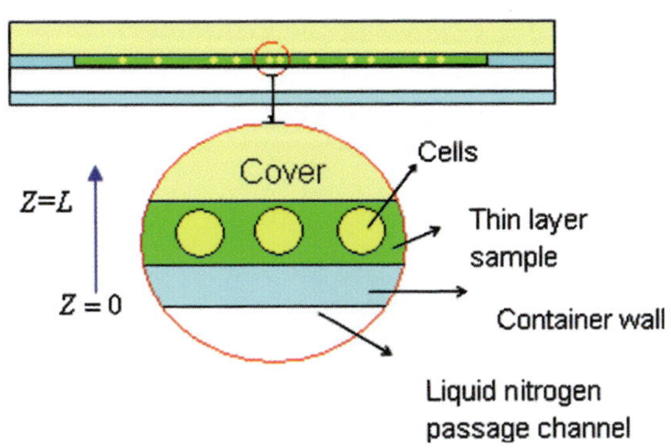

Figure 7. The cross-sectional view of the sample holder for cryogenic osilating heating pipe (COHP) shows the spatial relationship between the thin layer sample of cell suspension, the container's cover and wall, and the liquid nitrogen passage channel: Z is the thickness coordinate, and Z = 0 is the location where COHP working fluid is in contact with the container wall, and Z = L is the location where the thin sample is in contact with the cover. Figure reprinted and redrawn from [Han *et al.*, 2008] with permission from Elsevier.

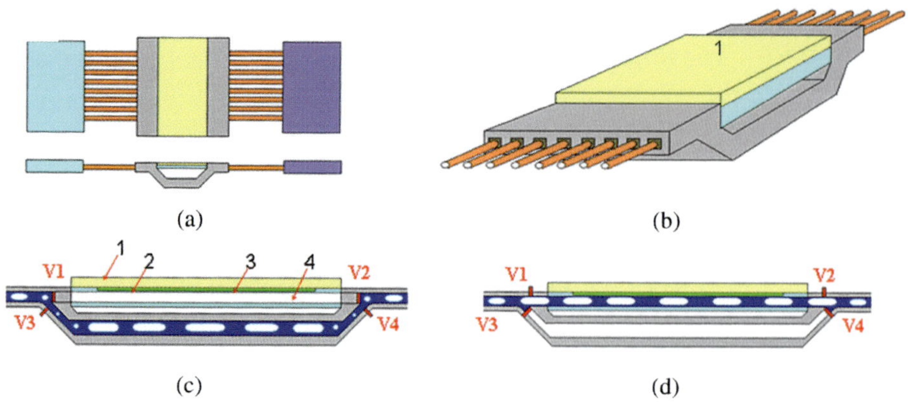

Figure 8. The cooling procedure with the COHP device. (a) and (b) The appearance when the working section is loaded with a sample container. (c) The cross-sectional view: 1, the cover; 2, the thin layer sample (in green color); 3, the thin container wall under the sample; and 4, the liquid nitrogen passage channel under the sample container. (d) To cool the thin film sample, valves (V) 3 and V4 are closed and V1 and V2 are open so that the liquid nitrogen slugs and nitrogen bubbles will flow through the new passages under the thin container wall, and the thin film evaporation will absorb the major heat release. Figure reprinted and redrawn from [Han *et al.*, 2008] with permission from Elsevier.

10^7 °C/min) in the cryopreservation sample [Han *et al.*, 2008]. Under this condition, almost all the biological materials can be vitrified.

In addition, the sample chamber can be changed on the same working platform as illustrated in Figure 8 [Han *et al.*, 2008]. By switching off the valves V1 and V2, the vitrified sample chamber is isolated from the oscillating flow of cryogenic fluid and can be transferred into liquid nitrogen for storage and future use. Meanwhile, a new sample container can be connected to the platform. By turning on the valves V1 and V2 and turning off the valves V3 and V4, the new sample could be vitrified by the circulating motion of vapor and liquid nitrogen within a fraction of seconds. Furthermore, the working platforms can be arranged into a matrix pattern with common evaporator and condenser for mass production [Han *et al.*, 2008]. However, this theoretical analyses need further experimental validation.

8. Quartz Microcapillary (QMC)

The cooling rate of a biological sample contained in a straw is not only dependent on the size of the straw but also the thermal properties of the straw material. Given the high thermal conductivity of quartz (8 W/m/K, compared to 0.2 W/m/K for plastics and 0.8 W/m/K for glass), the quartz microcapillary (QMC) with an outer diameter of 200 μm can be used to

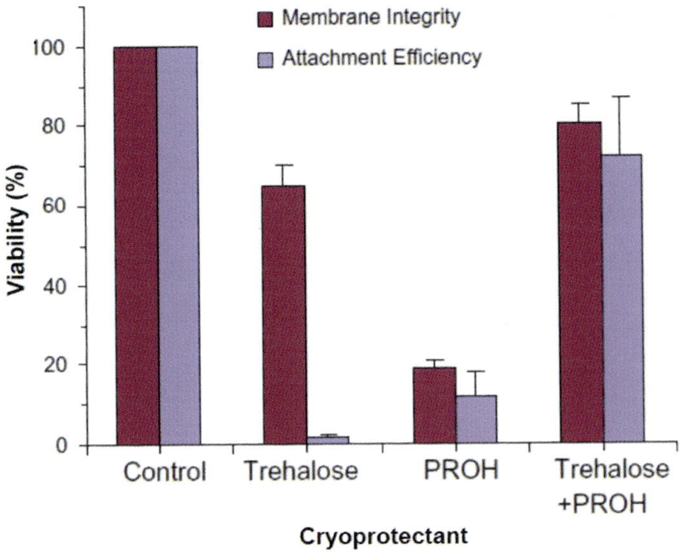

Figure 9. Immediate cell viability (membrane integrity) and attachment efficiency of mESCs post-cryopreservation using three different CPAs. The concentrations of trehalose and 1,2-propanediol (PROH) are 0.5 and 2 M for all the conditions, respectively. Values represent the mean ± standard deviation of at least three independent experiments. Figure reprinted and redrawn from [He *et al.*, 2008] with permission from Elsevier.

achieve a cooling rate as high as 100,000 °C/min by simply plunging it into liquid nitrogen [He *et al.*, 2008]. The thickness of the wall is 10 μm, rendering an inner diameter of 180 μm, which is slightly bigger than the diameter of human oocytes (~150 μm). Under this ultrafast cooling rate, a solution with 2 M PROH can be apparently vitrified.

For cell experiment, murine embryonic stem cells (mESCs) were tested. As shown in Figure 9 [He *et al.*, 2008], 2 M PROH alone could not protect cells well from cryoinjuries during vitrification (viability <20%) although apparent vitrification can be achieved. On the other hand, 0.5M trehalose alone can maintain membrane integrity for many cells (viability: ~65% in terms of membrane integrity), but the attachment efficiency is very low (~12%). When 2 M PROH and 0.5 M trehalose are combined together, the cell attachment efficiency can increase to ~72%. These results indicate that PROH and trehalose can work synergistically to protect mESCs during vitrification. The exact mechanism for this synergism needs further investigation.

Furthermore, it is shown that the use of slush nitrogen, instead of liquid nitrogen, can increase the cooling rate to 250,000 °C/min [Risco *et al.*, 2007]. This is because the super cooled liquid nitrogen can reduce the amount of vaporization, which forms a "vapor coat" over the capillary surface that induces the major heat resistance for cooling. As a result, a lower concentration

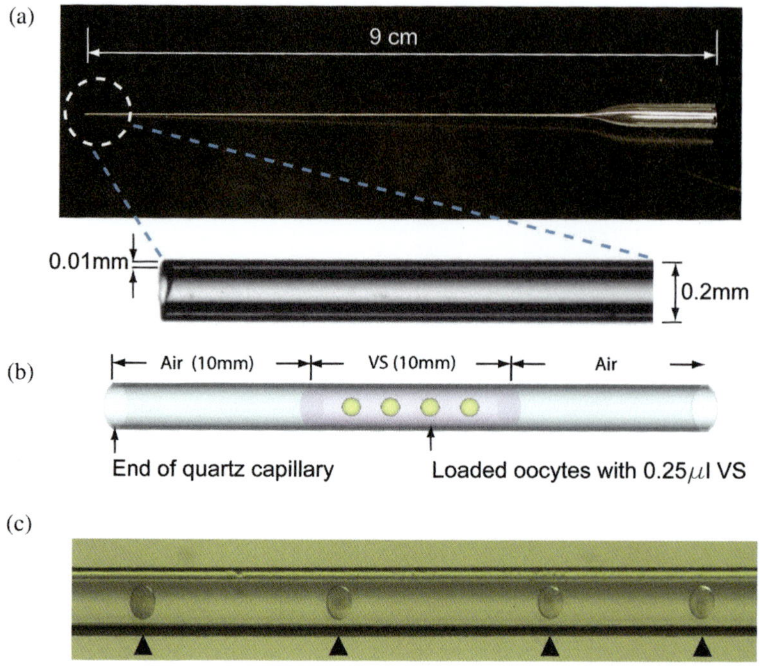

Figure 10. QMC vitrification of oocytes. (a) Typical pictures of the QMC; (b) a schematic illustration of the QMC loaded with oocytes; and (c) photomicrograph of the oocytes (arrowheads) loaded in QMC. VS, vitrification solution. Figure reprinted and redrawn from [Lee *et al.*, 2010a] with permission from Elsevier.

of CPA (1.5 M PROH) can be used to achieve apparent vitrification. Of note, the warming rate of QMC or open pulled straw (OPS) is much lower than their cooling rate [Risco *et al.*, 2007], which implies improving the warming rate is crucial for successful vitrification.

As a result, QMC vitrification in slush nitrogen can be used to cryopreserve mouse oocytes with only 1.5 M PROH as the penetrating CPA [Lee, *et al.*, 2010a]. As shown in Figure 10 [Lee, *et al.*, 2010a], the outer diameter is 200 μm with a wall thickness of 10 μm, resulting a cooling rate as high as 250,000 °C/min in nitrogen slush. Several mouse oocytes were loaded near the tip inside the QMC in a very small volume of vitrification solution (0.25 μl). The results show that the survival rate of oocytes post-vitrification is up to 90% and the fertilization and blastocyst formation rates are similar to that of fresh oocytes [Lee, *et al.*, 2010a]. Even though the CPA concentration is considered low here, the CPA was loaded before cooling and unloaded after warming by multiple steps [Lee *et al.*, 2010a]. However, this method can be improved by using 1.4 M EG and 1.1M DMSO as the penetrating CPAs and 1 M trehalose as the non-penetrating CPA. After warming at 37 °C with one-step CPA unloading, the viability of oocytes can be as high as 94%

post-vitrification with uncompromised embryonic development to the blastocyst stage [Choi *et al.*, 2015a].

9. Hydrogel Microencapsulation

Hydrogel microencapsulation originally used for 3D cell culture and drug delivery [Chang, 1964; Agarwal *et al.*, 2013] has been proposed as another promising strategy to achieve low-CPA vitrification. It has been found that the water in alginate hydrogel microcapsules can be preferentially vitrified compared to the water in bulk solution with 1.5 M or more DMSO [Zhang *et al.*, 2010]. As a result, mouse mesenchymal stromal cells encapsulated in the alginate microcapsules can be vitrified in quartz microcapillary at an outer diameter of 400 μm (10 μm in wall thickness) at a cooling rate of ~50,000 °C/min. The viability of the encapsulated cells post-vitrification in QMC can reach 88.9% while that of non-encapsulated cells is only 42.0% [Zhang *et al.*, 2010].

The hydrogel microcapsules can be produced by electrospray or microfluidics and the diameter of those microcapsules is ~100 μm or less [Zhang and He, 2009]. Figure 11 [Zhang *et al.*, 2010] shows the amount of ice

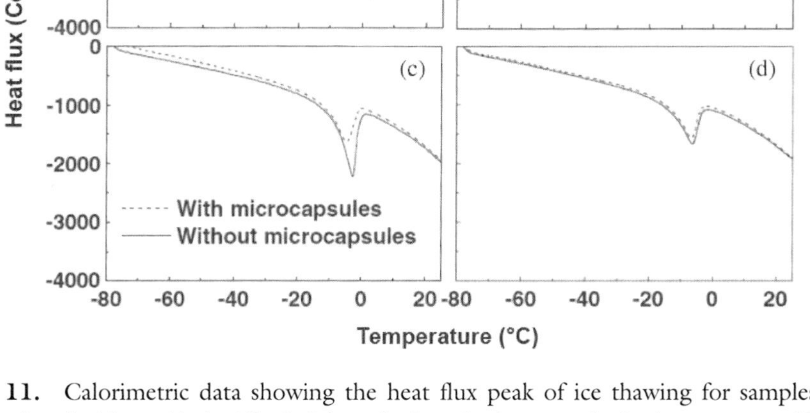

Figure 11. Calorimetric data showing the heat flux peak of ice thawing for samples with (solid line) and without (dashed line) alginate hydrogel microcapsules in the presence of 0% (a), 5% (b), 10% (c), and 15% (v/v) (d) DMSO. Figure reprinted and redrawn from [Zhang *et al.*, 2010] with permission from Springer.

formation within the samples with various concentrations of DMSO. As the concentration of DMSO increases, the amount of ice formation (indicated by the peak area) monotonically decreases due to two reasons: the decrease of water volume and the decrease of ice-formation extension in the sample during cooling. In addition, there is little difference of ice formation between the samples with and without microcapsules when the DMSO concentration is below or above 1.5 M, which indicates that the water in alginate hydrogel microcapsules can be vitrified in the presence of 1.5 M DMSO. As a result, the microcapsules become wrinkled post-vitrification without DMSO and stay intact with 1.5 M DMSO [Zhang *et al.*, 2010]. Moreover, although the treatment with $CaCl_2$ solution (0.15 M) cannot further promote the vitrification of water within hydrogel, it can enhance the mechanical strength of the alginate matrix and help to maintain the integrity of microcapsules cryopreserved with a low concentration of DMSO (1 M) [Zhang *et al.*, 2010].

More recently, it is found that the alginate hydrogel microencapsulation can inhibit devitrification of ice recrystallization during warming (Figure 12)

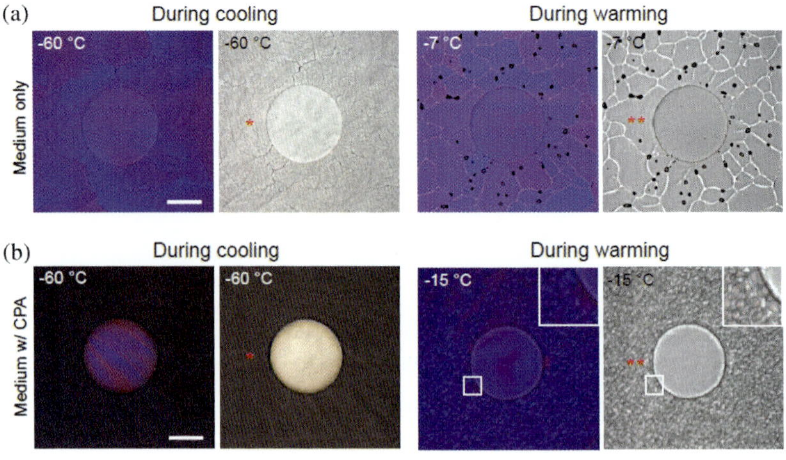

Figure 12. Inhibition of devitrification/ice recrystallization by alginate hydrogel microcapsules during warming. Images of both polarized and phase contrast microscopy are shown at –60 °C at the end of cooling and during warming with annealing for 20 min at –7 and –15 °C for an alginate hydrogel microcapsule in mESC medium without (a) and with (b) CPA, respectively. Ice recrystallization during warming indicated by the formation of large ice crystals (LICs, **) at –7 and –15 °C from the fine ice crystals (FICs, *) at –60 °C is visible outside the microcapsule while it is minimal inside the microcapsule for both solutions. The dark dots in the images during warming are precipitated salts after ice recrystallization. Scale bar: 100 μm. The alginate hydrogel microcapsule can effectively inhibit devitrification/ice recrystallization even in the absence of any CPA (a). Figure reprinted and redrawn from [Huang *et al.*, 2015] with permission from John Wiley and Sons.

[Huang *et al.*, 2015]. This is probably because the viscosity of the hydrogel microcapsules is immensely high at the devitrification or ice recrystallization temperature (range from –20 °C to –7 °C depending on the CPA concentration), which can effectively inhibit the diffusion of water molecules to rearrange into ice crystals, or the matrix of the calcium alginate chains can attach to the surface of the fine ice crystals to hinder their growth. This finding is significant for improving cell cryopreservation by vitrification since warming rate is generally much lower than cooling rate with the same cooling/warming device, and devitrification/ice recrystallization during warming often induces IIF that is a lethal event to cells. As demonstrated in Figure 13a, apparent vitrification in the PS can be achieved with VSP solution containing only 2 M penetrating CPA (PROH) in the presence of 1.3M trehalose (non-penetrating CPA). However, significant ice formation due to devitrification and possible ice recrystallization is observable in the bulk solution during warming. However, high cell viability (~80%) is achieved for the cells encapsulated in alginate hydrogel microcapsule while almost all cells without alginate hydrogel microencapsulation are dead (Figure 13b). These data clearly demonstrate that alginate hydrogel encapsulation can protect cells from the cryoinjury associated with devitrification and ice recrystallization during the warming step of a vitrification procedure.

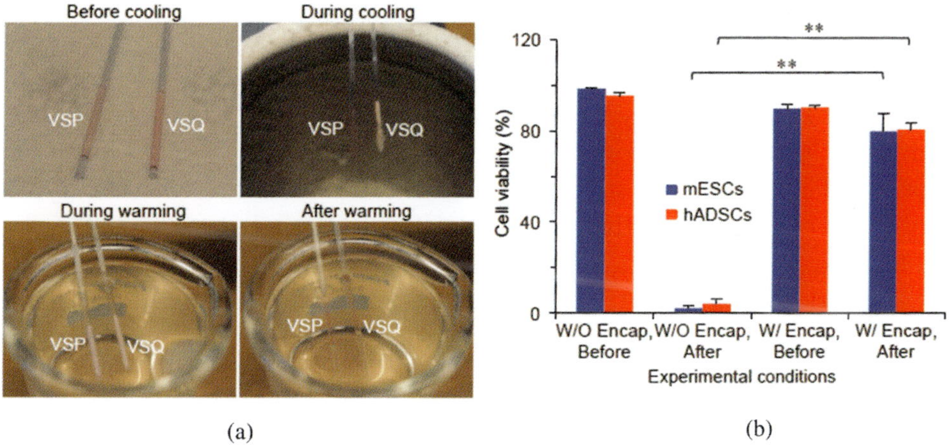

Figure 13. Low-CPA vitrification of stem cells with plastic straw (PS). (a) Typical images showing the vitrification of two different cell culture medium-based solutions in PS. VSP includes 2 M PROH and 1.3 M trehalose in the culture medium of mESCs or human adipose-derived stem cells (hADSCs), while VSQ contains 1.5 M PROH and 0.5 M trehalose in cell culture medium. (b) Viability of mESCs and hADSCs either with or without alginate hydrogel microencapsulation before and after vitrification in VSP using PS. The double asterisks indicate $p < 0.01$. Figure reprinted and redrawn from [Huang *et al.*, 2015] with permission from John Wiley and Sons.

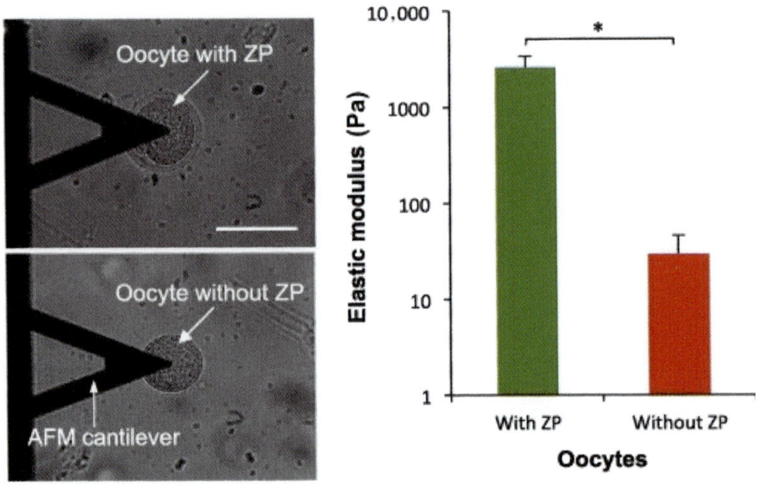

(a) (b)

Figure 14. Mechanical property of oocytes with and without zona pellucida (ZP): (a) Typical micrographs showing the setup for measuring the mechanical properties of oocytes with and without ZP using atomic force microscopy (AFM). (b) Quantitative data of the elastic modulus of oocytes with and without ZP determined by AFM. The asterisk indicates $p < 0.05$. Figure reprinted and redrawn from [Choi *et al.*, 2015b] with permission from Elsevier.

Similar effect has also been found for the zona pellucida (ZP) of mouse oocytes during QMC-based low-CPA vitrification [Choi *et al.*, 2015b]. The ZP consists of hydrogel of glycoproteins and renders the intact oocytes at least eighty-five times stronger than oocytes without a ZP (in terms of elastic modulus, Figure 14). The strong shell of ZP can protect oocytes from cryoinjury by not only providing the resistance to the mechanical stress due to possible ice formation during vitrification, but also by damping ice formation during cooling and inhibiting devitrification and ice recrystallization during warming.

A summary of the various methods aforementioned for cell cryopreservation by vitrification is given in Table 1. The general doctrine is that as the size of the cooling/warming device or cryopreservation sample decreases, the cooling and warming rates increase and the CPA concentration required for vitrification decreases. Moreover, the use of hydrogel microencapsulation for improving cell vitrification is a paradigm-shifting strategy as it enables low-CPA vitrification in a large sample volume with PS. With the cell-based medicine and assisted reproduction industry burgeoning in recent years, the low-CPA vitrification techniques should find widespread applications in the near future.

Table 1. A summary of materials and devices used for cell vitrification.

Materials/devices	Maximum volume (μl)	Cooling rate (°C/min)	CPA (M)	Cells studied	Comments
Plastic Straw (PS)	250–500	2000–3000	5–8	embryos, preantral follicles, nucellar cells, oocytes, tissues	multiple steps of CPA loading and unloading
Open Pulled Straw (OPS)	20–50	~20,000	4–6	stem cells, embryos, ova	size and wall thickness depends on manufactural pulling process
Electron Microscopy (EM) Grids			5–7	Oocytes, multipronuclear zygotes, blastocysts	short exposure time with artificial shrinkage of blastocoele
Cryoloop	5		1.5–2	sperm	with precooling in liquid nitrogen vapor
Droplet Vitrification			1.5	hepatocytes, fibroblasts, cardiomyocytes, stem cells, RAJI cells	direct cooling of the droplets with diameter of ~100 μm
Cryogenic Oscillation Heat Pipe (COHP)	50–100	~10,000,000		none	achieve highest cooling rate but without cell experiments
Quartz Microcapillary (QMC)	2–5	50,000–100,000	1.5–2	stem cells, oocytes	warming rate is usually lower than cooling rate
Hydrogel Microencapsulation	5–250	2,000–50,000	1.5–2	mesenchymal stem cells, embryonic stem cells	hydrogel encapsulation can preferentially vitrify water during cooling and inhibit devitrification/ice recrystallization during warming

References

Agarwal, P., S. Zhao, P. Bielecki, W. Rao, J. K. Choi, Y. Zhao, J. Yu, W. Zhang and X. He (2013). One-step microfluidic generation of pre-hatching embryo-like core-shell microcapsules for miniaturized 3D culture of pluripotent stem cells. *Lab Chip*, 13(23), pp. 4525-4533.

Akachi, H. (1990). Structure of a heat pipe. Google Patent US4921041.

Bastings, L., C. C. Beerendonk, J. R. Westphal, L. F. Massuger, S. E. Kaal, F. E. van Leeuwen, D. D. Braat and R. Peek (2013). Autotransplantation of cryopreserved ovarian tissue in cancer survivors and the risk of reintroducing malignancy: a systematic review. *Human Reproduction Update*, 19(5), pp. 483–506.

Bhat, S. N., A. Sharma and S. V. Bhat (2005). Vitrification and glass transition of water: insights from spin probe ESR. *Physical Review Letters*, 95(23), pp. 235702.

Bischof, J. C. (2000). Quantitative measurement and prediction of biophysical response during freezing in tissues. *Annual Review of Biomedical Engineering*, 2, pp. 257–288.

Bischof, J. C. (2006). Micro and nanoscale phenomenon in bioheat transfer. *Heat and Mass Transfer*, 42(10), pp. 955–966.

Bischof, J. C. and X. M. He (2005). Thermal stability of proteins. *Cell Injury: Mechanisms, Responses, and Repair*, 1066, pp. 12–33.

Bischof, J. C., W. F. Wolkers, N. M. Tsvetkova, A. E. Oliver and J. H. Crowe (2002). Lipid and protein changes due to freezing in dunning AT-1 cells. *Cryobiology*, 45(1), pp. 22–32.

Boldt, J., D. Cline and D. McLaughlin (2003). Human oocyte cryopreservation as an adjunct to IVF-embryo transfer cycles. *Human Reproduction*, 18(6), pp. 1250–1255.

Boutron, P. (1986). Comparison with the theory of the kinetics and extent of ice crystallization and of the glass-forming tendency in aqueous cryoprotective solutions. *Cryobiology*, 23(1), pp. 88–102.

Carvalho, A. A., L. R. Faustino, C. M. Silva, S. V. Castro, H. K. Luz, R. Rossetto, C. A. Lopes, C. C. Campello, J. R. Figueiredo, A. P. Rodrigues and A. P. Costa (2011). Influence of vitrification techniques and solutions on the morphology and survival of preantral follicles after in vitro culture of caprine ovarian tissue. *Theriogenology*, 76(5), pp. 933–941.

Chang, T. M. (1964). Semipermeable Microcapsules. *Science*, 146(3643), pp. 524–525.

Chao, B. H., X. He and J. C. Bischof (2004). Pre-treatment inflammation induced by TNF-alpha augments cryosurgical injury on human prostate cancer. *Cryobiology*, 49(1), pp. 10–27.

Chen, C. (1986). Pregnancy after human oocyte cryopreservation. *Lancet*, 1(8486), pp. 884–886.

Chen, S. U., Y. R. Lien, H. F. Chen, K. H. Chao, H. N. Ho and Y. S. Yang (2000). Open pulled straws for vitrification of mature mouse oocytes preserve patterns of meiotic spindles and chromosomes better than conventional straws. *Human Reproduction*, 15(12), pp. 2598–2603.

Choi, J. K., H. Huang and X. He (2015a). Improved low-CPA vitrification of mouse oocytes using quartz microcapillary. *Cryobiology*, 70, pp. 269–272.

Choi, J. K., T. Yue, H. Huang, G. Zhao, M. Zhang and X. He (2015b). The crucial role of zona pellucida in cryopreservation of oocytes by vitrification. *Cryobiology*, 71, pp. 350–355.

Coad, J. E. and J. C. Bischof (2003). Histologic differences between cryothermic and hyperthermic therapies. *Thermal Treatment of Tissue: Energy Delivery and Assessment II*, 4954, pp. 27–36.

Demirci, U. and G. Montesano (2007). Cell encapsulating droplet vitrification. *Lab Chip*, 7(11), pp. 1428–1433.

Desai, N. N., H. Blackmon and J. Goldfarb (2004). Single sperm cryopreservation on cryoloops: an alternative to hamster zona for freezing individual spermatozoa. *Reproductive Biomedicine Online*, 9(1), pp. 47–53.

Edgar, D. H. and D. A. Gook (2012). A critical appraisal of cryopreservation (slow cooling versus vitrification) of human oocytes and embryos. *Human Reproduction Update*, 18(5), pp. 536–554.

Fahy, G. M., D. R. MacFarlane, C. A. Angell and H. T. Meryman (1984). Vitrification as an approach to cryopreservation. *Cryobiology*, 21(4), pp. 407–426.

Fahy, G. M., B. Wowk, J. Wu and S. Paynter (2004a). Improved vitrification solutions based on the predictability of vitrification solution toxicity. *Cryobiology*, 48(1), pp. 22–35.

Fahy, G. M., B. Wowk, J. Wu, J. Phan, C. Rasch, A. Chang and E. Zendejas (2004b). Cryopreservation of organs by vitrification: perspectives and recent advances. *Cryobiology*, 48(2), pp. 157–178.

Fowler, A. and M. Toner (2005). Cryo-injury and biopreservation. *Annals of the New York Academy of Sciences*, 1066, pp. 119–135.

Gao, D. and J. K. Critser (2000). Mechanisms of cryoinjury in living cells. *Institute for Laboratory Animal Research Journal*, 41(4), pp. 187–196.

Han, X., H. Ma, A. Jiao and J. K. Critser (2008). Investigations on the heat transport capability of a cryogenic oscillating heat pipe and its application in achieving ultra-fast cooling rates for cell vitrification cryopreservation. *Cryobiology*, 56(3), pp. 195–203.

He, X. (2011a). Preservation of embryonic stem cells. In C. Atwood (ed.), *Manipulation and Utilization of Embryonic Stem Cells for Basic and Practical Applications* (Vienna, Austria, InTech Publisher), pp. 113–138.

He, X. (2011b). Thermostability of biological systems: fundamentals, challenges, and quantification. *The Open Biomedical Engineering Journal*, 5, pp. 47–73.

He, X. and J. C. Bischof (2005). Analysis of thermal stress in cryosurgery of kidneys. *Journal of Biomechanical Engineering*, 127(4), pp. 656–661.

He, X., E. Y. Park, A. Fowler, M. L. Yarmush and M. Toner (2008). Vitrification by ultra-fast cooling at a low concentration of cryoprotectants in a quartz micro-capillary: a study using murine embryonic stem cells. *Cryobiology*, 56(3), pp. 223–232.

Heng, B. C., L. L. Kuleshova, S. M. Bested, H. Liu and T. Cao (2005). The cryopreservation of human embryonic stem cells. *Biotechnology and Applied Biochemistry*, 41(Part 2), pp. 97–104.

Hoffmann, N. E. and J. C. Bischof (2001). Cryosurgery of normal and tumor tissue in the dorsal skin flap chamber: part II — injury response. *Journal of Biomechanical Engineering*, 123(4), pp. 310–316.

Huang, H., J. K. Choi, W. Rao, S. Zhao, P. Agarwal, G. Zhao and X. He (2015). Alginate hydrogel microencapsulation inhibits devitrification and ice recrystallization and enables large-volume low-CPA cell vitrification. *Advanced Functional Materials*, 25, pp. 6839–6850.

Hunt, C. J. (2011). Cryopreservation of human stem cells for clinical application: a review. *Transfusion Medicine and Hemotherapy*, 38(2), pp. 107–123.

Hunt, C. J., D. E. Pegg and S. E. Armitage (2006). Optimising cryopreservation protocols for haematopoietic progenitor cells: a methodological approach for umbilical cord blood. *CryoLetters*, 27(2), pp. 73–86.

Isachenko, V., I. Lapidus, E. Isachenko, A. Krivokharchenko, R. Kreienberg, M. Woriedh, M. Bader and J. M. Weiss (2009). Human ovarian tissue vitrification versus conventional freezing: morphological, endocrinological, and molecular biological evaluation. *Reproduction*, 138(2), pp. 319–327.

Jiao, A., X. Han, J. K. Critser and H. Ma (2006). Numerical investigations of transient heat transfer characteristics and vitrification tendencies in ultra-fast cell cooling processes. *Cryobiology*, 52(3), pp. 386–392.

Karlsson, J. O. M., E. G. Cravalho and M. Toner (1994). A model of diffusion-limited ice growth inside biological cells during freezing. *Journal of Applied Physics*, 75(9), pp. 4442–4455.

Karlsson, J. O. and M. Toner (1996). Long-term storage of tissues by cryopreservation: critical issues. *Biomaterials*, 17(3), pp. 243–256.

Lee, H. J., H. Elmoazzen, D. Wright, J. Biggers, B. R. Rueda, Y. S. Heo, M. Toner and T. L. Toth (2010a). Ultra-rapid vitrification of mouse oocytes in low cryoprotectant concentrations. *Reproductive BioMedicine Online*, 20(2), pp. 201–208.

Lee, J. Y., J. E. Lee, D. K. Kim, T. K. Yoon, H. M. Chung and D. R. Lee (2010b). High concentration of synthetic serum, stepwise equilibration and slow cooling as an efficient technique for large-scale cryopreservation of human embryonic stem cells. *Fertility and Sterility*, 93(3), pp. 976–985.

Liu, J., T. Viverette, M. Virgin, M. Anderson and D. Paresh (2005). A study of the impact of freezing on the lyophilization of a concentrated formulation with a high fill depth. *Pharmceutical Development and Technology*, 10(2), pp. 261–272.

Mazur, P. (1970). Cryobiology: the freezing of biological systems. *Science*, 168(3934), pp. 939–949.

Mazur, P. (1984). Freezing of living cells: mechanisms and implications. *American Journal of Physiology*, 247(3 Part 1), pp. C125–142.

Mazur, P., S. P. Leibo and E. H. Chu (1972). A two-factor hypothesis of freezing injury. Evidence from Chinese hamster tissue-culture cells. *Experimental Cell Research*, 71(2), pp. 345–355.

Menezo, Y. (2004). Cryopreservation of IVF embryos: which stage? *European Journal of Obstetrics, Gynecology and Reproductive Biology*, 113 Supplement 1, pp. S28–32.

Natsume, K., T. Mito, N. Yanagi, H. Tamura, T. Tamada, K. Shikimachi, N. Hirano and S. Nagaya (2011). Heat transfer performance of cryogenic oscillating heat pipes for effective cooling of superconducting magnets. *Cryogenics*, 51(6), pp. 309–314.

Niwa, H., J. Miyazaki and A. G. Smith (2000). Quantitative expression of Oct-3/4 defines differentiation, dedifferentiation or self-renewal of ES cells. *Nature Genetics*, 24(4), pp. 372–376.

Park, S. P., E. Y. Kim, D. I. Kim, N. H. Park, Y. S. Won, S. H. Yoon, K. S. Chung and J. H. Lim (1999). Simple, efficient and successful vitrification of bovine blastocysts using electron microscope grids. *Human Reproduction*, 14(11), pp. 2838–2843.

Park, S. P., E. Y. Kim, J. H. Oh, H. K. Nam, K. S. Lee, S. Y. Park, E. M. Park, S. H. Yoon, K. S. Chung and J. H. Lim (2000). Ultra-rapid freezing of human multipronuclear zygotes using electron microscope grids. *Human Reproduction*, 15(8), pp. 1787–1790.

Rabin, Y., P. S. Steif, M. J. Taylor, T. B. Julian and N. Wolmark (1996). An experimental study of the mechanical response of frozen biological tissues at cryogenic temperatures. *Cryobiology*, 33(4), pp. 472–482.

Rall, W. F. and G. M. Fahy (1985). Ice-free cryopreservation of mouse embryos at −196 degrees C by vitrification. *Nature*, 313(6003), pp. 573–575.

Rambhatla, S., R. Ramot, C. Bhugra and M. J. Pikal (2004). Heat and mass transfer scale-up issues during freeze drying: II. Control and characterization of the degree of supercooling. *AAPS PharmSciTech*, 5(4), pp. e58.

Reubinoff, B. E., M. F. Pera, G. Vajta and A. O. Trounson (2001). Effective cryopreservation of human embryonic stem cells by the open pulled straw vitrification method. *Human Reproduction*, 16(10), pp. 2187–2194.

Riggs, R., J. Mayer, D. Dowling-Lacey, T. F. Chi, E. Jones and S. Oehninger (2010). Does storage time influence postthaw survival and pregnancy outcome? An analysis of 11,768 cryopreserved human embryos. *Fertility and Sterility*, 93(1), pp. 109–115.

Risco, R., H. Elmoazzen, M. Doughty, X. He and M. Toner (2007). Thermal performance of quartz capillaries for vitrification. *Cryobiology*, 55(3), pp. 222–229.

Rizzino, A. (2010). Stimulating progress in regenerative medicine: improving the cloning and recovery of cryopreserved human pluripotent stem cells with ROCK inhibitors. *Regenerative Medicine*, 5(5), pp. 799–807.

Rupp, C. C., N. E. Hoffmann, F. R. Schmidlin, D. J. Swanlund, J. C. Bischof and J. E. Coad (2002). Cryosurgical changes in the porcine kidney: histologic analysis with thermal history correlation. *Cryobiology*, 45(2), pp. 167–182.

Sakai, A., S. Kobayashi and I. Oiyama (1990). Cryopreservation of nucellar cells of navel orange (Citrus sinensis Osb. var. brasiliensis Tanaka) by vitrification. *Plant Cell Reproduction*, 9(1), pp. 30–33.

Schuster, T. G., L. M. Keller, R. L. Dunn, D. A. Ohl and G. D. Smith (2003). Ultra-rapid freezing of very low numbers of sperm using cryoloops. *Human Reproduction*, 18(4), pp. 788–795.

Searles, J. A., J. F. Carpenter and T. W. Randolph (2001). The ice nucleation temperature determines the primary drying rate of lyophilization for samples frozen on a temperature-controlled shelf. *Journal of Pharmceutical Science*, 90(7), pp. 860–871.

Son, W. Y., S. H. Yoon, H. J. Yoon, S. M. Lee and J. H. Lim (2003). Pregnancy outcome following transfer of human blastocysts vitrified on electron microscopy grids after induced collapse of the blastocoele. *Human Reproduction*, 18(1), pp. 137–139.

Steponkus, P. L. and D. V. Lynch (1989). Freeze/thaw-induced destabilization of the plasma membrane and the effects of cold acclimation. *Journal of Bioenergetics and Biomembranes*, 21(1), pp. 21–41.

Vajta, G., P. Holm, M. Kuwayama, P. J. Booth, H. Jacobsen, T. Greve and H. Callesen (1998). Open pulled straw (OPS) vitrification: a new way to reduce cryoinjuries of bovine ova and embryos. *Molecular Reproduction and Development*, 51(1), pp. 53–58.

Vutyavanich, T., W. Piromlertamorn and S. Nunta (2010). Rapid freezing versus slow programmable freezing of human spermatozoa. *Fertility and Sterility*, 93(6), pp. 1921–1928.

Yamada, T., K. Kuroda, Y. Jitsuyama, D. Takezawa, K. Arakawa and S. Fujikawa (2002). Roles of the plasma membrane and the cell wall in the responses of plant cells to freezing. *Planta*, 215(5), pp. 770–778.

Yang, G., A. Zhang, L. X. Xu and X. He (2009). Modeling the cell-type dependence of diffusion-limited intracellular ice nucleation and growth during both vitrification and slow freezing. *Journal of Applied Physics*, 105(11), pp. 114701.

Zhang, W. and X. He (2009). Encapsulation of living cells in small (approximately 100 microm) alginate microcapsules by electrostatic spraying: a parametric study. *Journal of Biomechanical Engineering*, 131(7), pp. 074515.

Zhang, W., G. Yang, A. Zhang, L. X. Xu and X. He (2010). Preferential vitrification of water in small alginate microcapsules significantly augments cell cryopreservation by vitrification. *Biomedical Microdevices*, 12(1), pp. 89–96.

Zhou, G. B., S. E. Zhu, Y. P. Hou, F. Jin, Q. E. Yang, Z. Q. Yang, G. B. Quan and H. M. Tan (2005). Vitrification of mouse embryos at various stages by open-pulled straw (OPS) method. *Animal Biotechnology*, 16(2), pp. 153–163.

Zuo, Z. J., M. T. North and K. L. Wert (2001). High heat flux heat pipe mechanism for cooling of electronics. *IEEE Transactions on Components and Packaging Technologies*, 24(2), pp. 220–225.

Chapter 5

Multi-scale Thermal Conductivity Measurements for Cryobiological Applications

Harishankar Natesan, Jeunghwan Choi, Sean Lubner, Chris Dames and John Bischof

1. Introduction

Regenerative medicine and therapeutic approaches using heat sinks or sources are driving the need for improved understanding of heat transfer at micro-and nano-scales. Of the various property values related to heat transfer, thermal conductivity is a measure of how readily heat can be transferred across a medium and will be the focus of this chapter particularly in the cryogenic (<–40°C), sub-zero (<0°C) and hypothermic(<10°C) regimes where water, protein and lipid phase changes occur. At the microscale, we review the use of three omega and laser flash techniques for measurement of thermal conductivity in thin (1–1000 μm), anisotropic, and multi-layered tissues. At the nanoscale, pump-probe and scanning probe methods can be used to measure thermal conductivity at short time scales (10 ps–100 ns) and length scales (10 nm–1 μm), particularly in the coating and surrounding medium around metallic nanoparticles (1 nm–100 nm). Here we present these new thermal conductivity measurement techniques and briefly discuss their existing or potential use in cryobiological applications at the micro/nanoscale.

2. Cryobiological Applications

Cryobiology encompasses several applications (see Table 1) that either preserve or destroy biomaterials by freezing. This is achieved in both cases by cooling the material to hypothermic (<10°C), subzero (<0°C) and/or cryogenic temperatures (<–40°C). Examples of how these applications have been deployed for various biomaterials at the molecular, cellular and tissue scale are shown in Table 2. Importantly, these biomaterials are themselves made up of a wide range of constituents, with the possibility of both crystalline and amorphous (vitrified) phases, whose thermal conductivity ranges from 0.1 W/m.K (lipids) to 2.3 W/m.K (ice) [Owen, 2006]. Thermal conductivity is a critical determinant in how fast these materials achieve low/ high temperatures and as such is a critical determinant of either life or death after the process. A more precise understanding of thermal conductivity in micro and nanoscale biological systems is just now being determined in part due to the advent of new metrology techniques, the topic of this review.

Table 1. Overview of cryobiological applications.

	Application	Definition
Biopreservation	Cryopreservation	Preservation by freezing [Mazur, 1984; Karlsson and Toner, 1996]
	Lyophilization/Freeze drying	Preservation by drying after freezing [Crowe *et al.*, 1992; Carpenter *et al.*, 1997]
	Vitrification	Preservation by attaining a glassy state [Fahy *et al.*, 1984; Song *et al.*, 2000]
	Hypothermic preservation	Preservation at temperatures between 0°C and 37°C [Belzer and Southard, 1988; Steponkus, 1996]
Biodestruction	Cryosurgery/Cryotherapy/ Cryoablation	Destruction of tissue by freezing [Gage and Baust, 1998; Hoffmann and Bischof, 2002]

Table 2(a). Applications of biopreservation of biomaterials at different length scales.

Bio-material	Length scale	Bio-preservation	Rep. ref.
Tissues and Organs	>10 mm	Liver	[de Kanter *et al.*, 1998; Maas *et al.*, 2000]
		Kidney	[Fahy and Ali, 1997; de Kanter *et al.*, 1998]
		Heart valves	[Crescenzo *et al.*, 1993; Niwaya *et al.*, 1995; Gall *et al.*, 1998]
Thin Tissues	100 μm–10 mm	Skin	[Wainwright, 1995; Bello *et al.*, 2001]
		Ovarian tissue	[Hovatta *et al.*, 1996; Donnez *et al.*, 2004; Oktay *et al.*, 2004]
		Cartilage	[Muldrew *et al.*, 1994; Pegg *et al.*, 2006a; Pegg *et al.*, 2006b]
Cells	1–100 μm	Red and white blood cells	[Kleinschuster *et al.*, 1979; Meryman, 1989; Fuller and Grout, 1991; Rowe, 1994; Rowe, 1995]
		Reproductive cells (sperm, oocytes, Embryos)	[Hagedorn *et al.*, 1997; Paynter, 2000; Suquet *et al.*, 2000; O'Connell *et al.*, 2002; Purdy, 2006; Guan *et al.*, 2008; Barbas and Mascarenhas, 2009]
		CHO cells	[Kleman *et al.*, 2008]
Molecules	1–50 nm	Protein and liposomes	[Crowe *et al.*, 1985; Arvinte and Steponkus, 1988; Rudolph, 1988; Talsma *et al.*, 1991; Carpenter *et al.*, 1993; Carpenter *et al.*, 1997]
		Drugs	[Tang and Pikal, 2004; Abdelwahed *et al.*, 2006; Steven and Larry, 2010]
		DNA	[Ando *et al.*, 1999]

Table 2(b). Applications of biodestruction of biomaterials at different length scales.

Bio-material	Length scale	Biodestruction	Rep. ref.
Tissues and Organs	>10 mm	Prostrate tumor	[Onik *et al.*, 1993; Bischof *et al.*, 1997; Donnelly *et al.*, 2002]
		Lung tumor	[Homasson *et al.*, 1986; Homasson, 1989; Mo and G, 2004]
		Breast tumor	[Staren *et al.*, 1997; Morin *et al.*, 2004; Sabel *et al.*, 2005]
Thin Tissues	100 μm–10 mm	Pulmonary vein	[Avitall *et al.*, 2003; Sarabanda *et al.*, 2005]
		Renal tumor	[Uchida *et al.*, 1995]
		Subcutaneous fat	[Manstein *et al.*, 2008; Klein *et al.*, 2009]
Cells	1–100 μm	Cancer carcinoma cells	[Bischof *et al.*, 1997; Larson *et al.*, 2000; Clarke *et al.*, 2001; Hanai *et al.*, 2001; Yang *et al.*, 2003]
		Cardiovascular (smooth muscle cells)	[Balasubramanian *et al.*, 2008]
		Neural cells	[Silani *et al.*, 1988]
Molecules	1–50 nm	Cryosurgical adjuvants + liposome	[Goel *et al.*, 2009; Shenoi *et al.*, 2009; Zhang *et al.*, 2010; Shenoi *et al.*, 2011]
		DNA	[Lyscov and Moshkovsky, 1969; Meryman, 1971; Pegg, 1972; Fahy, 1986; Davis *et al.*, 2000]
		Eutectic agents	[Lyscov and Moshkovsky, 1969; Davis *et al.*, 2000; Han and Bischof, 2003]

3. Thermal Conductivity at Short Length and Time Scales

One measure of the ability of a material to conduct heat is the property entitled thermal conductivity. At the microscale, the physics of thermal conductivity is readily understood as fundamentally arising from the random walk of individual energy carriers that carry the heat. For large samples at bulk length scales, these energy carriers in solids and gases obey the kinetic theory of gases that defines the thermal conductivity contribution of each type of energy carrier as follows [Toberer *et al.*, 2012],

$$k_{carrier} = \frac{1}{3} c \rho v \lambda \tag{1}$$

where c is specific heat capacity, ρ is density, v is the average velocity of the energy carriers, and λ is their mean free path between collisions. The most important energy carriers in solids are free electrons and phonons (lattice vibrations). These energy carriers contribute to the overall thermal conductivity as shown below [Incropera, 2011],

$$k = k_{el} + k_{ph} \tag{2}$$

Where k_{el} is the electronic contribution and k_{ph} is the phononic contribution to thermal conductivity. Metals, in general, have free valence electrons that contribute to the majority of heat transfer. On the other hand, dielectrics do not have free electrons as they are electrical insulators. In this case, phonon is the major energy carrier.

In the biological materials of interest in this chapter, the dominant heat carrier is atomic vibrations. In crystalline materials these are universally discussed as phonons. In amorphous materials, the corresponding transfer of vibrational energy is sometimes discussed in terms of related entities known as diffusons and propagons [Allen *et al.*, 1999]. A simpler "minimum k" approximation whereby all phonons are treated with a mean free path equal to half of their wavelength has also proven remarkably successful at explaining the thermal conductivity of amorphous solids [Slack, 1979; Cahill and Pohl, 1987]. The individual contributions depend upon the atomic and molecular structure of the material.

At steady state, the familiar continuum Fourier's Law of heat conduction can break down whenever the forcing and/or boundary conditions vary substantially at length scales comparable to or smaller than the *bulk* mean free path ($L \leq \lambda$). Often, we can still determine an effective thermal conductivity with a different λ (i.e. λ_{eff}) while still using the same values of ρ, c, and v from the bulk regime ($L > \lambda$). For many idealized geometries (cylinders, sheets), the exact expressions for the effective λ are obtainable from the Boltzmann transport equation as given by Equation (3)

$$\lambda_{eff} = f\left(\lambda_{bulk}, L\right). \tag{3}$$

Equation (1) implies that k is directly proportional to 'λ' of the energy carrier. λ can change for several reasons. In an infinite single-crystal sample, the intrinsic, or "bulk," phonon λ is typically limited by phonon scattering with point defects and with other phonons. In real samples the additional scattering by physical boundaries can also be important whether external (sample surfaces) or internal (grain boundaries). This can be profound when the characteristic length scale between boundaries, L, is comparable to 'λ', as represented by the Knudsen number,

$$Kn = \frac{\lambda}{L}. \tag{4}$$

For most geometries, the thermal conductivity reduction due to boundary scattering begins to become observable for $Kn \geq 0.1$, and typically scales as $k \propto k_{Bulk}/Kn$ for $Kn >> 1$ [Toberer *et al.*, 2012]. If $Kn \geq 0.1$, thermal conductivity is appreciably reduced due to boundary scattering. For instance, at 37°C, thermal conductivity of bulk crystalline silicon is 148 W/m.K, whereas the thermal conductivity of 56 nm diameter Si nanowire is ~25 W/m. K [Touloukian *et al.*, 1970; Li *et al.*, 2003]. This dramatic reduction arises because the bulk λ in Si at 37°C is estimated from Equation (1) to be ~8×10^{-8} m, so in the 56 nm wire the corresponding Kn is only ~1.3, indicating very strong reductions.

Thus, a material's bulk λ is the critical parameter for identifying the breakdown of Fourier's law at steady state due to the effects of microscopic sample sizes. For liquid water at 5°C, taking the bulk values of c as 4.2kJ/kg. K, ρ as 1000 kg/m^3, k as 0.58 W/m.K [Owen, 2006], v as velocity of sound in water to be ~1430 m/s [Vogt *et al.*, 2008], the mean free path crudely determined from Equation (1) is ~0.3 nm. Similarly, for ice at −100°C, taking c as 1.4kJ/kg.m^3, ρ as 920 kg/m^3, k as 3.48 W/m.K [Owen, 2006], v as ~4200 m/s [Vogt *et al.*, 2008], the mean free path is ~1.9 nm. These estimates show that microscopic size effects are more likely to be important in ice than in liquid water, though nanostructures of characteristic length ~20 nm or smaller (i.e. $Kn \geq 0.1$) would be required to see an appreciable effect.

Even in a large bulk sample, Fourier's Law can also break down due to small time scale effects. Assuming continuum exists, and that Fourier's law holds, conduction is governed by the heat diffusion equation, a parabolic partial differential equation [Incropera, 2011], which implies that the information of temperature disturbances travel at infinite speed. This assumes that the effect of boundary conditions is felt instantaneously throughout the material. However, this assumption neglects the finite time it takes for an energy carrier to travel between collisions to exchange energy with a neighboring energy carrier. In other words, Fourier's law will break down whenever the forcing and/or boundary conditions vary substantially over *time* scales comparable to or smaller than the bulk *relaxation time* , $\tau = \lambda/v$ [Toberer *et al.*, 2012]. Similar to small length scales, the apparent thermal conductivity of a material can also be understood as being reduced at small time scales i.e. $t < \tau$ where t is the time scale of the heat conduction transient in the sample. Here however it is less common to use a modified τ (i.e. τ_{eff}) or a modified k_{eff}, [Yang and Dames, 2013], with most theoretical treatments moving directly to the more fundamental Boltzmann transport equation or its radiation counterpart [Collins *et al.*, 2013; Regner *et al.*, 2014]. Based on such analyses, the nearly universal expectation is that ultrafast phenomena should only cause a substantial breakdown in the parabolic heat diffusion equation when $t < \tau$. Thus estimates of a material's τ are a powerful guideline to the onset of such sub-continuum regimes. Continuing the estimates for bulk λ given above, for liquid water at 5°C the expected bulk τ is ~0.21 ps, and for ice at –100°C the expected τ is ~0.46 ps. On the contrary, there has been a claim that thermal relaxation time becomes important in heterogeneous biological tissues even at macroscale length and time scales ($L > \lambda$ and $t > \tau$)[Mitra *et al.*, 1995], but this has been challenged by others as issues with their experimental protocol or instrumental errors [Scott *et al.*, 2009].

There are several biomedical applications at short time scales, as summarized in Table 3. For instance, there is increasing interest in understanding thermal transport in biomaterials irradiated by short duration (~ fs) laser pulses [Huettmann *et al.*, 2003; Csaki *et al.*, 2007]. If this is used on a nanoparticle, the extent of heating could stay within the nanoparticle at short time scales. This is referred to as thermal confinement [Pissuwan *et al.*, 2006; Roper *et al.*, 2007]. The time scales depend on the material property and size. A continuum estimate for the thermal confinement timescale is given in Equation (5).

$$t_{confine} = \frac{d^2}{27\left(\frac{k}{\rho c}\right)}. \tag{5}$$

Here, d is the particle diameter (m), of the nanoparticle. This equation is derived from Fourier heat conduction for a semi-infinite solid. For the thermal penetration depth to be within a gold nanoparticle, the calculated t time duration of the pulse should be of the order of tens of fs [Qin and Bischof, 2012]. However, this time scale determined by Fourier heat equation is comparable to the relaxation time of gold nanoparticles ($t_{confine} \sim \tau$). Hence the Fourier law is no longer valid and nanoscale heat transfer effects need to be incorporated

Table 3. Ultrafast-thermal processes in biomedical applications.

System	Length scale	Application	Time scale	Reference
Metal nanoparticle	10–100 nm	Nanoparticle heating, nanosurgery	~100 fs	[Leary *et al.*, 2006; Uchugonova *et al.*, 2012]
Nanoparticle in bio-environment	10–100 nm	Heat transport across nanoparticle interface into the medium	~100–1000 ps	[Ge *et al.*, 2005; Schmidt *et al.*, 2008a; Alper and Hamad-Schifferli, 2010]
Collagen	10 nm	Laser processing of biomaterials (collagen, agar etc.)	~100 fs	[Zhang *et al.*, 2013]
Proteins	1–10 nm	Heat dissipation by proteins to the solvent	~10 ps	[Lian *et al.*, 1994]
Biomolecules	1–50 nm	Structural conformation of individual biomolecules, biochemical interactions in individual biomolecules	~100 μs	[Neuweiler *et al.*, 2009; Kapanidis and Strick, 2009]

to explain the heat conduction within a nanoparticle [Chen, 1996]. However, the most important time scale would be the diffusion of heat from the nanoparticles to the surrounding matrix ($t > t_{confine}$). This phenomenon is discussed more in Section 8.2 Molecular and Nanoscale. Thus, thermal conductivity and thermal relaxation are important material properties of heat conduction at short length scales and time scales [Ge *et al.*, 2005; Schmidt *et al.*, 2008a; Alper and Hamad-Schifferli, 2010]. As time and distance scales continue to reduce in biomaterial research, these topics will require deeper study.

4. Thermal Conductivity Measurement Techniques (Macro To Nano)

Whenever valid, Fourier's law can be used to measure thermal conductivity by first imposing a known temperature gradient ∇T(K/m) across an isotropic material and then measuring the ensuing heat flux Q (W/m^2). Alternatively, the heat flux can be defined and the thermal gradient calculated. In either case, thermal conductivity, the material constant in Fourier's Law, can then be obtained as, $k = -Q/\Delta T$.

The measurement of thermal conductivity in biomaterials has been of great interest for more than half a century, especially in the field of food refrigeration [Hill *et al.*, 1967; Poppendiek *et al.*, 1967; Bowman *et al.*, 1975; Chato, 1985; Sweat, 1995]. Subsequently, several models were developed to predict the thermal conductivity of various foods based on weight and/or volume averaged composition. Some of the most important models are tabulated in Table 4. In these models, the continuous phase is usually water and/or ice and/or fat (i.e. lipid) and the dispersed or discontinuous phase is comprised

Table 4. Thermal conductivity models for composite media.

Thermal conductivity model	Characteristics	Equation
Eucken [1940]	A modification of the Maxwell equation in which one component is dispersed in a spherical bubble within a continuous surrounding component (i.e. water/fat).	$k = k_c \dfrac{1-\left[1-a\left(\dfrac{k_d}{k_c}\right)\right]b}{1+(a-1)b}$ where k: thermal conductivity. $a = 3k_c/(2k_c + k_d)$ $b = V_d/(V_c + V_d)$ V: volume
Kopelman [1966]	Isotropic, two component system.	$k = k_c \dfrac{1-M}{1+M(1-L)}$ where L^3 = volume fraction of the discontinuous phase $M = L^2(1-k_d/k_c)$
	Anisotropic, two component system used for homogenous fibrous foods.	$k = k_c\left[1-N^2\left(1-\dfrac{k_d}{k_c}\right)\right]$ $k_\perp = k_c \dfrac{1-P}{1-P(1-N)}$ where N^2 = volume fraction of the discontinuous phase $P = N(1-k_d/k_c)$

(*Continued*)

Table 4. (*Continued*)

Thermal conductivity model	Characteristics	Equation
Levy [1981]	Modified version of Maxwell–Eucken model.	$$k = \frac{k_2\left[(2+\gamma)+2(\gamma-1)F_1\right]}{(2+\gamma)-(\gamma-1)F_1}$$ where subscript 1, 2 refer to first and second component $\gamma = k_1 / k_2$ F_1 is a function of γ and is not expanded here for brevity. Please refer to the literature for details [Levy, 1981].
Murakami and Okos [1989]	Based on the electrical resistance analogy. The parallel model and the perpendicular model form the upper and lower limit of thermal conductivity of a particular mixture of food	Parallel model: $$k = \sum x_i^v k_i$$ where subscript 'i' refers to individual constituent $$x_i^v = \frac{x_i/\rho_i}{\sum\left(\frac{x_i}{\rho_i}\right)}$$ Perpendicular model: $$k = \frac{1}{\sum x_i^v / k_i}$$

of protein, lipid and mineral. These models were validated for biomaterials by several macroscale measurement techniques. The main limitation with these models is that they were developed based on '*k*' measurement under equilibrium conditions in the biomaterial. However, cryopreservation process involves non-equilibrium thermodynamics (during vitrification or supercooling). For instance, a biomaterial will undergo vitrification with greater probability at higher cooling rates. Thus, *k* of the biomaterial will be closer to ice at lower cooling rates due to crystallization and *k* may remain low at higher cooling rates due to vitrification. The issue can be solved by measuring *k* as a function of temperature and thermal history (cooling rate) [Han and Bischof, 2004; Choi and Bischof, 2010] The next section will briefly discuss macroscale thermal conductivity measurement techniques and elaborate in detail on the microscale and nanoscale thermal conductivity measurement techniques.

5. Macroscale Thermal Conductivity Measurement Techniques

The pursuit of thermal conductivity measurement of macroscale biomaterials has led to several methods as noted in Table 5. Each of these measurements establishes either the heat flux or the thermal gradient in order to extract thermal conductivity as already noted and shown pictorially in Figure 1(a). Each of these methods will be briefly reviewed with differences in heat source, steady state vs. transience nature, geometry and directionality noted. These have previously been reviewed for macroscale thermal property measurements in cryobiology [Choi and Bischof, 2010].

Table 5. List of macroscale thermal conductivity techniques for biomaterials.

Meas. tech.	Heat source	Temp. meas.	Heat transfer model	Directionality	Rep. ref.
Guarded hot plate	Thermal*	Thermocouple (TC)	Steady state 1-D longitudinal	Cross-plane	[Hill *et al.*, 1967; Poppendiek *et al.*, 1967; Committee, 2011]
Thermal comparator	Thermal*	Probe (effective TC)	Transient 1-D cylindrical	Not sensitive	[Vendrik and Vos, 1957; Morley, 1966; Vachon *et al.*, 1967]
Radial heat flow method	Thermal*	Thermocouple	Steady state 1-D cylindrical	In-plane	[Glassbrenner and Slack, 1964]
Heated thermocouple	Electrical**	Thermocouple	Transient 1-D spherical	Not sensitive	[Grayson, 1952]
Transient hot wire	Electrical**	Thermocouple	Transient 1-D cylindrical	Not sensitive	[Blackwell, 1954; Andersson and Backstrom, 1976; Bhattacharya and Mahajan, 2003; Assael *et al.*, 2004]
Chato's probe	Electrical**	Thermocouple (constant temperature)	Transient 1-D spherical	Not sensitive	[Chato, 1968; Balasubramaniam and Bowman, 1977; Chato, 1980]
Pulse decay method	Electrical Single pulse**	Thermocouple	Transient 1-D spherical	Not sensitive	[Chen *et al.*, 1981; Valvano *et al.*, 1985; Patel *et al.*, 1987; Zhang *et al.*, 2002; Choi *et al.*, 2013]

Notes: *Thermal — Heat source heated by secondary heating through heat exchanger.
**Electrical — Heat source heated by Joule heating.

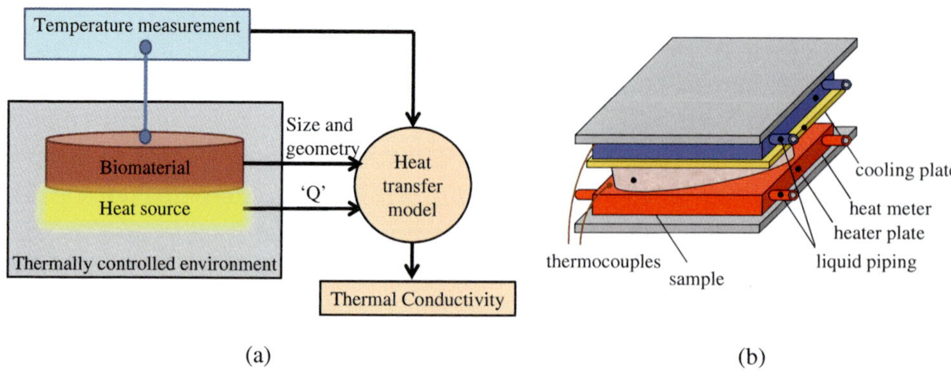

Figure 1. (a) A pictorial representation of most thermal conductivity measurement techniques; (b) schematic of guarded hot plate method (the guard is concentric with the sample but is omitted for clarity).

These macroscale methods can be broadly classified into steady state and transient methods. Steady state methods use the knowledge of heat flux and temperature response from the sample in a one-dimensional heat transfer model to determine thermal conductivity. These methods require long durations in order to reach steady state (could be several hours depending on the size of the sample) and proper "guarding" to avoid parasitic heat losses thereby supporting the necessary one-dimensional heat flow. Further, these methods suffer from contact resistance between the sample and the heat source [Hill *et al.*, 1967; Poppendiek *et al.*, 1967; Committee, 2011]. The main advantage of the steady state methods is that they can directly measure thermal conductivity and anisotropy. Examples of steady state method are guarded hot plate [Hill *et al.*, 1967; Poppendiek *et al.*, 1967; Committee, 2011] and the radial heat flow method [Glassbrenner and Slack, 1964]. In guarded hot plate as shown in Figure 1(b), the sample is sandwiched between a heater plate on one side and a cooling plate on the other side, which heats and cools the sample at a constant rate. The heat meter and the thermocouples placed in contact with the sample measures heat flow and temperature response respectively. Once steady state is achieved, the temperature response in the sample is measured and applied into one-dimensional heat transfer models to determine thermal conductivity.

In contrast, transient methods are typically much faster (<1 minute) to perform and have a simpler hardware design. In these cases, a spherical or a cylindrical probe is inserted into the tissue and can simultaneously apply heat and measure temperature. The heat flow is in the form of either a pulse [Chen *et al.*, 1981; Valvano *et al.*, 1985; Patel *et al.*, 1987; Zhang *et al.*, 2002; Choi *et al.*, 2013] or continuous heating [Chato, 1968; Balasubramaniam and Bowman, 1977; Chato, 1980; Blackwell, 1954; Andersson and Backstrom, 1976; Bhattacharya and Mahajan, 2003; Assael *et al.*, 2004]. While typically faster than steady state approaches, they also suffer from contact resistance between the sample and the probe. Hence, the heat flow from the probe needs to penetrate deeper into the sample so that contact resistance becomes negligible at long time scales of heating. This requires the size of the samples to be generally greater than ~10 mm thereby limiting their application to thick tissues and organs. Also, they are invasive, leading to tissue damage.

6. Microscale Thermal Conductivity Measurement Techniques

Macroscale techniques cannot be used for thin tissues or cellular systems (<10 mm). In order to address this need, there are now various techniques that can be adapted, such as the two shown in Table 6. Specifically, thermal conductivity of films (<1 mm) has been extensively studied in the field of microelectronics and thermo-electrics, for which a variety of techniques has been developed in the past two decades [Tritt, 2004]. These techniques will each be discussed in more detail below.

6.1. *3ω Method*

One commonly used thin film thermal conductivity measurement technique for inorganic materials is called the 3ω method [Cahill and Pohl, 1987; Dames and Chen, 2005]. The first application of this technique to measure thermal conductivity of amorphous solids was developed by Cahill and Pohl in 1987 [Cahill and Pohl, 1987]. In the last three decades, this technique was modified to measure thin films [Cahill *et al.*, 1994; Borca-Tasciuc *et al.*, 2001], multilayered films [Olson *et al.*, 2005], anisotropic solids [Kim *et al.*, 1999; Tong and Majumdar, 2006], suspended wires [Lu *et al.*, 2001]

Table 6. List of microscale thermal conductivity measurement techniques.

Meas. tech.	Heat source	Temp. meas.	Heat transfer model	Directionality	Sample size (thickness)	Rep. ref.
3 omega method	Electrical** (1–1000 Hz)	Third harmonic voltage response	1-D Quasi-steady cylindrical	Radial (bulk samples); cross-plane (thin films)	0.1 μm to 2mm	[Cahill and Pohl, 1987]
Laser flash method	Optical*** (Single laser pulse)	Infrared detection of the opposite side	Transient cylindrical	Cross-plane and in-plane	100 μm to 2mm	[Parker *et al.*, 1961; Ohta *et al.*, 1989]

Notes: *Thermal — Heat source heated by secondary heating through heat exchanger.
**Electrical — Heat source heated by Joule heating.
***Optical — Heat source heated by photothermal effect.

and liquids [Choi and Kim, 2008]. This method is unique compared to other techniques because it is neither a steady state nor a transient technique. It is a quasi-steady technique wherein the sample is periodically heated. Essential to the technique is a thin metallic line that acts as heater and thermometer in intimate contact with a sample as shown in Figure 2(a).

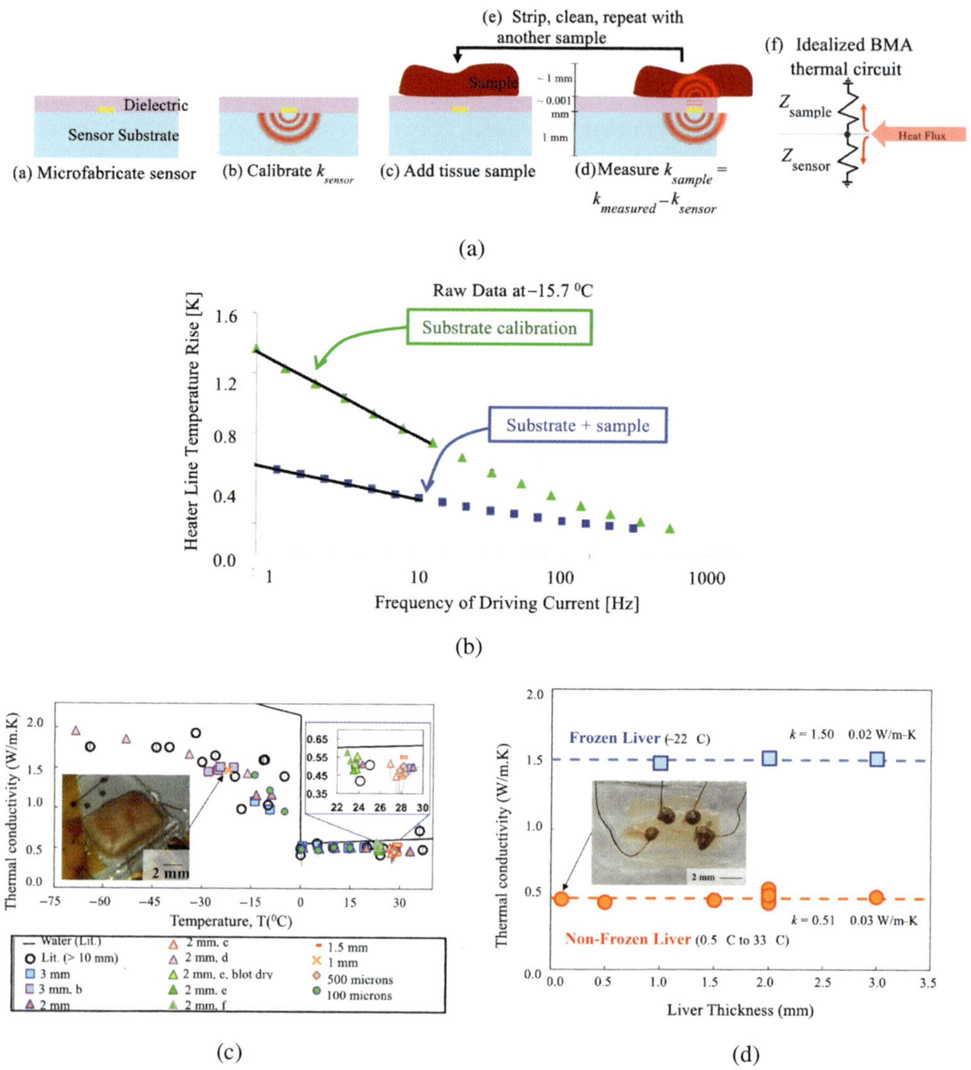

Figure 2. (a) Cross sectional view. Layout, process and thermal circuit for supported 3 omega method; (b) temperature rise 'ΔT_{AC}' at the heater line as a function of frequency of driving current 'ω'; (c) measurement of thermal conductivity of mouse liver; (d) thermal conductivity is independent of thickness in mouse liver sliced to varying thickness between ~100 μm and ~3 mm. Reprinted with permission from Lubner *et al.*, Copyright 2015, AIP Publishing LLC.

By exciting the heater line with an alternating current ($I_0 \sin(\omega t)$), Joule heating results as shown below:

$$Q = \frac{I_0^2 R_0}{2} - \frac{I_0^2 R_0}{2}(\cos 2\omega t). \quad (6)$$

Equation (6) comprises a steady DC term and an AC term oscillating at '2ω'. The resulting temperature rise in the heater line would similarly comprise a DC and an AC term oscillating at '2ω'.

$$\Delta T = (\Delta T)_{DC} + (\Delta T)_{AC} \cos(2\omega t + \varphi). \quad (7)$$

However, there is an additional 'φ' term in the AC component, which is due to phase lag between Joule heating and subsequent surface temperature rise. This is indicative of both the thermal inertia and thermal impedance in the surrounding medium (sample and substrate). These temperature perturbations are further transferred to the temperature-dependent electrical resistance of the heater line due to the linear relationship between them.

$$R = R_0 + R_0 \alpha \Delta T_{DC} + R_0 \alpha \left|\Delta T_{AC}\right| \cos(2\omega t + \varphi) \quad (8)$$

where I, R, ω, t, α, φ are current, electrical resistance, angular frequency (rad/sec), time, temperature coefficient of resistance of the heater line and phase lag respectively. Following Ohm's law, the '2ω' oscillations in resistance will combine with the applied '1ω' oscillations in alternating current to produce '3ω' oscillations in voltage as shown below.

$$V = I_0 R_0 \left[\sin(\omega t(1 + \alpha \Delta T_{DC})) + \frac{\alpha \Delta T_{AC}}{2} \left[\sin(3\omega t + \varphi) - \sin(\omega t + \varphi) \right] \right] \quad (9)$$

$$V = V_{1\omega} + V_{3\omega}. \quad (10)$$

Equations (9) and (10), in which the voltage comprises of a '1ω' and a '3ω' component, shows that the '3ω' voltage contains information of the AC temperature rise (ΔT_{AC}) in the heater line. In short, sinusoidal Joule heating of known quantity is applied and the resulting temperature rise in the heater line is determined from the '3ω' voltage. Hence, this technique is called the 3 omega method [Cahill and Pohl, 1987].

A semi-infinite solid with a sinusoidal line heat source term is the preferred heat transfer model. The readers are directed to another reference where the mathematical procedure to obtain thermal conductivity from the temperature rise based on the model is discussed [Cahill and Pohl, 1987].

Considering only the real part (in phase) of temperature oscillation, the equation for thermal conductivity is given in Equation (11).

$$k_{\text{in phase}} = \left[\frac{V_{1\omega}^3}{4\pi LR^2}\frac{dR}{dT}\right]\frac{\ln(\omega)}{V_{3\omega,\,\text{in phase}}}. \tag{11}$$

In Equation (11), R and dR/dT are obtained by measuring the resistance of the sensor at different temperatures. The intercept and slope of the resulting R vs T curve would give R_0 and dR/dT respectively. Also R can be determined from R_0 and dR/dT for the temperature of measurement. Then $V_{3\omega,\text{in phase}}$ is measured for different frequencies of electrical current and plotted as shown in Figure 2(b). The slope of the line can then be used to determine thermal conductivity. This method is referred to as the slope method. A major advantage of slope method is that it is not sensitive to contact resistance.

Recently, the 3 omega method has been used for measurement of thermal conductivity of biological solutions [Park *et al.*, 2012a], biological tissues such as mouse liver [Lubner *et al.*, 2015] as shown in Figure 2(c–d) and even biological cells [Park *et al.*, 2013]. These studies use a "supported 3ω method", whereby the heater line is micro-fabricated onto a substrate of known thermal properties and the biomaterial is placed on top of the heater line as shown in Figure 2(a). To an excellent approximation, the resulting measured k value is simply the sum of that of the biomaterial and the substrate [Lubner *et al.*, 2015]. The thermal conductivity of the substrate is known from calibration by measuring without any sample on top of the sensor as shown in Figure 2(b). In addition, water layer(s) and/or evaporation around the sample need to be carefully monitored to ensure reproducible measurement. The sample sizes of interest for this measurement are typically 10's of μm to mm. The accuracy achieved is 6%–7% for fresh tissue/water and 2%–5% for frozen tissue/ ice [Lubner *et al.*, 2015].

In principle, by adapting known single-sided 3ω solutions for multilayer systems [Feldman, 1999; Borca-Tasciuc *et al.*, 2001], the supported 3 omega method should also be capable of measuring the thermal conductivity of composite tissues and contact resistance between tissues as well. One attractive aspect of this method is that the amount of the sample measured can be varied. Specifically, the frequency of the driving current can be changed to control the penetration depth, defined as $(D/2\omega)^{0.5}$. In fact, this technique has been used with low frequency (<1 Hz) electric current to sense the backside boundary of the sample (~2 mm thick) [Natesan *et al.*, 2016]. Future work in this area can also address anisotropic biomaterial properties (i.e. along and across dendritic ice fronts, of ice formation within tissues).

6.2 *Laser Flash Method*

Laser flash method is an ASTM standardized optical technique that has been used primarily in the measurement of thermal conductivity of inorganic wafers and films (100 μm–1 mm) [Committee, 2013]. It is a non-invasive non-contact method and thus has the potential to offer several advantages for biomedical applications. Laser flash method was first introduced by Parker *et al.* in 1961 [Parker *et al.*, 1961]. Figure 3(a) shows a high intensity short duration laser pulse heating one side of a sample. The resulting temperature response on the rear side is recorded as a function of time by either a thermocouple or an infrared (IR) detector. This temperature versus time curve can then be used to extract thermal diffusivity from a heat transfer model, as the diffusion length $\sqrt{\alpha t}$ is directly related to the thickness of the sample. From thermal diffusivity, thermal conductivity is then obtained if volumetric heat capacity (ρc) is known. The heat transfer model as developed by Parker *et al.*, [1961] is explained in detail. The temperature distribution for a plane wall of finite thickness 'L' is given by Equation (12).

$$T(x,t)=\frac{1}{L}\int_0^L T(x,0)\,dx+\frac{2}{L}\sum_{n=1}^{\infty} exp\left(-n^2\pi^2\frac{\alpha t}{L^2}\right) \times\cos\frac{n\pi x}{L}\int_0^L T(x,0)\,\cos\frac{n\pi x}{L}\,dx \tag{12}$$

where α thermal diffusivity (m^2/s), x distance from the front surface exposed to the heat source, t time. Initially, the incident light is assumed to instantaneously heat a sample uniformly up to a depth of 'g'. For depth 'x' greater than 'g', the sample is at a uniform temperature T_0 (x, $t = 0$). Thus the initial conditions are summarized below,

$$T(x,0)=\frac{Q}{\rho c g} \quad \text{for } 0 < x < g \tag{13}$$

$$T(x,0)=0 \quad \text{for } g < x < L \tag{14}$$

where Q is incident radiant energy (J/m^2). The temperature of the back surface is independent of 'g' as long as $g \ll L$ which is true for most cases.

$$T(L,t)=\frac{Q}{\rho c L}\left(1+2\sum_{n=1}^{\infty}(-1)^n exp\left(-n^2\pi^2\frac{\alpha t}{L^2}\right)\right). \tag{15}$$

By reducing Equation (15), an expression for thermal diffusivity, volumetric heat capacity and hence thermal conductivity could be determined. Thermal diffusivity is given by

$$\alpha = \frac{0.48L^2}{\pi^2 t_{1/2}} \tag{16}$$

where $t_{1/2}$ is time for the back surface to reach half the maximum temperature rise. Volumetric heat capacity could be determined from the incident radiant energy and temperature vs time curve as shown below,

$$\rho c = \frac{Q}{LT_m} \tag{17}$$

where T_m is maximum temperature rise at the rear surface (K). Thus thermal conductivity can be calculated from thermal diffusivity and heat capacity. As this is an absolute measurement technique, calibration is not needed. The major limitations of this model are the assumptions of absence of parasitic heat loss from the sample, zero pulse duration as shown in Figure 3(b) and one-dimensional heat flow (i.e. cross plane). Finally, the model assumes that the sample is a black body (i.e. maximum absorption and emission at a given temperature). This can be approximated by coating the front and rear sample surfaces with material of low transmittance and reflectance and high absorptance. An example would be black coatings made of dry graphite spray, however, this might lead to poor repeatability in coating and non-homogenous coating. This would reduce the accuracy of the measurement by 2% to 5% [Akoshima *et al.*, 2013]. Later, several models came up to accurately model the process and are tabulated in Table 7.

When laser flash is applied to thinner samples, there are several challenges. For instance, a main challenge is the need to capture the temperature decay

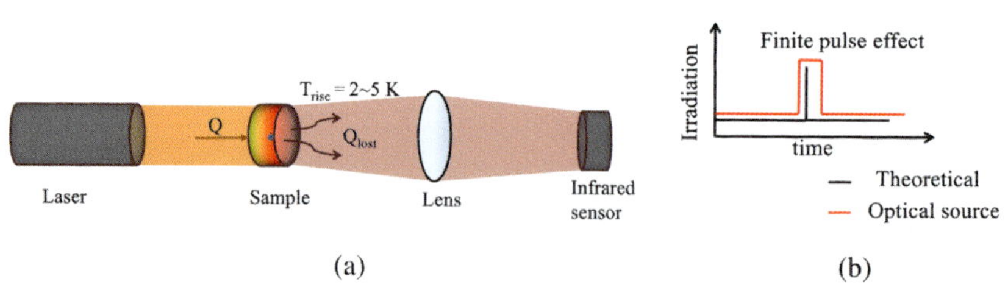

Figure 3. (a) Schematic of Laser flash method. Q: heat input into the sample on the front side. Q_{lost}: backside heat loss. T_{rise}: Temperature rise on the backside of the sample. (b) Illustration of finite pulse effect vs instantaneous heating.

Table 7. Thermal models of laser flash method.

Model	Major characteristics
Parker *et al.*, [1961]	• Infrared radiation of negligible pulse duration • 1-D heat transfer • Ignores convection/radiation heat loss from both the surfaces. • Cannot be applied to very high temperature (i.e. requires <1000 K)
Parker *et al.*, [1962]	• Includes convective heat loss from one surface • Accounts for internal heat generation within the sample as a function of depth if present during heat diffusion
Rudkin *et al.*, [1962]	• Radio waves as a heat source • Can be applied to higher temperature (<2000 K) measurements
Cowan [1963]	• Can be applied to high temperature (>2500 K) measurements • Surface heat loss (convection/ radiation) is included
Cape and Lehman [1963]	• Accounts for finite duration of the laser pulse in the model • Includes heat loss due to radiation at both the facial and radial surfaces

that occurs with increasing rapidity in thin samples. For thin samples, the laser pulse duration must be shorter than the heat diffusion time across the sample thickness. Otherwise, the heat transfer model presented above, which assumes that the heat source is instantaneous, would fail or require careful modification. Further, the sampling rate of the temperature detector must also be high enough to capture the transient temperature decay at the rear end. While thermocouples were initially used for this purpose, infrared (IR) detectors were employed to detect the temperature response [Ohta *et al.*, 1989]. The main advantages of IR detectors are that they are non-contact type, accurate and work better for large temperature changes. Also IR detectors can measure temperature of a specific area compared to a point measurement by thermocouples [Moffat and Kim, 2004].

Laser flash method has also been used to measure thermal conductivity of polymers of thickness 14–250 μm [Choy *et al.*, 1987]. The major challenge specifically for biomaterials, however, is the evaporative heat loss that occurs at small length scales, requiring careful experimental methods. For instance, Tada *et al.* [1978] developed an experimental setup to measure the thermal conductivity of liquids, such as water, by placing roughly 1 mm of the fluid between metal disks (one of which is heated by laser). This modification also allows measurement of molten salt [Ohta *et al.*, 1990]. Though the traditional laser flash technique is generally non-contact, this modification would require contact between disks and the sample introducing the issue of contact resistance. While few biomaterials have been measured with this approach, the thermal diffusivity of 1.2–1.5 mm thick human tooth is one example [Panas *et al.*, 2003]. Thus this technique may be able to measure thermal conductivity within sub-millimetre thick tissues. The uncertainty of this technique varies from ±3% to 5% depending on material and temperature [Hay *et al.*, 2005].

7. Nanoscale Thermal Conductivity Measurement Techniques

Nanoscale thermal conductivity measurement techniques have been recently developed to probe materials of dimension in the order of tens of nm to hundreds of nm, thereby revealing the mechanism of heat transport in both crystalline and amorphous solids [Toberer *et al.*, 2012]. Some of the representative techniques are listed in Table 8 and discussed in detail in the next section.

7.1. *Scanning Thermal Microscopy (SThM) Method*

SThM can be used to map thermal conductivity [Nonnenmacher and Wickramasinghe, 1992; Pylkki *et al.*, 1994; Florescu *et al.*, 2002; Gu *et al.*,

Table 8. List of nanoscale thermal conductivity measurement techniques.

Measurement technique	Heat source	Temperature measurement	Heat transfer model	Directionality	Spatial resolution	Rep. ref.
Scanning Thermal Microscopy (SThM)	Electrical/ Optical	Thermocouple, thermistor or RTD	Quasi-steady model	Spherical radial	100–1000 nm	[Majumdar *et al.*, 1993; Pylkki *et al.*, 1994]
Time domain thermo reflectance (TDTR) microscopy	Optical*** — Laser (>1 MHz pulsed)	Using surface reflectivity dependence on temperature	Transient	In-plane and cross-plane	10–600 nm	[Paddock and Eesley, 1986; Cahill, 2004]
Transient Absorption Method (TAM)	Optical*** — Laser (>1 MHz pulsed)	Using transmitted signal to measure absorbance	Transient	Spherical radial	1–20 nm	[Huang *et al.*, 2013]

Notes: *Thermal — Heat source heated by secondary heating through heat exchanger.
**Electrical — Heat source heated by Joule heating.

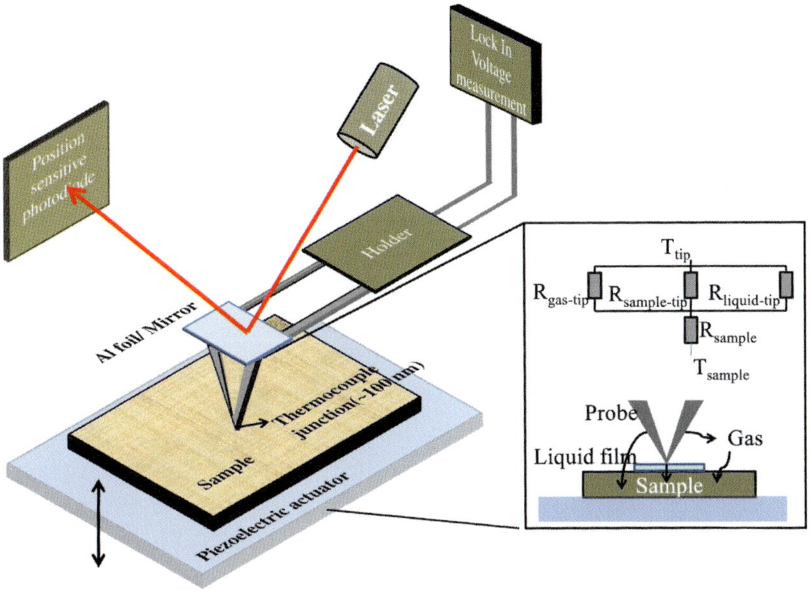

Figure 4. Scanning thermal microscopy. The inset shows the heat transfer mechanism between the heated tip and the sample.

2002; Assael *et al.*, 2004; Gomès *et al.*, 2013] or temperature [Williams and Wickramasinghe, 1988; Majumdar *et al.*, 1993; Kim *et al.*, 2012; Lee *et al.*, 2013] on the surface of a sample or its sub-surface [Hammiche *et al.*, 1996a; Hammiche *et al.*, 1996b; Hammiche *et al.*, 2002] at scales between the nm to μm range. SThM consists of a micro-fabricated probe that is moved across the sample surface. The probe is typically a pointed tip made of a temperature sensor, such as thermocouple or thermistor, and has a size in the range of 100 nm. Nanofabrication of probes has been described elsewhere in the literature [Luo *et al.*, 1996]. Figure 4 shows the schematic of the scanning thermal microscopy. The probe is held in position by a cantilever or shank, and a reflecting mirror is placed on top of the cantilever in order to reflect an incident laser onto an optical sensor. As the probe is scanned across the surface of the sample, the cantilever is deflected due to any irregularity on the sample surface, thereby causing a corresponding change in the direction of the reflected laser beam. The displacement of the reflected laser on the optical sensor can therefore be used to determine the surface topology of the sample. More commonly, it is used as part of a feedback loop which moves the stage to enforce zero displacement of the laser. By heating the tip and recording the amount of heat loss from the tip, a thermal conductivity map of the sample surface can be obtained.

Historically, the scanning probe microscopy was modified to probe surface topography based on temperature as a feedback [Williams and Wickramasinghe, 1986]. In 1988; a dedicated instrument to map temperature was developed with a resolution of less than 1 mK and a spatial resolution of around 100 nm [Williams and Wickramasinghe, 1988]. This was then modified to map thermal conductivity based on the principle of scanning tunneling microscopy. However, this method was limited in application to only electrically conducting surfaces [Nonnenmacher and Wickramasinghe, 1992].

Majumdar and colleagues developed a breakthrough measurement technique that for the first time simultaneously maps surface topography and surface temperature of the electrically insulating and conducting sample. A cantilever acting as a thermocouple measures temperature and uses interatomic force between the tip and the sample as a feedback signal to maintain a constant tip-sample distance. [Majumdar *et al.*, 1993]. However, the thermocouple tip could not be heated, limiting its application only to temperature measurement. In response, Pylkki and colleagues developed a resistive thermal probe that could be electrically heated [Pylkki *et al.*, 1994]. Thus, a constant current could be used to electrically heat the resistive probe, which cools down due to neighbouring sample surface. By monitoring the temperature of the resistive probe, a qualitative map of thermal conductivity was obtained.

This method has been applied to measure thermal conductivity of carbon nanotubes [Shen *et al.*, 2010] and polyethylene nanofibers [Shi *et al.*, 2009] using a modified bi-metallic cantilever. Though very little work is published, SThM could have a significant impact in application for nanoscale biomaterials [Nikiforov *et al.*, 2011]. Similar techniques, such as Atomic Force Microscopy and Scanning Tunnelling Microscopy, have been applied to map surface topography of protein, lipids, DNA and intracellular structures [Baró *et al.*, 1985; Allison *et al.*, 2010]. Since nanoscale biomaterials, such as cells and biomolecules, are often suspended in an aqueous solution, the probe has to be electrically insulated and protected from the enhanced parasitic thermal losses in an aqueous environment. Li and Gianchandani (2003) made the first attempt to design a SThM probe for use in aqueous environment. In this work, both the substrate and the cantilever were immersed in an aqueous environment to avoid surface tension effects on the cantilever. However, they mapped the surface thermal conductance of HeLa tumour cells in a non-aqueous environment as shown in Figure 5 [Li and Gianchandani, 2003], several measurements were made by this technique proving its potential to measure biomaterials in a liquid environment including even a lab on a chip

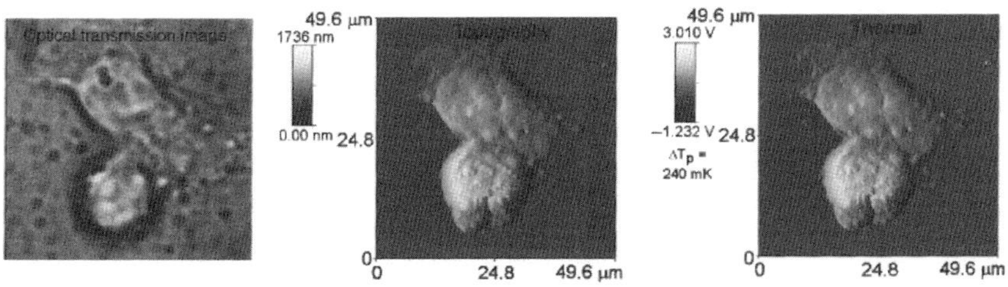

(a) (b) (c)

Figure 5. (a) Optical Transmission image; (b) topography; (c) thermal image of nucleus of HeLa cells attached to a glass substrate undergoing mitosis. Reprinted with permission from Li and Gianchandani, Copyright 2003, Elsevier.

[Nikiforov *et al.*, 2011; Tovee and Kolosov, 2013]. Another advantage of this technique is that it can provide additional information on latent heat during phase transition and specific heat [Hammiche *et al.*, 1996a; Shnyrov *et al.*, 1997; Price *et al.*, 1999; Craig *et al.*, 2002; Zhang *et al.*, 2009].

Although powerful, SThM faces several limitations, first and foremost being the complex heat transfer mechanism between the heated probe and the surface. At the nanoscale, liquid film formation on the surface due to humidity in the environment enhances the possible number of pathways for heat transfer. As shown in Figure 4 (inset), heat transfer occurs through tip-gas-sample, tip-liquid-sample and tip-sample [Shi and Majumdar, 2001; Lefèvre *et al.*, 2006; Park *et al.*, 2008]. Secondly, the thermal conductivity is mostly qualitative due to lack of proper calibration techniques and heat transfer models. As shown in Figure 5, the interplay between surface roughness and temperature measurement causes error in thermal conductivity measurement. Nevertheless, this system provides an opportunity to physically probe biomaterials with a nanoscale resolution. This provides exciting opportunities to probe thermal conductivity of individual cell components and even study the effect of macromolecules, nanoparticles, cryoprotective agents (CPAs) and ice crystals on them. Though SThM has not been widely used to probe temperature and thermal conductivity in biomaterials, Atomic Force Microscopy, a related technique, has been used extensively to probe topography in live cells. For instance, it has been used to image cells as a function of temperature [Espenel *et al.*, 2008]. Also, a biological AFM has been developed to probe biomaterials at cryogenic temperatures [Han *et al.*, 1995]. Thus these measurements provide an opportunity to repeat similar measurement on thermal conductivity using SThM. For instance, imaging using FTIR and Raman Spectroscopy in the past has provided valuable information on protein and lipid phase change at the cellular level [Bischof *et al.*, 2002;

Wolkers *et al.*, 2007; Dong *et al.*, 2009]. Similarly, thermal measurements inside cells during phase change could provide new information at molecular scale, potentially changing our understanding of the micro-environments inside cells and their influence on molecular phase change.

7.2. *Time Domain Thermoreflectance (TDTR) Method*

TDTR is a transient technique that utilizes the dependence of reflectance of materials such as gold on their surface temperature to determine thermal properties of thin films [Paddock and Eesley, 1986; Capinski, 1999; Huxtable *et al.*, 2004], amorphous [Wang *et al.*, 2013a; Wang *et al.*, 2013b], anisotropic [Schmidt *et al.*, 2008a; Feser and Cahill, 2012], multilayered structures [Cahill, 2004] and microscale liquid droplets [Schmidt *et al.*, 2008c]. In this technique, the surface of the sample is coated with a metallic transducer with high temperature coefficient of reflectivity. Figure 6 shows the schematic of the time-domain thermal reflectance technique. Two lasers, a pump and probe beam, are employed as a heater and temperature sensor respectively. The pump beam heats the surface with a femtosecond duration Ti-Sapphire laser pulse. This results in very fast heating of the thin metallic (transducer) surface. Subsequently, the surface is cooled by the underlying sample by conduction. The higher the thermal conductivity of the sample, the faster the surface temperature decay, which correspondingly causes a change in the surface reflectivity of the sample. The time delayed probe beam can then measure this reflectivity and therefore surface temperature change. By fitting

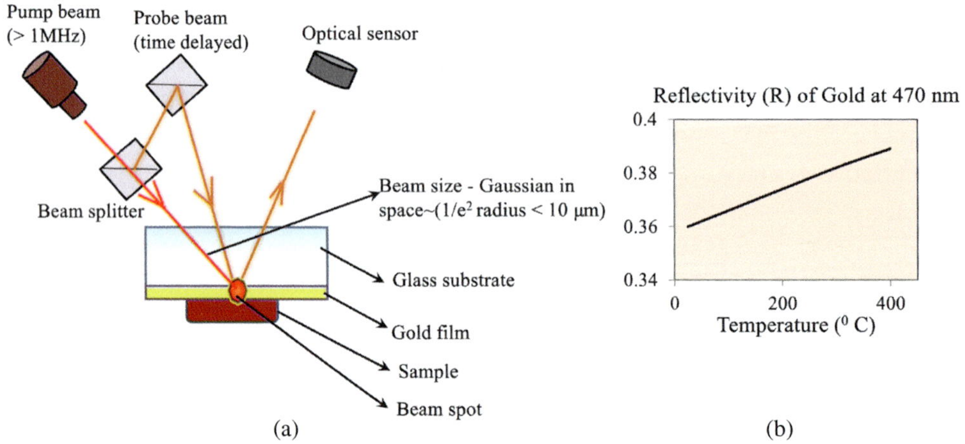

Figure 6. (a) Time-Domain Thermoreflectance Method; (b) plot shows the temperature dependence of reflectivity of gold film. Figure reprinted from Beran [1985].

this behavior with a suitable heat transfer model, the thermal properties of the sample can be extracted. In general, negligible lateral heat spreading, due to smaller thermal penetration depth (<100 nm) than the beam spot size (~10 μm–100 μm), enables the use of one-dimensional heat flow model [Paddock and Eesley, 1986; Capinski, 1999; Cahill, 2004]. In Figure 6, the gold film on the glass substrate is the transducer that dissipates heat to both the substrate and the sample. For this case, the analytical solution for the bi-directional, one-dimensional heat transfer is available in the literature [Schmidt *et al.*, 2008b].

A major difference between laser flash method and time domain TDTR is that the former uses an IR detector/ thermocouple (~1 kHz sampling frequency) whereas the latter uses a high frequency laser beam (>10 MHz sampling frequency) to detect the temperature change which can yield improved spatial and temporal resolution. For instance, laser flash method detects temperature on the opposite side of the sample from the heated side, whereas TDTR detects temperature on the same side of the sample that is heated, consequently, heat need not propagate through the sample for TDTR, resulting in faster more confined measurements. Specifically, this translates into the ability of TDTR to measure temperature decay at much shorter time scales (~100–1000 ps), scaling in biological samples to diffusion lengths ($\sqrt{at}$) on the order of 100–600 nm. This technique has been used to understand heat transport phenomena at very short time scales (~several hundred ps or less) such as thermal conductance across solid-solid interface [Costescu *et al.*, 2003], solid-liquid interface [Ge *et al.*, 2006], chemical bonds [Losego *et al.*, 2012] and even across molecular chains [Wang *et al.*, 2007]. It has also been used to provide insights into heat transfer in protein films [Foley *et al.*, 2014] (Figure 7). This is an absolute measurement and does not require calibration. The accuracy of the technique is limited by the uncertainty in thickness of the transducer film and is usually about 7% at frequency of 10 MHz [Koh *et al.*, 2009]. Though very limited work has been done in experimentally understanding the deviations in Fourier law at short length and time scales effects (as discussed in Section), TDTR has been instrumental in observing the effects in Si and Mgo and Al/ $Si_{0.99}Ge_{0.01}$ alloys [Wilson *et al.*, 2013; Wilson and Cahill, 2014]. There are several challenges in TDTR application to biomaterials. First the sample needs to be heated by an order of 1–10°C to achieve appreciable response in the output signal to temperature change. For instance, in Figure 6(b), the graph between reflectivity of gold and temperature shows that temperature rise should be high enough to cause detectable change in reflectivity, which is usually around 1–10°C [Ujihara, 2003]. This value is higher compared to other techniques

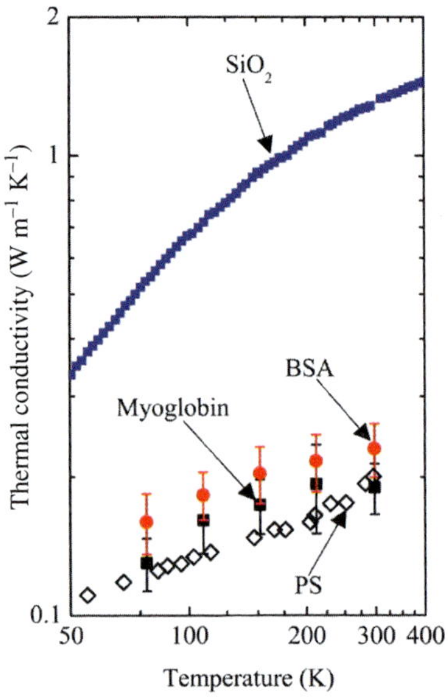

Figure 7. Thermal conductivity of protein films using TDTR. Reprinted with permission from Foley *et al.*, Copyright 2014, American Chemical Society.

such as 3 omega method at ~0.1–1°C[Dames and Chen, 2005]. Next, the minimum beam size of a 100 X objective lens that can be used in TDTR is >1 μm for a wavelength of 785 nm. Hence, it would be impossible to probe heat transport between individual nanoparticles and surrounding medium (<100 nm), due to the inability of localizing the heat source to the nanoparticles. A variation of this method has been developed to address this limitation as described in the next section.

7.3. *Transient Absorption Method*

This new technique referred to as transient absorption method (TAM) has been modified to use nanoparticles as transducers of laser heat energy, and can be used to measure changes in absorption instead of reflection at short time scales (>100 ps) [Wilson *et al.*, 2002; Ge *et al.*, 2004; Park *et al.*, 2012b]. Figure 8(a) shows the schematic of the transient absorption method. A tuned laser pump beam optically heats gold nanorods dispersed in a solution. Once heated by a pulse, the nanorods are cooled by the surrounding medium. The resulting temperature decay leads to change in optical absorbance (Δ_{Abs}) of the nanorods with time. A probe beam sends signals to a detector on the opposite side of the sample which measures the transmitted signal and therefore the change in absorption (Δ_{Abs}) by the nanorods as shown in Figure 8(b).

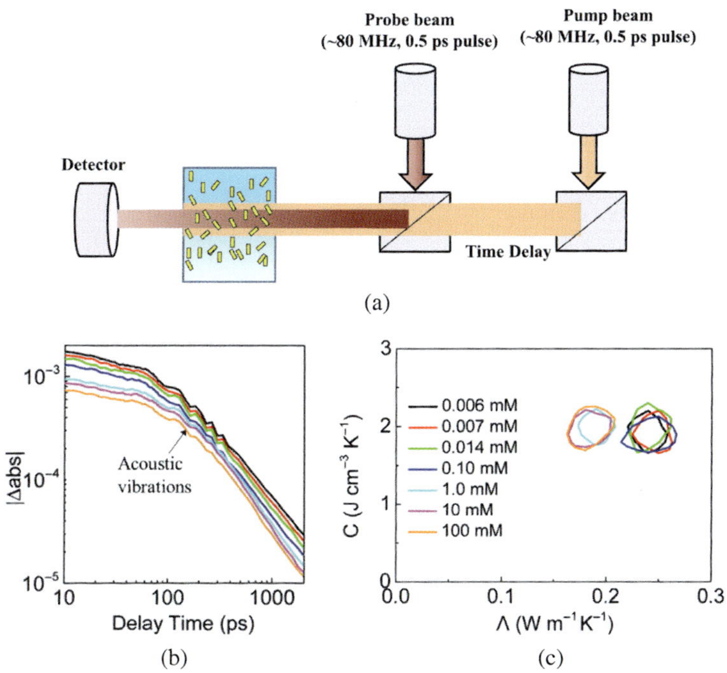

Figure 8. (a) Schematic of the transient absorption spectroscopy; (b) transient absorption signal as function of time; (c) volumetric specific heat and thermal conductance of CTAB coated on the gold nanorod surface. Figures from panels (b) and (c) are reprinted with permission from Huang *et al.*, Copyright 2013, American Chemical Society.

Time below 10 ps is not plotted as this captures non-equilibrium electron-phonon coupling events within the nanoparticle. The disturbance in Δ_{Abs} between 100 and 500 ps indicate acoustic vibrations in the nanoparticle. Above 500 ps, the signal is in the diffusion regime and the slope is limited by cooling of the nanoparticles at the interface. This is proportional to the thermal conductance of the interface between the nanoparticles and the medium. Therefore, thermal conductance of the interface could be obtained from the slope of the signal ($\alpha\Delta_{Abs}$) vs time plot above 500 ps in the diffusion regime. From the thermal conductance and geometry of the interface coating and the nanorods obtained by imaging techniques, thermal conductivity of the interface layer can be obtained. This technique has been applied to measure thermal conductivity and specific heat capacity of coatings on the gold nanorods [Huang *et al.*, 2013]. The results are plotted in Figure 8(c). The technique is currently limited to measure thermal properties of only gold nanorods due to their unique optical properties. The instrument provides absolute measurement of thermal conductivity and does not need calibration. This device is currently being used to answer several exciting questions, such as the physics of heat transfer across the nanoparticle-medium interface and the rate of nanoparticle heat dissipation [Cahill *et al.*, 2014]. With this new ability to probe molecular heat transfer and heat transport at the nanoparticle-polymer-solvent interface, new research topics are opening up in the rational thermal design and analysis of nanoparticle composites for biomedical applications.

8. Measurement Opportunities at Different Scales in Cryobiology

8.1. *Tissue and Cellular — Microscale*

A recent review of biomaterial properties in the cryogenic regime gives a reasonably complete picture of the thermal properties available at the tissue level [Choi and Bischof, 2010]. What is clear from this review and recent work [Lubner *et al.*, 2012] is that there are limited measurements of thermal properties of thin tissues (<1mm) such as pulmonary vein, esophagus, phrenic nerve, small diameter arteries, and cells (<100 μm) for application in thermal therapies and cryopreservation. Anisotropy and composite structures are very common in thin tissues where contact resistance may play a role in heat diffusion during various energy-based surgeries. In addition, upon tissue or bulk system freezing the ice crystals change the thermal conductivity based on the directionality of crystal growth, crystallized fraction and morphology which are often determined by the freeze front (planar, dendritic and epitaxial). One thermal analysis approach to study this was presented by Rubinsky and Ikeda using a directional solidification stage [Rubinsky and Ikeda, 1985]. This stage applies the Bridgman technique by propagating the sample at a known velocity through a large temperature gradient ($\Delta T > 20°\text{C}$ /mm) spanning the phase change, thereby creating directional crystal growth. They observed that the ice front interface changed from an initial planar front in pure water to a dendritic structure in solution due to non-equilibrium cooling effects as shown in Figure 9(a). The resulting ice crystal morphology consists of cell compartments separated by non-frozen freeze concentrated mushy zones with typical dendritic structures on the 10s of μm scale, that eventually undergo eutectic phase change below the eutectic temperature. This could result in a lamellar structure of ice and eutectic mixture. Consequently, anisotropy in thermal conductivity based on the direction of crystal growth can be expected especially in cryosurgical applications as shown in Figure 9 (b–c) and hence needs further study. Experiments conducted in liver tissue [Rubinsky *et al.*, 1990; Bischof *et al.*, 1993] showed that the ice crystals grow along the vascular structure that is the path of least resistance to

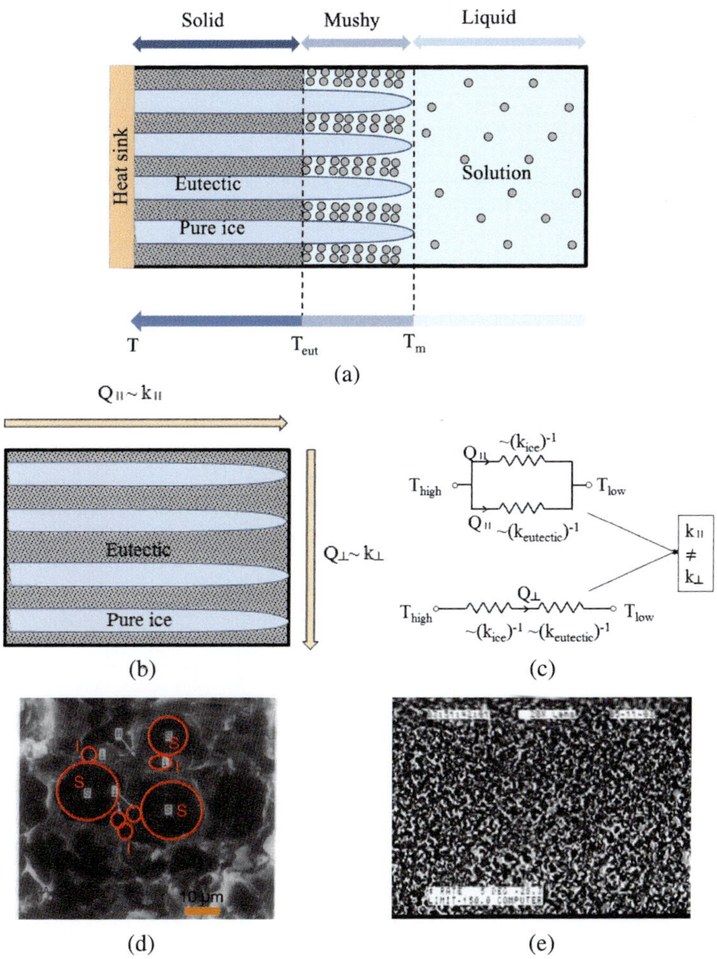

Figure 9. (a) Schematic of the directional freezing with three zones; Liquid above melting point (T_m), mushy above eutectic point (T_{eut}) and solid below T_{eut}; (b) resulting ice crystal morphology has anisotropy in structure; (c) electrical analogy of heat transfer shows anisotropy in thermal conductivity; (d) cross section of ice crystal morphology in liver tissue frozen at 220 °C/min. S — large frozen crystals in sinusoids (blood vessels), i — smaller ice crystals in hepatocytes. Scale bar — 10 μm [Rubinsky *et al.*, 1990]; (e) ice morphology in pre-nucleated 1X PBS + 6M glycerol solution. Black dots represents ice and white dots represents the solute. Figures from panels (d) and (e) are reprinted with permission from Choi and Bischof, Copyright 2010, Elsevier.

crystallization (Figure 9(d)). However, in the case of cryopreservation, the crystallization is generally based on ice nucleation sites thereby resulting in ice crystals growing randomly in all directions as shown in Figure 9(e) [Choi and Bischof, 2010]. Thus, we do not expect any anisotropy in the thermal conductivity for this case.

Heat transfer at the cellular scale, with and without crystalline, amorphous or nanoparticle inclusions, are forefront areas of research in cryobiology. For instance, high speed cryomicroscopy has recently been developed to understand freezing inside cells at rates of 130°C/min to visualize the precise moment of inception and growth of ice crystals within and between cells [Stott and Karlsson, 2009]. Further, pulsed lasers are now being used to rapidly thaw vitrified embryos (10^5–10^7°C/min), thereby avoiding devitrification and cell death [Jin *et al.*, 2014]. These heat transfer processes are critical to maintaining cellular viability, but precise understanding becomes complicated due to the short length and time scales involved. For instance, complications can occur due to co-existence of amorphous and crystalline phase inclusions in these systems. Artificial inclusions such as nanoparticles can also change thermal conductivity and heat generation [Ge *et al.*, 2004]. The interface between the nanoparticle and the biological environment are often comprised of polymeric layers such as PEG (polyethylene glycol), which help increase the circulation time in blood. Eventually, proteins opsonize or attach to the particle from the blood [Shah *et al.*, 2012; Walkey and Chan, 2012], leading to a nm-sized coating [Lacerda *et al.*, 2010]. The heat transfer in and around these natural and artificial (i.e. nanoparticle) inclusions are active areas of research [Wang *et al.*, 2007; Carter *et al.*, 2009] at the nanoscale.

In the case of vitrification, thermal conductivity measurement could be a viable quality control tool to assess the morphology and quality of vitrified biomaterials. For instance, in the case of vitrification of thin tissues such as arteries and veins, the formation of cracks and ice crystals in normal, vitrified and thawed tissues is a major hurdle for successful preservation. Thus detection of cracks and ice crystals at the microscale would be helpful for design and optimization of vitrification protocols. One way to ascertain the presence of cracks and ice crystals is by using the fact that thermal conductivity increases by 50–100% in crystallized vs. vitrified tissues and solutions as shown in Figure 10. Similarly, a reduction in thermal conductivity due to cracks (and hence contact resistance) would be expected. Similar concepts have been successfully applied in the past to perform non-destructive testing to identify internal cracks and flaws in inorganic materials [Thomas *et al.*, 1986; Sakagami and Kubo, 2002].

8.2. *Molecular and Nanoscale*

Increasingly technologies that manipulate biological systems at small length scales (<100 nm) and short time scales (<100 ns) are being developed. For instance, nanoparticles (1–100 nm) as illustrated in Figure 11(a) are being

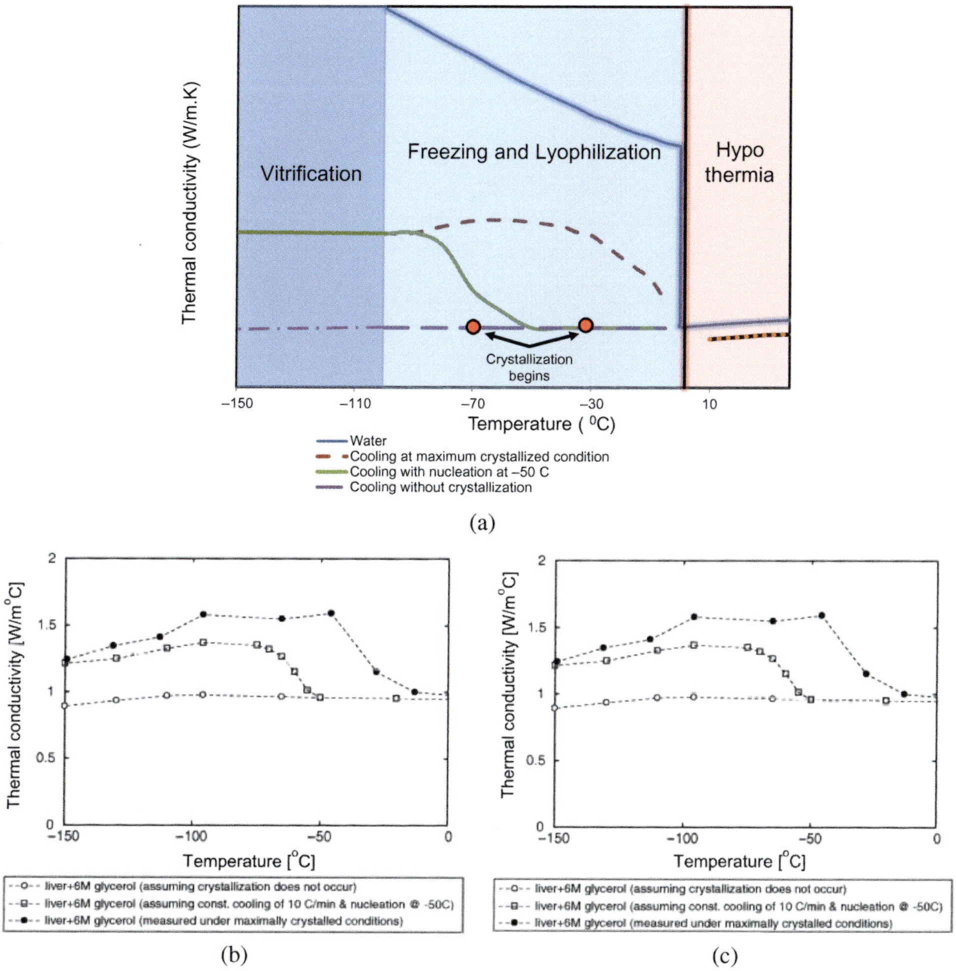

Figure 10. (a) Schematic of thermal conductivity of biomaterial as a function of temperature. Biomaterial is cooled at different rates from 37°C to –150°C. Temperature regime of various biomedical applications is labelled at the top; (b) thermal conductivity of PBS + 6M glycerol; (c) thermal conductivity of porcine liver + 6 M glycerol determined during cooling from 0°C to –150°C. Figures from panels (b) and (c) are reprinted with permission from Choi and Bischof, Copyright 2010, Elsevier.

used to carry molecular adjuvants to "pre-condition" tumors so that cryosurgical iceballs can become more uniformly destructive to cancer within the 0°C isotherm [Goel *et al.*, 2007]. In cryopreservation, iron oxide nanoparticles have shown the ability to uniformly and rapidly heat and thaw vitrified biomaterials, thereby possibly addressing a major bottleneck for the vitrification of larger biomaterials [Etheridge *et al.*, 2014]. While the two examples above may not dramatically change the bulk thermal conductivity, they can

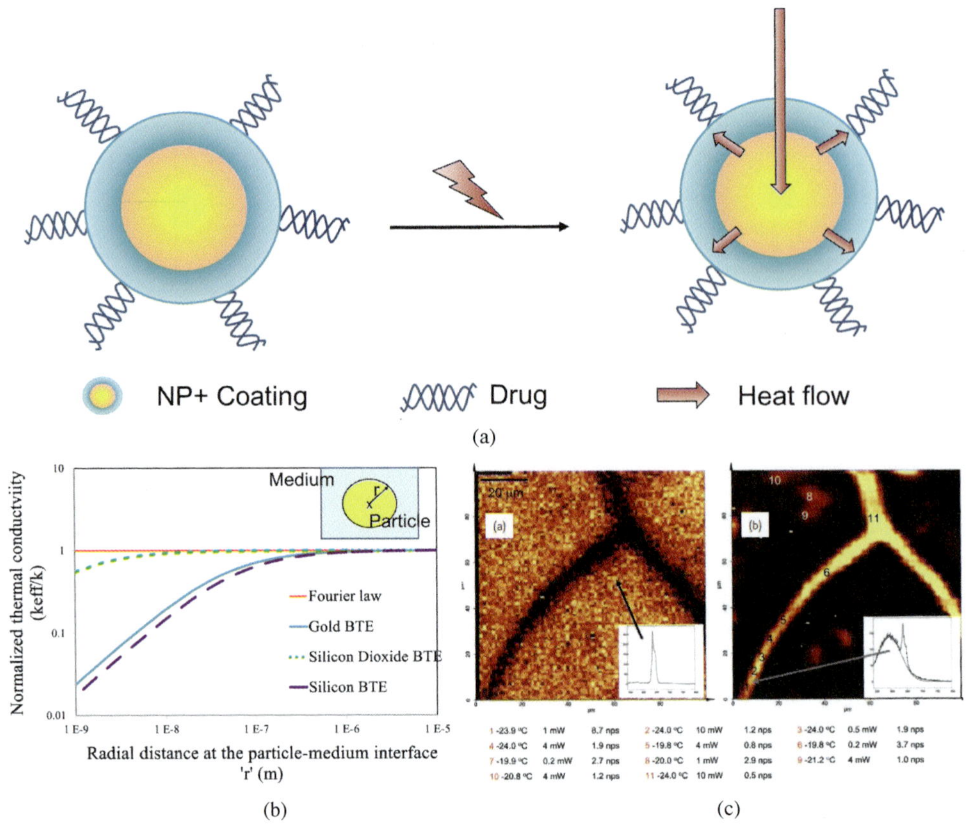

Figure 11. (a) Schematic of a typical nanoparticle heating in biomedical applications; BTE-refers to values determined based on Boltzmann transport equation; (b) deviation of effective thermal conductivity of the medium around the nanoparticle from Fourier law at small length scales [Chen, 1996; Espenel *et al.*, 2008]; (c) heating of 50 nm Au nanoparticles embedded in 100 μm × 100 μm cross section of ice. Figure from panel (c) is reprinted with permission from Govorov and Richardson, Copyright 2007, Elsevier.

reduce blood flow (i.e. perfusion) *in vivo* [Shenoi *et al.*, 2013]. This can allow an ice ball to grow larger and create heat generation for improved thawing. Further approaches that directly change the thermal properties of the tissue are cryo-nanosurgery in which larger amounts of nanoparticles are injected directly into tissues in order to increase the thermal conductivity [Liu and Deng, 2009]. Other approaches are to use nanoparticles to change the nucleation or devitrification temperatures of CPAs to increase the efficacy as cryopreservatives [Han *et al.*, 2008], or to manipulate cellular transport activity by nanoparticles to improve either cryosurgery or cryopreservation [Yue *et al.*, 2014] .

Clearly understanding heat transfer in nanoparticle-laden biomaterials, and specifically the nanoscale heat transfer from nanoparticles into the biomaterials when the nanoparticles are used as a heat source, is increasingly important in cryobiology. For instance, Chen theoretically predicted that thermal transport between heated nanoparticles and the surrounding medium decreases for very small nanoparticle length scales [Chen, 1996]. Figure 11(b) theoretically demonstrates that the thermal conductivity predicted by microscale heat transfer theories departs significantly from Fourier law at the nanoparticle-medium interface for nanoparticles of different materials. For details on the theory behind the plot, readers should refer to Chen [1996]. This drop in apparent thermal conductivity near the interface would result in higher nanoparticle temperatures but less rate of heat gain in the medium. One reason for this is that small length scales ($Kn \geq 0.1$) result in scattering of heat carriers by the nanoparticle boundary (thermal contact resistance effects). Thus, the geometry and internal structure of the nanoparticle play an important role in determining the size limit of application of bulk heat transfer behavior (i.e. Fourier's law).

At the micro/nanoscale, there have been some efforts to quantify heating of gold nanoparticles in cryogenic applications. In a proof of concept micro/nanoscale study, Govorov and colleagues were able to detect and quantify ice melting around laser heated nanoparticles (aggregated vs non-aggregated) using confocal Raman micro-spectroscopy, as shown in Figure 11(c) [Govorov and Richardson, 2007]. The need for experimental techniques that allow us to probe thermal conductivity and heat transfer at these short length (<100 nm) and time scales (<100 ns) are clear. They will ultimately help design the most efficient heat transfer processes for cryobiological application.

9. Conclusion

This review has discussed needs and opportunities for thermal conductivity measurements at micro and nano length scales in cryobiology. After briefly introducing cryobiological applications, we discussed the theoretical understanding and measurement of thermal conductivity at short length and time scales. Specifically, many new techniques have been developed to understand thermal conductivity of micro and nanoscale inorganic materials for applications in waste energy scavenging and cooling of electronics. Though most of the techniques discussed in this chapter have not been applied to cryogenic temperatures (e.g. < –40°C), there is no inherent limitation in the technique to measure thermal conductivity at such low temperatures. In the future these techniques can be used in cryobiology thereby facilitating manipulation of phase and thermal behaviour and ultimately life and death at the molecular, membrane, cellular and tissue level.

References

Abdelwahed, W., G. Degobert, S. Stainmesse and H. Fessi (2006). Freeze-drying of nanoparticles: formulation, process and storage considerations. *Advanced Drug Delivery Reviews*, 58(15), pp. 1688–1713.

Akoshima, M., M. Neda and T. Baba (2013). *Quantitative Evaluation of the Effect of Black-Coating for Laser Flash Measurement*, DEStech Publications, Inc.

Allen, P. B., J. L. Feldman, J. Fabian and F. Wooten (1999). Diffusons, locons and propagons: character of atomie yibrations in amorphous Si. *Philosophical Magazine Part B*, 79(11–12), pp. 1715–1731.

Allison, D. P., N. P. Mortensen, C. J. Sullivan and M. J. Doktycz (2010). Atomic force microscopy of biological samples. *Wiley Interdisciplinary Reviews, Nanomedicine and Nanobiotechnology*, 2(6), pp. 618–634.

Alper, J. and K. Hamad-Schifferli (2010). Effect of ligands on thermal dissipation from gold nanorods. *Langmuir*, 26(6), pp. 3786–3789.

Andersson, P. and G. Backstrom (1976). Thermal conductivity of solids under pressure by the transient hot wire method. *Review of Scientific Instruments*, 47(2), pp. 205–209.

Ando, S., D. Putnam, D. W. Pack and R. Langer (1999). PLGA microspheres containing plasmid DNA: preservation of supercoiled DNA via cryopreparation and carbohydrate stabilization. *Journal of Pharmaceutical Sciences*, 88(1), pp. 126–130.

Arvinte, T. and P. L. Steponkus (1988). Characterization of the pH-induced fusion of liposomes with the plasma membrane of rye protoplasts. *Biochemistry*, 27(15), pp. 5671–5677.

Assael, M. J., C. F. Chen, I. Metaxa and W. A. Wakeham (2004). Thermal conductivity of suspensions of carbon nanotubes in water. *International Journal of Thermophysics*, 25(4), pp. 971–985.

Avitall, B., D. Urboniene, G. Rozmus, D. Lafontaine, R. Helms and A. Urbonas (2003). New cryotechnology for electrical isolation of the pulmonary veins. *Journal of Cardiovascular Electrophysiology*, 14(3), pp. 281–286.

Balasubramaniam, T. A. and H. F. Bowman (1977). Thermal conductivity and thermal diffusivity of biomaterials: a simultaneous measurement technique. *Transactions of the ASME. Series K, Journal of Biomechanical Engineering*, 99(3), pp. 148–154.

Balasubramanian, S. K., R. T. Venkatasubramanian, A. Menon and J. C. Bischof (2008). Thermal injury prediction during cryoplasty through in vitro characterization of smooth muscle cell biophysics and viability. *Annals of Biomedical Engineering*, 36(1), pp. 86–101.

Barbas, J. P. and R. D. Mascarenhas (2009). Cryopreservation of domestic animal sperm cells. *Cell and Tissue Banking*, 10(1), pp. 49–62.

Baró, A. M., R. Miranda, J. Alamán, N. García, G. Binnig, H. Rohrer, C. Gerber and J. L. Carrascosa (1985). Determination of surface topography of biological specimens at high resolution by scanning tunnelling microscopy. *Nature*, 315(6016), pp. 253–254.

Bello, Y. M., D. A. F. Falabella and W. H. Eaglstein (2001). Tissue-engineered skin. *American Journal of Clinical Dermatology*, 2(5), pp. 305–313.

Belzer, F. O. and J. H. Southard (1988). Principles of solid-organ preservation by cold storage. *Transplantation*, 45(4), pp. 673–676.

Beran, U.-D. D. A. (1985). The reflectance behaviour of gold at temperatures up to 500°C. *Tschermaks mineralogische und petrographische Mitteilungen*, 34(3–4), pp. 211–215.

Bhattacharya, A. and R. L. Mahajan (2003). Temperature dependence of thermal conductivity of biological tissues. *Physiological Measurement*, 24(3).

Bischof, J., K. Christov and B. Rubinsky (1993). A morphological study of cooling rate response in normal and neoplastic human liver tissue: cryosurgical implications. *Cryobiology*, 30(5), pp. 482–492.

Bischof, J. C., D. Smith, P. V. Pazhayannur, C. Manivel, J. Hulbert and K. P. Roberts (1997). Cryosurgery of Dunning AT-1 rat prostate tumor: thermal, biophysical, and viability response at the cellular and tissue level. *Cryobiology*, 34(1), pp. 42–69.

Bischof, J. C., W. F. Wolkers, N. M. Tsvetkova, A. E. Oliver and J. H. Crowe (2002). Lipid and protein changes due to freezing in Dunning AT-1 cells. *Cryobiology*, 45(1), pp. 22–32.

Blackwell, J. H. (1954). A transient flow method for determination of thermal constants of insulating materials in bulk. Part I—theory. *Journal of Applied Physics*, 25(2).

Borca-Tasciuc, T., A. R. Kumar and G. Chen (2001). Data reduction in 3omega method for thin-film thermal conductivity determination. *Review of Scientific Instruments*, 72(4), pp. 2139–2147.

Bowman, H. F., E. G. Cravalho and M. Woods (1975). Theory, measurement, and application of thermal properties of biomaterials. *Annual Review of Biophysics and Bioengineering*, 4(1), pp. 43–80.

Cahill, D. G. (2004). Analysis of heat flow in layered structures for time-domain thermoreflectance. *Review of Scientific Instruments*, 75(12), pp. 5119–5122.

Cahill, D. G., G. M. Choi, J. Park, J. Huang, W. Wang and R. Wilson. "Ultrafast Heat Transfer in Nanoscale Materials." *http://users.mrl.illinois.edu/cahill/upenn_sep14.pdf.*

Cahill, D. G., M. Katiyar and J. R. Abelson (1994). Thermal conductivity of a-Si: H thin films. *Physical Review B*, 50(9).

Cahill, D. G. and R. O. Pohl (1987). Thermal conductivity of amorphous solids above the plateau. *Physical Review B (Condensed Matter)*, 35(8), pp. 4067–4073.

Cape, J. A. and G. W. Lehman (1963). Temperature and finite pulse-time effects in the flash method for measuring thermal diffusivity. *Journal of Applied Physics*, 34(7), pp. 1909–1913.

Capinski, W. (1999). Thermal-conductivity measurements of GaAs/AlAs superlattices using a picosecond optical pump-and-probe technique. *Physical Review B*, 59(12), pp. 8105–8113.

Carpenter, J. F., M. J. Pikal, B. S. Chang and T. W. Randolph (1997). Rational design of stable lyophilized protein formulations: some practical advice. *Pharmaceutical Research*, 14(8), pp. 969–975.

Carpenter, J. F., S. J. Prestrelski and T. Arakawa (1993). Separation of freezing- and drying-induced denaturation of lyophilized proteins using stress-specific stabilization: I. enzyme activity and calorimetric studies. *Archives of Biochemistry and Biophysics*, 303(2), pp. 456–464.

Carter, J. A., Z. Wang and D. D. Dlott (2009). Ultrafast nonlinear coherent vibrational sum-frequency spectroscopy methods to study thermal conductance of molecules at interfaces. *Accounts of Chemical Research*, 42(9), pp. 1343–1351.

Chato, J. C. (1968). A method for the measurement of the thermal properties of biological materials. *Symposium on Thermal Problems in Biotechnology* (The American Society of Mechanical Engineers, New York).

Chato, J. C. (1980). Measurement of thermal properties of growing tumors. *Annals of the New York Academy of Sciences*, 335(1), pp. 67–85.

Chato, J. C. (1985). Measurement of thermal properties of biological materials. *Heat transfer in Medicine and Biology*, 1, pp. 167–173.

Chen, G. (1996). Nonlocal and nonequilibrium heat conduction in the vicinity of nanoparticles. *Journal of Heat Transfer*, 118(3), pp. 539–545.

Chen, M. M., K. R. Holmes and V. Rupinskas (1981). Pulse-decay method for measuring the thermal conductivity of living tissues. *Journal of Biomechanical Engineering*, 103, pp. 253–260.

Choi, J. and J. C. Bischof (2010). Review of biomaterial thermal property measurements in the cryogenic regime and their use for prediction of equilibrium and non-equilibrium freezing applications in cryobiology. *Cryobiology*, 60(1), pp. 52–70.

Choi, J., M. Morrissey and J. Bischof (2013). Thermal processing of biological tissue at high temperatures: impact of protein denaturation and water loss on the thermal properties of human and porcine liver in the range 25–80 °C. *Journal of Heat Transfer*, 135(6), p. 061302.

Choi, S. R. and D. Kim (2008). Real-time thermal characterization of 12 nl fluid samples in a microchannel. *Review of Scientific Instruments*, 79(6), pp. 064901.

Choy, C. L., W. P. Leung and Y. K. Ng (1987). Thermal diffusivity of polymer films by the flash radiometry method. *Journal of Polymer Science Part B: Polymer Physics*, 25(9), pp. 1779–1799.

Clarke, D. M., J. M. Baust, R. G. Van Buskirk and J. G. Baust (2001). Chemo-cryo combination therapy: an adjunctive model for the treatment of prostate cancer. *Cryobiology*, 42(4), pp. 274–285.

Collins, K. C., A. A. Maznev, Z. Tian, K. Esfarjani, K. A. Nelson and G. Chen (2013). Non-diffusive relaxation of a transient thermal grating analyzed with the Boltzmann transport equation. *Journal of Applied Physics*, 114(10).

Committee, E. (2011). Test method for evaluating the resistance to thermal transmission of materials by the guarded heat flow meter technique, ASTM International.

Committee, E. (2013). Test method for thermal diffusivity by the flash method, ASTM International.

Costescu, R. M., M. A. Wall and D. G. Cahill (2003). Thermal conductance of epitaxial interfaces. *Physical Review B*, 67(5).

Cowan, R. D. (1963). Pulse Method of Measuring Thermal Diffusivity at High Temperatures. *Journal of Applied Physics*, 34(4), pp. 926–927.

Craig, D. Q. M., V. L. Kett, C. S. Andrews and P. G. Royall (2002). Pharmaceutical applications of micro-thermal analysis. *Journal of Pharmaceutical Sciences*, 91(5), pp. 1201–1213.

Crescenzo, D. G., S. L. Hilbert, R. H. Messier Jr, P. W. Domkowski, M. K. Barrick, P. L. Lange, V. J. Ferrans, R. B. Wallace and R. A. Hopkins (1993). Human cryopreserved homografts: electron microscopic analysis of cellular injury. *The Annals of Thoracic Surgery*, 55(1), pp. 25–31.

Crowe, J. H., F. A. Hoekstra and L. M. Crowe (1992). Anhydrobiosis. *Annual Review of Physiology*, 54(1), pp. 579–599.

Crowe, L. M., J. H. Crowe, A. Rudolph, C. Womersley and L. Appel (1985). Preservation of freeze-dried liposomes by trehalose. *Archives of Biochemistry and Biophysics*, 242(1), pp. 240–247.

Csaki, A., F. Garwe, A. Steinbrück, G. Maubach, G. Festag, A. Weise, I. Riemann, K. König and W. Fritzsche (2007). A parallel approach for subwavelength molecular surgery using gene-specific positioned metal nanoparticles as laser light antennas. *Nano Letters*, 7(2), pp. 247–253.

Dames, C. and G. Chen (2005). 1ω, 2ω, and 3ω methods for measurements of thermal properties. *Review of Scientific Instruments*, 76(12).

Davis, D. L., E. P. O'Brie and C. M. Bentzley (2000). Analysis of the degradation of oligonucleotide strands during the freezing/thawing processes using MALDI-MS. *Analytical Chemistry*, 72(20), pp. 5092–5096.

de Kanter, R., P. Olinga, I. Hof, M. de Jager, W. A. Verwillegen, M. J. H. Slooff, H. J. Koster, D. K. F. Meijer and G. M. M. Groothuis (1998). A rapid and simple method for cryopreservation of human liver slices. *Xenobiotica*, 28(3), pp. 225–234.

Dong, J., A. Hubel, J. C. Bischof and A. Aksan (2009). Freezing-induced phase separation and spatial microheterogeneity in protein solutions. *The Journal of Physical Chemistry B*, 113(30), pp. 10081–10087.

Donnelly, B. J., J. C. Saliken, D. S. Ernst, N. Ali-Ridha, P. M. A. Brasher, J. W. Robinson and J. C. Rewcastle (2002). Prospective trial of cryosurgical ablation of the prostate: five-year results. *Urology*, 60(4), pp. 645–649.

Donnez, J., M. M. Dolmans, D. Demylle, P. Jadoul, C. Pirard, J. Squifflet, B. Martinez-Madrid and A. Van Langendonckt (2004). Livebirth after orthotopic transplantation of cryopreserved ovarian tissue. *The Lancet*, 364(9443), pp. 1405–1410.

Espenel, C., M.-C. Giocondi, B. Seantier, P. Dosset, P.-E. Milhiet and C. Le Grimellec (2008). Temperature-dependent imaging of living cells by AFM. *Ultramicroscopy*, 108(10), pp. 1174–1180.

Etheridge, M. L., Y. Xu, L. Rott, J. Choi, B. Glasmacher and J. C. Bischof (2014). RF heating of magnetic nanoparticles improves the thawing of cryopreserved biomaterials. *Technology*, 2(3), pp. 229–242.

Eucken, A. (1940). Allgemeine gesetzmäßigkeiten für das wärmeleitvermögen verschiedener stoffarten und aggregatzustände. *Forschung auf dem Gebiet des Ingenieurwesens A*, 11(1), pp. 6–20.

Fahy, G. M. (1986). The relevance of cryoprotectant "toxicity" to cryobiology. *Cryobiology*, 23(1), pp. 1–13.

Fahy, G. M. and S. E. Ali (1997). Cryopreservation of the mammalian kidney. II. Demonstration of immediate ex vivo function after introduction and removal of 7.5M cryoprotectant. *Cryobiology*, 35(2), pp. 114–131.

Fahy, G. M., D. R. MacFarlane, C. A. Angell and H. T. Meryman (1984). Vitrification as an approach to cryopreservation. *Cryobiology*, 21, pp. 407–426.

Feldman, A. (1999). Algorithm for solutions of the thermal diffusion equation in a stratified medium with a modulated heating source. *High Temperatures–High Pressures*, 31(3).

Feser, J. P. and D. G. Cahill (2012). Probing anisotropic heat transport using time-domain thermoreflectance with offset laser spots. *Review of Scientific Instruments*, 83(10), pp. 104901–104907.

Florescu, D. I., L. G. Mourokh, F. H. Pollak, D. C. Look, G. Cantwell and X. Li (2002). High spatial resolution thermal conductivity of bulk ZnO (0001). *Journal of Applied Physics*, 91(2), pp. 890–892.

Foley, B. M., C. S. Gorham, J. C. Duda, R. Cheaito, C. J. Szwejkowski, C. Constantin, B. Kaehr and P. E. Hopkins (2014). Protein thermal conductivity measured in the solid state reveals anharmonic interactions of vibrations in a fractal structure. *The Journal of Physical Chemistry Letters*, 5(7), pp. 1077–1082.

Fuller, B. J. and B. W. W. Grout (1991). *Clinical Applications of Cryobiology*, CRC Press.

Gage, A. A. and J. Baust (1998). Mechanisms of tissue injury in cryosurgery. *Cryobiology*, 37(3), pp. 171–186.

Gall, K. L., S. E. Smith, C. A. Willmette and M. F. O'Brien (1998). Allograft heart valve viability and valve-processing variables. *The Annals of Thoracic Surgery*, 65(4), pp. 1032–1038.

Ge, Z., D. G. Cahill and P. V. Braun (2004). AuPd metal nanoparticles as probes of nanoscale thermal transport in aqueous solution. *The Journal of Physical Chemistry B*, 108(49), pp. 18870–18875.

Ge, Z., D. G. Cahill and P. V. Braun (2006). Thermal conductance of hydrophilic and hydrophobic interfaces. *Physical Review Letters*, 96(18).

Ge, Z., Y. Kang, T. A. Taton, P. V. Braun and D. G. Cahill (2005). Thermal transport in au-core polymer-shell nanoparticles. *NanoLetters*, 5(3), pp. 531–535.

Glassbrenner, C. J. and G. A. Slack (1964). Thermal conductivity of silicon and germanium from 3°K to the melting point. *Physical Review*, 134(4A), pp. A1058–A1069.

Goel, R., K. Anderson, F. Schmidlin, G. Vercellotti, J. C. Bischof, J. Belcher and J. Slaton (2009). Adjuvant approaches to enhance cryosurgery. *Journal of Biomechanical Engineering*, 131(7), pp. 074003–074003.

Goel, R., D. Swanlund, J. Coad, G. F. Paciotti and J. C. Bischof (2007). TNF-alpha-based accentuation in cryoinjury—dose, delivery, and response. *Molecular Cancer Therapeutics*, 6(7), pp. 2039–2047.

Gomès, S., P. Newby, B. Canut, K. Termentzidis, O. Marty, L. Fréchette, P. Chantrenne, V. Aimez, J.-M. Bluet and V. Lysenko (2013). Characterization of the thermal conductivity of insulating thin films by scanning thermal microscopy. *Microelectronics Journal*, 44(11), pp. 1029–1034.

Govorov, A. O. and H. H. Richardson (2007). Generating heat with metal nanoparticles. *Nano Today*, 2(1), pp. 30–38.

Grayson, J. (1952). Internal calorimetry in the determination of thermal conductivity and blood flow. *Journal of Physiology*, 118, pp. 54–72.

Gu, Y. Q., X. L. Ruan, L. Han, D. Z. Zhu and X. Y. Sun (2002). Imaging of thermal conductivity with sub-micrometer resolution using scanning thermal microscopy. *International Journal of Thermophysics*, 23(4), pp. 1115–1124.

Guan, M., D. M. Rawson and T. Zhang (2008). Cryopreservation of zebrafish (Danio rerio) oocytes using improved controlled slow cooling protocols. *Cryobiology*, 56(3), pp. 204–208.

Hagedorn, M., F. W. Kleinhans, D. E. Wildt and W. F. Rall (1997). Chill sensitivity and cryoprotectant permeability of dechorionated zebrafish embryos, Brachydanio rerio. *Cryobiology*, 34(3), pp. 251–263.

Hammiche, A., D. J. Hourston, H. M. Pollock, M. Reading and M. Song (1996a). Scanning thermal microscopy: subsurface imaging, thermal mapping of polymer blends, and localized calorimetry. *Journal of Vacuum Science Technology B: Microelectronics and Nanometer Structures*, 14(2), pp. 1486–1491.

Hammiche, A., H. M. Montagu-Pollock, M. Reading and M. Song (2002). Method and apparatus for performing localized thermal analysis and sub-surface imaging by scanning thermal microscopy. Patent W0 199704036941.

Hammiche, A., H. M. Pollock, M. Song and D. J. Hourston (1996b). Sub-surface imaging by scanning thermal microscopy. *Measurement Science and Technology*, 7(2).

Han, B. and J. C. Bischof (2003). Enhancement of cell and tissue destruction in cryosurgery by use of eutectic freezing. In T. P. Ryan (ed.), SPIE Proceedings Vol. 4954: *Thermal Treatment of Tissue: Energy Delivery and Assessment II.*

Han, B. and J. C. Bischof (2004). Thermodynamic nonequilibrium phase change behavior and thermal properties of biological solutins for cryobiology applications. *Journal of Biomechanical Engineering*, 126, pp. 196–203.

Han, W., J. Mou, J. Sheng, J. Yang and Z. Shao (1995). Cryo atomic force microscopy: a new approach for biological imaging at high resolution. *Biochemistry*, 34(26), pp. 8215–8220.

Han, X., H. B. Ma, C. Wilson and J. K. Critser (2008). Effects of nanoparticles on the nucleation and devitrification temperatures of polyol cryoprotectant solutions. *Microfluidics and Nanofluidics*, 4(4), pp. 357–361.

Hanai, A., W.-L. Yang and T. S. Ravikumar (2001). Induction of apoptosis in human colon carcinoma cells HT29 by sublethal cryo-injury: mediation by cytochrome c release. *International Journal of Cancer*, 93(4), pp. 526–533.

Hay, B., J. R. Filtz, J. Hameury and L. Rongione (2005). Uncertainty of thermal diffusivity measurements by laser flash method. *International Journal of Thermophysics*, 26(6), pp. 1883–1898.

Hey, J. and D. MacFarlane (1998). Crystallization of ice in aqueous solutions of glycerol and dimethyl sulfoxide 2: ice crystal growth kinetics. *Cryobiology*, 37(2), pp. 119–130.

Hill, J. E., J. D. Leitman and J. E. Sunderland (1967). Thermal conductivity of various meats. *Food Technology*, 21(1143), pp. 91–96.

Hoffmann, N. E. and J. C. Bischof (2002). The cryobiology of cryosurgical injury. *Urology*, 60(2, Supplement 1), pp. 40–49.

Homasson, J. P. (1989). Cryotherapy in pulmonology today and tomorrow. *European Respiratory Journal*, 2(9), pp. 799–801.

Homasson, J. P., P. Renault, M. Angebault, J. P. Bonniot and N. J. Bell (1986). Bronchoscopic cryotherapy for airway strictures caused by tumors. *Chest*, 90(2), pp. 159–164.

Hovatta, O., R. Silye, T. Krausz, R. Abir, R. Margara, G. Trew, A. Lass and R. M. L. Winston (1996). Cryopreservation of human ovarian tissue using dimethylsulphoxide and propanediol-sucrose as cryoprotectants. *Human Reproduction*, 11(6), pp. 1268–1272.

Huang, J., J. Park, W. Wang, C. J. Murphy and D. G. Cahill (2013). Ultrafast thermal analysis of surface functionalized gold nanorods in aqueous solution. *ACS Nano*, 7(1), pp. 589–597.

Huettmann, G., B. Radt, J. Serbin and R. Birngruber (2003). Inactivation of proteins by irradiation of gold nanoparticles with nano- and picosecond laser pulses. In R. Steiner (ed.), *Therapeutic Laser Applications and Laser-Tissue Interactions*, Vol. 5142 of Proceedings, SPIE (Optical Society of America), Paper 5142-88.

Huxtable, S., D. G. Cahill, V. Fauconnier, J. O. White and J.-C. Zhao (2004). Thermal conductivity imaging at micrometre-scale resolution for combinatorial studies of materials. *Nature Materials*, 3(5), pp. 298–301.

Incropera, F. P. (2011). *Fundamentals of Heat and Mass Transfer*, John Wiley & Sons.

Jin, B., F. W. Kleinhans and P. Mazur (2014). Survivals of mouse oocytes approach 100% after vitrification in 3-fold diluted media and ultra-rapid warming by an IR laser pulse. *Cryobiology*, 68(3), pp. 419–430.

Kapanidis, A. N. and T. Strick (2009). Biology, one molecule at a time. *Trends in Biochemical Sciences*, 34(5), pp. 234–243.

Karlsson, J. O. M. and M. Toner (1996). Long-term storage of tissues by cryopreservation: critical issues. *Biomaterials*, 17(3), pp. 243–256.

Kim, J. H., A. Feldman and D. Novotny (1999). Application of the three omega thermal conductivity measurement method to a film on a substrate of finite thickness. *Journal of Applied Physics*, 86(7), pp. 3959–3963.

Kim, K., W. Jeong, W. Lee and P. Reddy (2012). Ultra-high vacuum scanning thermal microscopy for nanometer resolution quantitative thermometry. *ACS Nano*, 6(5), pp. 4248–4257.

Klein, K. B., B. Zelickson, J. G. Riopelle, E. Okamoto, E. P. Bachelor, R. S. Harry and J. A. Preciado (2009). Non-invasive cryolipolysis™ for subcutaneous fat reduction does not affect serum lipid levels or liver function tests. *Lasers in Surgery and Medicine*, 41(10), pp. 785–790.

Kleinschuster, S. J., K. R. VanKampen, R. Spendlove, D. Atlum, M. L. Zupancic and C. C. Muscoplat (1979). Cryopreservation of bovine mononuclear leukocytes. *American Journal of Veterinary Research*, 40(11), pp. 1649–1651.

Kleman, M. I., K. Oellers and E. Lullau (2008). Optimal conditions for freezing CHO-S and HEK293-EBNA cell lines: influence of Me2SO, freeze density, and PEI-mediated transfection on revitalization and growth of cells, and expression of recombinant protein. *Biotechnology and Bioengineering*, 100(5), pp. 911–922.

Koh, Y. K., S. L. Singer, W. Kim, J. M. O. Zide, H. Lu, D. G. Cahill, A. Majumdar and A. C. Gossard (2009). Comparison of the 3omega method and time-domain thermoreflectance for measurements of the cross-plane thermal conductivity of epitaxial semiconductors. *Journal of Applied Physics*, 105(5), pp. 054303–054307.

Kopelman, I. J. (1966). *Transient Heat Transfer and Thermal Properties in Food Systems*, Michigan State University. Department of Food Science.

Lacerda, S. H. D. P., J. J. Park, C. Meuse, D. Pristinski, M. L. Becker, A. Karim and J. F. Douglas (2010). Interaction of gold nanoparticles with common human blood proteins. *ACS Nano*, 4(1), pp. 365–379.

Larson, T. R., D. W. Rrobertson, A. Corica and D. G. Bostwick (2000). In vivo interstitial temperature mapping of the human prostate during cryosurgery with correlation to histopathologic outcomes. *Urology*, 55(4), pp. 547–552.

Leary, S. P., C. Y. Liu and M. L. J. Apuzzo (2006). Toward the emergence of nanoneurosurgery: part iii nanomedicine: targeted nanotherapy, nanosurgery, and progress toward the realization of nanoneurosurgery. *Neurosurgery*, 58(6), pp. 1009–1026.

Lee, W., K. Kim, W. Jeong, L. A. Zotti, F. Pauly, J. C. Cuevas and P. Reddy (2013). Heat dissipation in atomic-scale junctions. *Nature*, 498(7453), pp. 209–212.

Lefèvre, S., S. Volz and P.-O. Chapuis (2006). Nanoscale heat transfer at contact between a hot tip and a substrate. *International Journal of Heat and Mass Transfer*, 49(1–2), pp. 251–258.

Levy, F. L. (1981). A modified Maxwell-Eucken equation for calculating the thermal conductivity of two-component solutions or mixtures. *International Journal of Refrigeration*, 4(4), pp. 223–225.

Li, D., Y. Wu, P. Kim, L. Shi, P. Yang and A. Majumdar (2003). Thermal conductivity of individual silicon nanowires. *Applied Physics Letters*, 83(14), pp. 2934–2936.

Li, M. H. and Y. B. Gianchandani (2003). Applications of a low contact force polyimide shank bolometer probe for chemical and biological diagnostics. *Sensors and Actuators A: Physical*, 104(3), pp. 236–245.

Lian, T., B. Locke, Y. Kholodenko and R. M. Hochstrasser (1994). Energy flow from solute to solvent probed by femtosecond ir spectroscopy: malachite green and heme protein solutions. *The Journal of Physical Chemistry*, 98(45), pp. 11648–11656.

Liu, J. and Z.-S. Deng (2009). Nano-cryosurgery: advances and challenges. *Journal of Nanoscience and Nanotechnology*, 9(8), pp. 4521–4542.

Losego, M. D., M. E. Grady, N. R. Sottos, D. G. Cahill and P. V. Braun (2012). Effects of chemical bonding on heat transport across interfaces. *Nature Materials*, 11(6), pp. 502–506.

Lu, L., W. Yi and D. L. Zhang (2001). 3omega method for specific heat and thermal conductivity measurements. *Review of Scientific Instruments*, 72(7), pp. 2996–3003.

Lubner, S. D., J. Choi, Y. Hasegawa, A. Fong, J. C. Bischof and C. Dames (2012). Measurements of the thermal conductivity of sub-millimeter biological tissues. *ASME 2012 International Mechanical Engineering Congress and Exposition*, American Society of Mechanical Engineers.

Lubner, S. D., J. Choi, G. Wehmeyer, B. Waag, V. Mishra, H. Natesan, J. C. Bischof and C. Dames (2015). Reusable bi-directional 3ω sensor to measure thermal conductivity of 100 μm thick biological tissues. *Review of Scientific Instruments*, 86(1).

Luo, K., Z. Shi, J. Lai and A. Majumdar (1996). Nanofabrication of sensors on cantilever probe tips for scanning multiprobe microscopy. *Applied Physics Letters*, 68(3), pp. 325–327.

Lyscov, V. N. and Y. S. Moshkovsky (1969). DNA cryolysis. *Biochimica et Biophysica Acta (BBA) — Nucleic Acids and Protein Synthesis*, 190(1), pp. 101–110.

Maas, W. J. M., W. R. Leeman, J. P. Groten and J. J. M. van de Sandt (2000). Cryopreservation of precision-cut rat liver slices using a computer-controlled freezer. *Toxicology in Vitro*, 14(6), pp. 523–530.

Majumdar, A., J. P. Carrejo and J. Lai (1993). Thermal imaging using the atomic force microscope. *Applied Physics Letters*, 62(20), pp. 2501–2503.

Manstein, D., H. Laubach, K. Watanabe, W. Farinelli, D. Zurakowski and R. R. Anderson (2008). Selective cryolysis: a novel method of non-invasive fat removal. *Lasers in Surgery and Medicine*, 40(9), pp. 595–604.

Mazur, P. (1984). Freezing of living cells: mechanisms and implications. *The American Journal of Physiology*, 247(3 Pt 1), pp. C125–142.

Meryman, H. T. (1971). Cryoprotective agents. *Cryobiology*, 8(2), pp. 173–183.

Meryman, H. T. (1989). Frozen red cells. *Transfusion Medicine Reviews*, 3(2), pp. 121–127.

Mitra, K., S. Kumar, A. Vedevarz and M. K. Moallemi (1995). Experimental evidence of hyperbolic heat conduction in processed meat. *Journal of Heat Transfer*, 117(3), pp. 568–573.

Mo, M. and A. G (2004). Cryosurgery for lung cancer: clinical results and technical aspects. *Technology in Cancer Research & Treatment*, 3(2), pp. 143–150.

Moffat, R. J. and J. Kim (2004). *Temperature and Heat Transfer Measurements*, CRC Press.

Morin, J., A. Traore, G. Dionne, M. Dumont, B. Fouquette, M. Dufour, S. Cloutier and C. Moisan (2004). Magnetic resonance-guided percutaneous cryosurgery of breast carcinoma: technique and early clinical results. *Canadian Journal of Surgery*, 47(5), pp. 347–351.

Morley, M. J. (1966). Thermal conductivities of muscles, fats and bones. *International Journal of Food Science & Technology*, 1(4), pp. 303–311.

Muldrew, K., M. Hurtig, K. Novak, N. Schachar and L. E. McGann (1994). Localization of Freezing injury in articular cartilage. *Cryobiology*, 31(1), pp. 31–38.

Murakami, E. G. and M. R. Okos (1989). Measurement and prediction of thermal properties of foods, *Food Properties and Computer-Aided Engineering of Food Processing Systems.* In R. P. Singh and A. G. Medina (eds.) (Springer Netherlands), pp. 3–48.

Natesan, H., W. Hodges, J. Choi, S. Lubner, C. Dames and J. Bischof (2016). A microthermal sensor for focal therapy applications. *Scientific Reports*, 6, p. 21395.

Neuweiler, H., C. M. Johnson and A. R. Fersht (2009). Direct observation of ultrafast folding and denatured state dynamics in single protein molecules. *Proceedings of the National Academy of Sciences*, 106(44), pp. 18569–18574.

Nikiforov, M. P., S. Hohlbauch, W. P. King, K. Voïtchovsky, S. A. Contera, S. Jesse, S. V. Kalinin and R. Proksch (2011). Temperature-dependent phase transitions in zeptoliter volumes of a complex biological membrane. *Nanotechnology*, 22(5), 055709.

Niwaya, K., H. Sakaguchi, K. Kawachi and S. Kitamura (1995). Effect of warm ischemia and cryopreservation on cell viability of human allograft valves. *The Annals of Thoracic Surgery*, 60, Supplement 2, pp. S114–S117.

Nonnenmacher, M. and H. K. Wickramasinghe (1992). Scanning probe microscopy of thermal conductivity and subsurface properties. *Applied Physics Letters*, 61(2), pp. 168–170.

O'Connell, M., N. McClure and S. E. M. Lewis (2002). The effects of cryopreservation on sperm morphology, motility and mitochondrial function. *Human Reproduction*, 17(3), pp. 704–709.

Ohta, H., G. Ogura, Y. Waseda and M. Suzuki (1990). Thermal diffusivity measurements of molten salts using a three layered cell by the laser flash method. *Review of Scientific Instruments*, 61(10), pp. 2645–2649.

Ohta, H., H. Shibata and Y. Waseda (1989). New attempt for measuring thermal diffusivity of thin films by means of a laser flash method. *Review of Scientific Instruments*, 60(3), pp. 317–321.

Oktay, K., E. Buyuk, L. Veeck, N. Zaninovic, K. Xu, T. Takeuchi, M. Opsahl and Z. Rosenwaks (2004). Embryo development after heterotopic transplantation of cryopreserved ovarian tissue. *The Lancet*, 363(9412), pp. 837–840.

Olson, B. W., S. Graham and K. Chen (2005). A practical extension of the 3ω method to multilayer structures. *Review of Scientific Instruments*, 76(5).

Onik, G. M., J. K. Cohen, G. D. Reyes, B. Rubinsky, Z. Chang and J. Baust (1993). Transrectal ultrasound-guided percutaneous radical cryosurgical ablation of the prostate. *CANCER-PHILADELPHIA*, 72, pp. 1291–1299.

Owen, M. S. (2006). *2006 ASHRAE Handbook — Refrigeration*, "Thermal Properties of Foods," ASHRAE, Atlanta, Chap. 9, ISBN: 1-931862-87-7 (ISSN: 1930-7217).

Paddock, C. A. and G. L. Eesley (1986). Transient thermoreflectance from thin metal films. *Journal of Applied Physics*, 60(1), pp. 285–290.

Panas, A. J., S. Muda, J. Terpiłowski and M. Preiskorn (2003). Investigation of the thermal diffusivity of human tooth hard tissue. *International Journal of Thermophysics*, 24(3), pp. 837–848.

Park, B. K., N. Yi, J. Park and D. Kim (2012a). Development of a microfabricated sensor to measure thermal conductivity of picoliter scale liquid samples. *Review of Scientific Instruments*, 83(10), p. 106102.

Park, J., J. Huang, W. Wang, C. J. Murphy and D. G. Cahill (2012b). Heat transport between Au nanorods, surrounding liquids, and solid supports. *The Journal of Physical Chemistry C*, 116(50), pp. 26335–26341.

Park, B. K., N. Yi, J. Park and D. Kim (2013). Thermal conductivity of single biological cells and relation with cell viability. *Applied Physics Letters*, 102(20).

Park, K., G. L. Cross, Z. M. Zhang and W. P. King (2008). Experimental investigation on the heat transfer between a heated microcantilever and a substrate. *Journal of Heat Transfer*, 130(10), p. 102401.

Parker, W. J. and R. J. Jenkins (1962). Thermal conductivity measurements on bismuth telluride in the presence of a 2 MeV electron beam. *Advanced Energy Conversion*, 2, pp. 87–103.

Parker, W. J., R. J. Jenkins, C. P. Butler and G. L. Abbott (1961). Flash method of determining thermal diffusivity, heat capacity, and thermal conductivity. *Journal of Applied Physics*, 32(9), p. 1679.

Patel, P. A., J. W. Valvano, J. A. Pearce, S. A. Prahl and C. R. Denham (1987). A self-heated thermistor technique to measure effective thermal properties from the tissue surface. *Journal of Biomechanical Engineering*, 109(4).

Paynter, S. J. (2000). Current status of the cryopreservation of human unfertilized oocytes. *Human Reproduction Update*, 6(5), pp. 449–456.

Pegg, D. E. (1972). Perfusion of rabbit kidneys with cryoprotective agents. *Cryobiology*, 9(5), pp. 411–419.

Pegg, D. E., L. Wang, D. Vaughan and C. J. Hunt (2006a). Cryopreservation of articular cartilage. Part 2: Mechanisms of cryoinjury. *Cryobiology*, 52(3), pp. 347–359.

Pegg, D. E., M. C. Wusteman and L. Wang (2006b). Cryopreservation of articular cartilage. Part 1: Conventional cryopreservation methods. *Cryobiology*, 52(3), pp. 335–346.

Pissuwan, D., S. M. Valenzuela and M. B. Cortie (2006). Therapeutic possibilities of plasmonically heated gold nanoparticles. *Trends in Biotechnology*, 24(2), pp. 62–67.

Poppendiek, H. F., R. Randall, J. A. Breeden, J. E. Chambers and J. R. Murphy (1967). Thermal conductivity measurements and predictions for biological fluids and tissues. *Cryobiology*, 3(4), pp. 318–327.

Price, D. M., M. Reading, A. Hammiche and H. M. Pollock (1999). Micro-thermal analysis: scanning thermal microscopy and localised thermal analysis. *International Journal of Pharmaceutics*, 192(1), pp. 85–96.

Purdy, P. H. (2006). A review on goat sperm cryopreservation. *Small Ruminant Research*, 63(3), pp. 215–225.

Pylkki, R. J., P. J. Moyer and P. E. West (1994). Scanning near-field optical microscopy and scanning thermal microscopy. *Japanese Journal of Applied Physics*, 33(Part 1, No. 6B), pp. 3785–3790.

Qin, Z. and J. C. Bischof (2012). Thermophysical and biological responses of gold nanoparticle laser heating. *Chemistry Society Reviews*, 41(3), pp. 1191–1217.

Regner, K. T., A. J. H. McGaughey and J. A. Malen (2014). Analytical interpretation of nondiffusive phonon transport in thermoreflectance thermal conductivity measurements. *Physical Review B*, 90(6).

Roper, D. K., W. Ahn and M. Hoepfner (2007). Microscale heat transfer transduced by surface plasmon resonant gold nanoparticles. *The Journal of Physical Chemistry C*, 111(9), pp. 3636–3641.

Rowe, A. W. (1994). Cryopreservation of red blood cells. *Vox Sanguinis*, 67, pp. 201–206.

Rowe, A. W. (1995). Cryopreservation of blood: a historical perspective. *Transfusion Medicine and Hemotherapy*, 22(1), pp. 36–40.

Rubinsky, B. and M. Ikeda (1985). A cryomicroscope using directional solidification for the controlled freezing of biological material. *Cryobiology*, 22(1), pp. 55–68.

Rubinsky, B., C. Y. Lee, J. Bastacky and G. Onik (1990). The process of freezing and the mechanism of damage during hepatic cryosurgery. *Cryobiology*, 27(1), pp. 85–97.

Rudkin, R. L., R. J. Jenkins and W. J. Parker (1962). Thermal diffusivity measurements on metals at high temperatures. *Review of Scientific Instruments*, 33(1), pp. 21–24.

Rudolph, A. S. (1988). The freeze-dried preservation of liposome encapsulated hemoglobin: a potential blood substitute. *Cryobiology*, 25(4), pp. 277–284.

Sabel, M. S., M. A. Nehs, G. Su, K. P. Lowler, J. L. M. Ferrara and A. E. Chang (2005). Immunologic response to cryoablation of breast cancer. *Breast Cancer Research and Treatment*, 90(1), pp. 97–104.

Sakagami, T. and S. Kubo (2002). Applications of pulse heating thermography and lock-in thermography to quantitative nondestructive evaluations. *Infrared Physics & Technology*, 43(3–5), pp. 211–218.

Sarabanda, A. V., T. J. Bunch, S. B. Johnson, S. Mahapatra, M. A. Milton, L. R. Leite, G. K. Bruce and D. L. Packer (2005). Efficacy and safety of circumferential pulmonary vein isolation using a novel cryothermal balloon ablation system. *Journal of the American College of Cardiology*, 46(10), pp. 1902–1912.

Schmidt, A. J., J. D. Alper, M. Chiesa, G. Chen, S. K. Das and K. Hamad-Schifferli (2008a). Probing the gold nanorod–ligand–solvent interface by plasmonic absorption and thermal decay. *The Journal of Physical Chemistry C*, 112(35), pp. 13320–13323.

Schmidt, A. J., X. Chen and G. Chen (2008b). Pulse accumulation, radial heat conduction, and anisotropic thermal conductivity in pump-probe transient thermoreflectance. *Review of Scientific Instruments*, 79(11).

Schmidt, A., M. Chiesa, X. Chen and G. Chen (2008c). An optical pump-probe technique for measuring the thermal conductivity of liquids. *Review of Scientific Instruments*, 79(6).

Scott, E. P., M. Tilahun and B. Vick (2009). The question of thermal waves in heterogeneous and biological materials. *Journal of Biomechanical Engineering*, 131(7), p. 074518.

Shah, N. B., G. M. Vercellotti, J. G. White, A. Fegan, C. R. Wagner and J. C. Bischof (2012). Blood–nanoparticle interactions and in vivo biodistribution: impact of surface peg and ligand properties. *Molecular Pharmaceutics*, 9(8), pp. 2146–2155.

Shen, S., A. Henry, J. Tong, R. Zheng and G. Chen (2010). Polyethylene nanofibres with very high thermal conductivities. *Nature Nanotechnology*, 5(4), pp. 251–255.

Shenoi, M. M., J. K. Anderson and J. C. Bischof (2009). Nanoparticle enhanced thermal therapies. *Annual International Conference of the IEEE Engineering in Medicine and Biology Society, 2009. EMBC 2009.*

Shenoi, M. M., I. Iltis, J. Choi, N. A. Koonce, G. J. Metzger, R. J. Griffin and J. C. Bischof (2013). Nanoparticle delivered vascular disrupting agents (vdas): use of tnf-alpha conjugated gold nanoparticles for multimodal cancer therapy. *Molecular Pharmaceutics*, 10(5), pp. 1683–1694.

Shenoi, M. M., N. B. Shah, R. J. Griffin, G. M. Vercellotti and J. C. Bischof (2011). Nanoparticle preconditioning for enhanced thermal therapies in cancer. *Nanomedicine*, 6(3), pp. 545–563.

Shi, L. and A. Majumdar (2001). Thermal transport mechanisms at nanoscale point contacts. *Journal of Heat Transfer*, 124(2), pp. 329–337.

Shi, L., J. Zhou, P. Kim, A. Bachtold, A. Majumdar and P. L. McEuen (2009). Thermal probing of energy dissipation in current-carrying carbon nanotubes. *Journal of Applied Physics*, 105(10).

Shnyrov, V. L., J. M. Sanchez-Ruiz, B. N. Boiko, G. G. Zhadan and E. A. Permyakov (1997). Applications of scanning microcalorimetry in biophysics and biochemistry. *Thermochimica Acta*, 302(1), pp. 165–180.

Silani, V., A. Pizzuti, O. Strada, A. Falini, M. Buscaglia and G. Scarlato (1988). Human neuronal cell viability demonstrated in culture after cryopreservation. *Brain Research*, 473(1), pp. 169–174.

Slack, G. A. (1979). The thermal conductivity of nonmetallic crystals. *Solid State Physics*, 34, pp. 1–71.

Song, Y. C., B. S. Khirabadi, F. Lightfoot, K. G. M. Brockbank and M. J. Taylor (2000). Vitreous cryopreservation maintains the function of vascular grafts. *Nature Biotechnology*, 18, pp. 296–299.

Staren, E. D., M. S. Sabel, L. M. Gianakakis, G. A. Wiener, V. M. Hart, M. Gorski, K. Dowlatshahi, B. F. Corning, M. F. Haklin and G. Koukoulis (1997). Cryosurgery of breast cancer. *Archives of Surgery (Chicago, Ill.: 1960)*, 132(1), pp. 28–33; Discussion 34.

Steponkus, P. L. (1996). *Advances in Low-Temperature Biology*, Elsevier.

Steven, L. N. and A. G. Larry (2010). Freeze-drying: principles and practice, *Pharmaceutical Dosage Forms: Parenteral Medications*. Volume 2: Facility Design, Sterilization and Processing. In (Informa Healthcare), pp. 353–382.

Stott, S. L. and J. O. M. Karlsson (2009). Visualization of intracellular ice formation using high-speed video cryomicroscopy. *Cryobiology*, 58(1), pp. 84–95.

Suquet, M., C. Dreanno, C. Fauvel, J. Cosson and R. Billard (2000). Cryopreservation of sperm in marine fish. *Aquaculture Research*, 31(3), pp. 231–243.

Sweat, V. E. (1995). Thermal properties of foods. In M. A. Rao and S. S. H. Rizvi (eds.), *Engineering Properties of Food* (Marcel Dekker, New York), pp. 99–138.

Tada, Y., M. Harada, M. Tanigaki and W. Eguchi (1978). Laser flash method for measuring thermal conductivity of liquids; application to low thermal conductivity liquids. *Review of Scientific Instruments*, 49(9), pp. 1305–1314.

Talsma, H., M. J. v. Steenbergen, P. J. M. Salemink and D. J. A. Crommelin (1991). The cryopreservation of liposomes. 1. A differential scanning calorimetry study of the thermal behavior of a liposome dispersion containing mannitol during freezing/thawing. *Pharmaceutical Research*, 8(8), pp. 1021–1026.

Tang, X. and M. J. Pikal (2004). Design of freeze-drying processes for pharmaceuticals: practical advice. *Pharmaceutical Research*, 21(2), pp. 191–200.

Thomas, R. L., L. D. Favro and P. K. Kuo (1986). Thermal-wave imaging for nondestructive evaluation. *Canadian Journal of Physics*, 64(9), pp. 1234–1237.

Toberer, E. S., L. L. Baranowski and C. Dames (2012). Advances in thermal conductivity. *Annual Review of Materials Research*, 42(1), pp. 179–209.

Tong, T. and A. Majumdar (2006). Reexamining the 3-omega technique for thin film thermal characterization. *Review of Scientific Instruments*, 77(10), pp. 104902–104909.

Touloukian, Y. S., R. W. Powell, C. Y. Ho and P. G. Klemens (1970). Thermophysical Properties of Matter — The TPRC Data Series. Volume 1. Thermal Conductivity — Metallic Elements and Alloys.

Tovee, P. D. and O. V. Kolosov (2013). Mapping nanoscale thermal transfer in-liquid environment—immersion scanning thermal microscopy. *Nanotechnology*, 24(46).

Tritt, T. M. (2004). *Thermal Conductivity: Theory, Properties, and Applications*, Springer.

Uchida, M., Y. Imaide, K. Sugimoto, H. Uehara and H. Watanabe (1995). Percutaneous cryosurgery for renal tumours. *British Journal of Urology*, 75(2), pp. 132–137.

Uchugonova, A., M. Lessel, S. Nietzsche, C. Zeitz, K. Jacobs, C. Lemke and K. König (2012). Nanosurgery of cells and chromosomes using near-infrared twelve-femtosecond laser pulses. *Journal of Biomedical Optics*, 17(10), pp. 1015021–1015026.

Ujihara, K. (2003). Reflectivity of metals at high temperatures. *Journal of Applied Physics*, 43(5), pp. 2376–2383.

Vachon, R. I., F. J. Walker, D. F. Walker and G. H. Nix (1967). In vivo determination of thermal conductivity of bone using the thermal comparator technique. In *Digest of the 7th International Conference of Medical and Biological Engineering*, Stockholm, Sweden.

Valvano, J. W., J. R. Cochran and K. R. Diller (1985). Thermal conductivity and diffusivity of biomaterials measured with self-heated thermistors. *International Journal of Thermophysics*, 6(3), pp. 301–311.

Vendrik, A. J. H. and J. J. Vos (1957). A method for the measurement of the thermal conductivity of human skin. *Journal of Applied Physiology*, 11(2), pp. 211–215.

Vogt, C., K. Laihem and C. Wiebusch (2008). Speed of sound in bubble-free ice. *The Journal of the Acoustical Society of America*, 124(6), pp. 3613–3618.

Wainwright, D. J. (1995). Use of an acellular allograft dermal matrix (AlloDerm) in the management of full-thickness burns. *Burns*, 21(4), pp. 243–248.

Walkey, C. D. and W. C. W. Chan (2012). Understanding and controlling the interaction of nanomaterials with proteins in a physiological environment. *Chemical Society Reviews*, 41(7).

Wang, X., V. Ho, R. A. Segalman and D. G. Cahill (2013a). Thermal conductivity of high-modulus polymer fibers. *Macromolecules*, 46(12), pp. 4937–4943.

Wang, X., C. D. Liman, N. D. Treat, M. L. Chabinyc and D. G. Cahill (2013b). Ultralow thermal conductivity of fullerene derivatives. *Physical Review B*, 88(7).

Wang, Z., J. A. Carter, A. Lagutchev, Y. K. Koh, N.-H. Seong, D. G. Cahill and D. D. Dlott (2007). Ultrafast flash thermal conductance of molecular chains. *Science*, 317(5839), pp. 787–790.

Williams, C. C. and H. K. Wickramasinghe (1986). Scanning thermal profiler. *Applied Physics Letters*, 49(23), pp. 1587–1589.

Williams, C. C. and H. K. Wickramasinghe (1988). Photothermal imaging with sub-100-nm spatial resolution, *Photoacoustic and Photothermal Phenomena*. In P. D. P. Hess and P. D. J. Pelzl (eds.) (Springer Berlin Heidelberg), pp. 364–369.

Wilson, O. M., X. Hu, D. G. Cahill and P. V. Braun (2002). Colloidal metal particles as probes of nanoscale thermal transport in fluids. *Physical Review B*, 66(22).

Wilson, R. B. and D. G. Cahill (2014). Anisotropic failure of Fourier theory in time-domain thermoreflectance experiments. *Nature Communications*, 5.

Wilson, R. B., J. P. Feser, G. T. Hohensee and D. G. Cahill (2013). Two-channel model for nonequilibrium thermal transport in pump-probe experiments. *Physical Review B*, 88(14), p. 144305.

Wolkers, W. F., S. K. Balasubramanian, E. L. Ongstad, H. C. Zec and J. C. Bischof (2007). Effects of freezing on membranes and proteins in LNCaP prostate tumor cells. *Biochimica et Biophysica Acta (BBA)—Biomembranes*, 1768(3), pp. 728–736.

Yang, F. and C. Dames (2013). Mean free path spectra as a tool to understand thermal conductivity in bulk and nanostructures. *Physical Review B*, 87, pp. 035437.

Yang, W.-L., T. Addona, D. G. Nair, L. Qi and T. S. Ravikumar (2003). Apoptosis induced by cryo-injury in human colorectal cancer cells is associated with mitochondrial dysfunction. *International Journal of Cancer*, 103(3), pp. 360–369.

Yue, C., G. Zhao, J. Yi, C. Gao, L. Shen, Y. Zhang, Z. Wang and W. Liu (2014). Effect of hydroxyapatite nanoparticles on osmotic responses of pig iliac endothelial cells. *Cryobiology*, 69(2), pp. 273–280.

Zhang, H., S. Cheng, L. He, A. Zhang, Y. Zheng and D. Gao (2002). Determination of thermal conductivity of biomaterials in the temperature range 233–313K using a tiny detector made of a self-heated thermistor. *Cell Preservation Technology*, 1(2), pp. 141–147.

Zhang, J., M. Bunker, X. Chen, A. P. Parker, N. Patel and C. J. Roberts (2009). Nanoscale thermal analysis of pharmaceutical solid dispersions. *International Journal of Pharmaceutics*, 380(1–2), pp. 170–173.

Zhang, W., P. H. Chung, A. Zhang and S. Chen (2013). Laser processing of natural biomaterials, *Laser Technology in Biomimetics.* In V. Schmidt and M. R. Belegratis (eds.) (Springer Berlin Heidelberg), pp. 237–257.

Zhang, W., K. Gilstrap, L. Wu, R. B. K. C, M. A. Moss, Q. Wang, X. Lu and X. He (2010). Synthesis and characterization of thermally responsive pluronic f127–chitosan nanocapsules for controlled release and intracellular delivery of small molecules. *ACS Nano*, 4(11), pp. 6747–6759.

Chapter 6

Microscale Thermoelectric Devices for Use with Biosystems

Dinesh Pinisetty and Ram V. Devireddy

1. Introduction

This chapter outlines our recent efforts in modeling, fabricating and characterization of micro-scale array of thermal sensors and/or actuators (micro thermoelectric coolers/heaters) for use in biological systems. The micro-scale devices utilized thermoelectric fundamentals, i.e. the Peltier effect and the Seebeck effect. The fabrication of such devices necessitated the development of various thermodynamic models for pre-fabrication analysis and to optimize device dimensions and geometry. Prototype thermal sensor arrays were fabricated using a multi-step, UV lithography-based microfabrication process while prototype thermal actuators were fabricated using eletrodeposition techniques. To alleviate the problems associated with dense wiring and associated thermal sources/sinks within the devices, a technique was developed to fabricate a series of micro-bumps. These micro-bump structures were developed to assist in the isolation of the dense wiring layer, within a flip-chip bonded device. And finally, to improve the thermoelectric cooler device performance, we investigated template-based electrodeposition techniques to generate nanostructures (nanowires and nanotubes) of commonly used thermoelectric materials, like Bi_2Te_3 and Sb_2Te_3, for use in a micro-scale device.

2. Cryobiology Background

The use of low temperatures to store living systems (cryopreservation) has been successfully applied to a variety of mammalian systems [Mazur, 1970;

McGrath, 1988], including red blood cells, lymphocytes, gametes, embryos, cultured cells, hepatocytes, bone marrow stem cells, cornea, skin and pancreatic cells. Other instances of low temperature storage techniques which are currently being developed include preserving human and non-human mammalian oocytes [Rall, 1992; Bernard and Fuller, 1996], hematopoietic stem cells [Rowley, 1992; Hubel, 1997], cartilage/tissue engineered equivalents [Oegema *et al.*, 2000; Devireddy *et al.*, 2003; Neidert *et al.*, 2004], food substances [Fennema *et al.*, 1973], plant cold hardiness [Schrijnemakers and Van Iren, 1995], preservation of protozoa and parasites, insects, fish or aquatic species [Poleo *et al.*, 2001; Poleo *et al.*, 2005], microorganisms like algae [Walsh, 1998], canine liver and kidney slices [Day *et al.*, 1999], and spermatozoa from a variety of mammalian species, including human, mouse, equine, canine and boar [Devireddy and Bischof, 2003].

Reduced to its fundamentals, a typical cryopreservation protocol requires that biological systems tolerate four non-physiological conditions [Mazur, 1970]: (i) exposure to molar concentrations of cryoprotective additives (CPAs) causing "osmotic" injury; (ii) cooling to subzero temperatures, causing "chilling" injury; (iii) removal or conversion of almost all liquid cell water into the solid state, or the freezing process; and (iv) warming to room temperature, causing both "recrystallization" and other "thawing" injury. Of these four injury processes, a major hindrance is the one associated with freezing [Mazur, 1984; Rubinsky *et al.*, 1987; Rall, 1992]. When frozen in molar concentrations of CPAs, various types of cells exhibit survival that is strongly dependent on the cooling rate. The optimum-cooling rate that yields maximum survival is dependent both on the type and concentration of CPA. Clearly, minimizing the destructive effects of freezing is imperative for a freezing protocol to be effective and optimal. This objective can only be achieved by precise and quantitative knowledge of freezing processes in biological systems [Lovelock, 1953; Love, 1966; Diller, 1982; Toner, 1993].

Figure 1 shows a schematic of freezing response in a biological system. The two main consequences of freezing cells (isolated or embedded in tissue) in the presence of extracellular/vascular ice are (i) cellular dehydration or the loss of intracellular water and (ii) formation of intracellular ice (IIF). During freezing of a cell suspension and/or a tissue system, ice tends to form predominantly in the extracellular/vascular spaces [Lovelock, 1953; Love, 1966; Diller, 1982; Rubinsky *et al.*, 1987; Toner, 1993; Pazhayannur and Bischof, 1997]. The formation of extracellular ice increases the extracellular solute concentration and the intracellular fluid (water) permeates across the cell membrane to restore the chemical equilibrium. This process of "water

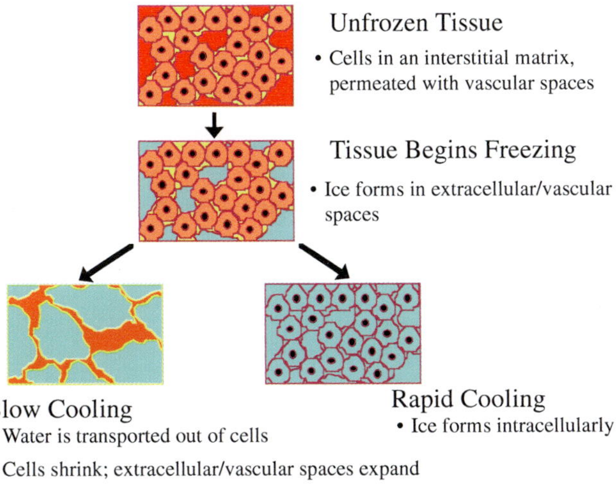

Figure 1. Freezing response in a tissue system. Upon freezing ice forms initially in the larger extracellular fluid/space. The consequent response is dependent on the imposed cooling rate, ice forms intracellularly at rapid cooling rates and water is transported out of the cells with the cell volume shrinking at slow cooling rates.

transport" or mass transfer across the cell membrane can only occur at relatively "slow" cooling rates and is governed by the membrane transport properties, reference hydraulic permeability of the plasma membrane (L_{pg}) and the activation energy of the hydraulic conductivity (*E*Lp) [Mansoori, 1975; Levin *et al.*, 1976]. The exact magnitude of the "slow" and "fast" cooling rates are specific to a cell type. For example in mouse embryos a cooling rate exceeding 2°C/min causes IIF while in red blood cells the same occurs for cooling rates greater than 850°C/min [Mazur, 1970; Mazur *et al.*, 1972]. Freezing injury may occur during this "slow" cooling process by the damaging effects of extracellular ice, time-dependent exposure to concentrated solutions and excessive cell shrinkage [Lovelock, 1953; Toner, 1993]. Conversely, at relatively "fast" cooling rates, heat transfer dominates over mass transfer and the extracellular ice formation occurs rapidly. This increases the likelihood that the under cooled intracellular fluid will form intracellular ice which is lethal to cells. Therefore, "too fast" or "too slow" cooling rates tend to kill cells and an "optimal" cooling rate will exist in the intermediate range. This has been confirmed experimentally for a variety of cells and the curve of cell survival plotted as a function of the cooling rate has a characteristic inverted U-shape [Mazur *et al.*, 1972].

Both water transport and IIF have been extensively studied in single and isolated cells using cryomicroscopy [Mazur, 1963; Diller, 1982; McGrath,

1988; Devireddy *et al.*, 1998]. In contrast, in tissue sections, two techniques are available to study water transport and none to study IIF. The two techniques to assess water transport out of cells in opaque tissue slices under dynamic freezing conditions are: (i) a low temperature microscopy method [Pazhayannur and Bischof, 1997] which uses a directional solidification stage [Rubinsky and Ikeda, 1985] and a two-step variation on the controlled freezing technique [Bischof *et al.*, 1993] and (ii) a differential scanning calorimetry (DSC) technique [Devireddy and Bischof, 1998; Devireddy *et al.*, 1998]. The water transport parameters (L_{pg} and E_{Lp}) are well characterized for a variety of mammalian tissue systems, including rat liver, prostrate tumor, the liver tissue of a freeze-tolerant wood frog and uterine fibroid tissue [Devireddy *et al.*, 1999a; Devireddy *et al.*, 1999b; Devireddy *et al.*, 1999c; Devireddy *et al.*, 2001]. The knowledge of *Lpg* and *ELp* allows a good biophysical understanding of cellular dehydration that occurs during freezing of tissue systems at "slow" cooling rates [Bischof, 2000; Devireddy and Bischof, 2003]. Unfortunately, similar knowledge of IIF, which occurs at "fast" cooling rates, is currently unattainable in tissue systems. This manuscript describes the fabrication of microscale thermoelectric sensors and actuators that could potentially be used to measure intracellular ice nucleation kinetics in artificial tissues. As a proof of concept for this application, prototype arrays of Type T micro-thermocouples, which have more stable characteristics at low temperatures [ASTM, 1970; Cygan, 2006], were designed, fabricated, and tested. Thermal modeling was used to characterize device performance and guide parameter selection. A multilayer UV-LIGA process was used to fabricate the prototype devices for preliminary evaluation at the cellular scale. Further sections of this chapter details our efforts to create a flip chip bonded integrated device [Kardak, 2008] along with our efforts to create microscale thermoelectric cooler with nanostructures [Prabhakar, 2006; Gupta *et al.*, 2010; Pinisetty and Devireddy, 2010; Pinisetty *et al.*, 2011a, Pinisetty *et al.*, 2011b].

3. Thermoelectric Fundamentals

Thermoelectric effects have been known since the early nineteenth century. The Seebeck effect is the underlying principle of the thermocouple, and it is also the basis for thermoelectric generators [Rosi *et al.*, 1959; Landecker, 1970; Rowe, 1996; Nuwayhid, 2002; Prabhakar *et al.*, 2007]. It states that, whenever there is a temperature difference between the two junctions of a circuit composed of two dissimilar conductors, a voltage can be observed in the circuit. The Seebeck coefficient is then defined as the ratio between the electric voltage and temperature difference. A second thermoelectric effect was identified by French scientist Jean Peltier. He observed that when an electric current passes through a junction between two different materials, there is a cooling or heating effect depending on the direction of the current. This thermoelectric cooling or heating phenomena is the basis for thermoelectric refrigeration. Lord Kelvin showed that, based on the principles of thermodynamics, there is a direct relationship between the Seebeck effect and the Peltier effect. He also predicted the existence of a third thermoelectric effect, the Thomson effect, which produces reversible heating or cooling when a current flows along a homogeneous conductor over which there is a temperature gradient, and must be taken into account whenever there is a substantial difference of temperature between the heat source and sink [Rowe, 1996; Rowe and Min, 1996].

The thermoelectric effects are generally not dependent on the physical size or the nature of the thermocouple junction. This is particularly true for thermocouples, i.e. as long as the two materials forming the thermocouple junction are in contact, the geometry of the junction does not affect temperature sensing based on the Seebeck effect. Primarily the materials comprising the junctions determine the performance of these devices. A micro-sized thermocouple can achieve the same level of efficiency and a faster response than a larger device because of the reduction in the probe heat capacity. In the case of thermoelectric cooling devices, the problems become more complicated [Prabhakar, 2006]. Both suitable materials and careful design are essential for achieving optimum performance.

3.1. *Thermocouples*

The most common application of the thermoelectric effect has been in thermocouples based on the Seebeck effect for temperature measurement. Miniaturization of thermocouples has advanced dramatically. Thermocouples with sizes down to 100 mm are commercially available. One research group has reported the fabrication of a thermocouple probe by electrochemical etching for use with an atomic force microscope [Majumdar *et al.*, 1993]. The resulting thermocouple junction was 25 micrometers in diameter. Thermal images obtained using this device indicated spatial resolution on a submicrometer level. Errors in the surface temperature measurement were estimated to be from 3% to 5%. The major sources of error in the technique were due to the size of the thermocouple probe and the probe fin effect.

The key to the development of the proposed instruments is the deposition of thermoelectric alloys in micromachined patterns. Electroplating has been demonstrated to be the best method for depositing metals and alloys in high aspect ratio microstructures [Panda and Podlaha, 2003; Lozano-Morales and Podlaha, 2008; Davis *et al.*, 2010; Gupta and Podlaha, 2010; Kim *et al.*, 2012]. Several standard thermocouple material pairs, with Seebeck coefficients ranging from 0.9 $\mu V/°C$ (at 100°C for Type B, with Pt and Rh) to 6.75 $\mu V/°C$ (at 100°C for Type E, with chromel (Ni 90% Cr) and constantan (Ni 45% Cu)). Not all thermocouple materials are easily electroplatable, but two of the combinations with the highest Seebeck coefficient, Type E and Type T (Cu and constantan, 46.8 $\mu V/°C$ at 100°C) are made up of metals and alloys that can be electrodeposited [Lozano-Morales and Podlaha, 2008; Gupta and Podlaha, 2008]. The Type T junction has a more limited temperature range (400°C max), but the Seebeck coefficient is more stable with temperature than for the Type E junction. Copper is easily electroplated and deposition of NiCu alloys in high aspect ratio patterns has recently been reported [Kim *et al.*, 2012].

Thermoelectric cooling depends on the electrodeposition of materials with good figure of merit. Three groups, one in France [Magri *et al.*, 1996a, 1996b] and another in Japan [Nagata *et al.*, 2000; Takahashi *et al.*, 2004] have reported the electrodeposition of amorphous bismuth telluride alloys. The approach employed by the French group was adapted and implemented at LSU. Uniform films of bismuth telluride were successfully deposited as thin films and in a lithographically patterned, 40 mm deep microstructure (see Figure 2(a); tilted for 60 degrees). The surface structure was found to be similar to the published results. The composition of the films was tested using EDAX and near stoichiometric ratios have been obtained (see Figure 2(b) and Table 1).

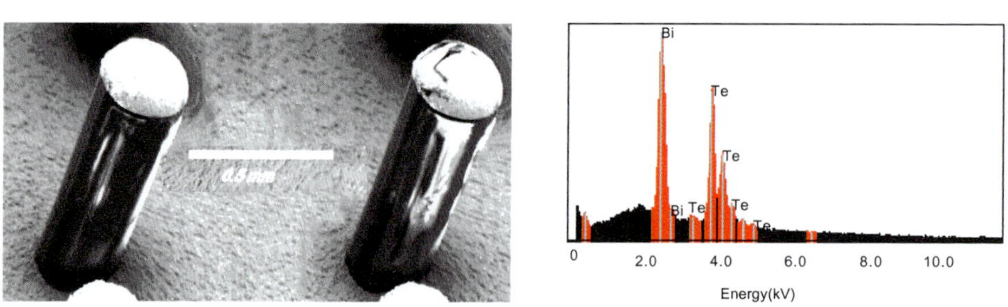

(a) (b)

Figure 2. Electrodeposition of amorphous bismuth telluride. (a) Deposit in lithographically patterned microstructures 40 mm deep; (b) EDAX analysis of the alloy composition showing the near stoichiometric ratio (see Table 1).

Table 1. EDAX analysis of the alloy composition showing the near stoichiometric ratio.

Element	Wt%	At%	X(mm)	Y(mm)
Bismuth	50.42	38.35	23.867	56.819
Telluride	49.5	61.65		
Total	99.92	100		

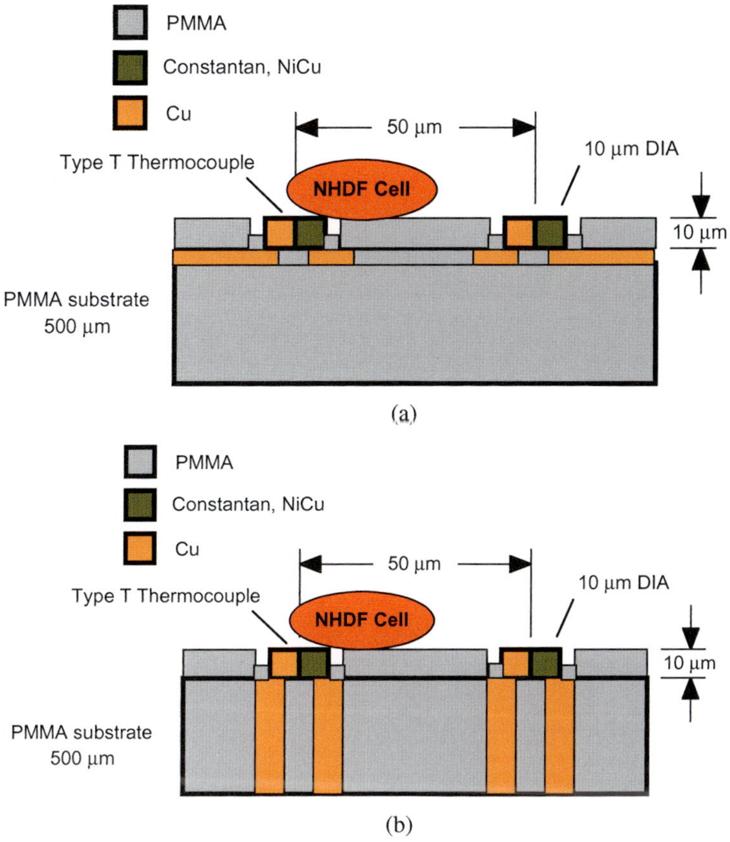

Figure 3. Schematics (not to scale) of two possible configurations for the thermocouple arrays. (a) Probes over a planar wiring layer on top of a substrate (PMMA or glass); (b) A 'vertical' wiring layer with vias for each half of the thermocouple junction passing through a polymer substrate. The contacts on the opposite side of the substrate can be bonded to contact pads on a PC board.

4. Thermoelectric Sensors

Two possible configurations for the thermocouple arrays are shown in Figure 3. The sensing area was a 3 × 3 array of Type-T thermocouples and a reference temperature thermocouple; one of the leads was copper, and the other was constantan. Wires of the same materials, micro-patterned on a silicon wafer, connected the junctions to the instrumentation. The wires and junctions were embedded in SU-8 photoresist (Microchem, Newton, MA) to create a level surface for the cells or tissue, which is placed at the center of the sensing area, above the thermocouple array.

4.1. *Numerical Modeling*

Suitable design and appropriate dimensions for the micro-patterned sections of the thermocouple device were determined parametrically through finite element modeling using ANSYS (Version 9.0, Canonsburg, PA). The modeling was used to study the influence of different factors on the response of the thermocouple to heat release by a cell, to aid in the refinement of the design of the instrument. Three design issues were considered: (i) The effect of the presence and dimensions of intermediate posts (Figure 4) on the thermocouple response; (ii) heat conduction through the thermocouples to the substrate; and (iii) thermal isolation and crosstalk between adjacent thermocouples array.

The system model consisted of an SU-8 base with embedded thermocouples, a single cell modeled as a cylinder, and the surrounding tissue. The cells were assumed to have a mean diameter of ~20 μm, a value that corresponds to the most commonly used cells (fibroblasts) in tissue engineering

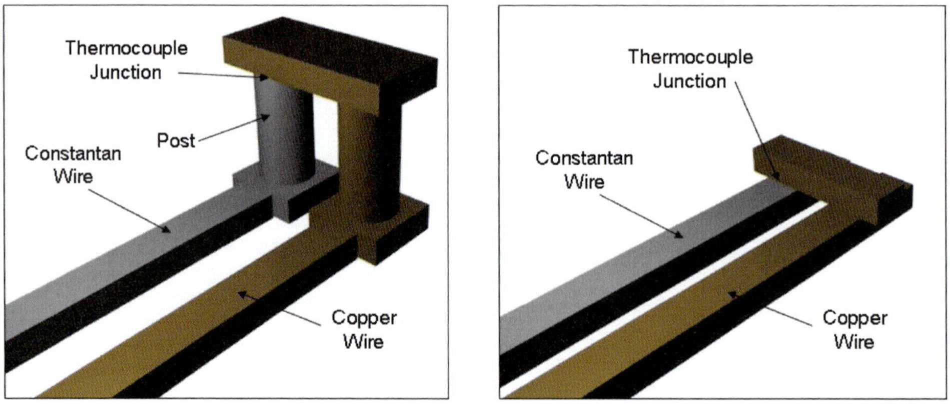

(a) (b)

Figure 4. Possible thermocouple junction geometries. (a) Junction arrangement with an intermediate post layer causing isolation of junction from wiring layer. (b) Junction arrangement without a post layer.

applications. Two cooling rates that bounded the conditions to be used in experiments were implemented in the simulations. A low and high cooling rate was examined: 5°C/min, and 500°C/min; reported in the literature as being relevant for measurement of intracellular ice formation in a wide range of cell types [28–30]. The temperature at which ice nucleates within the cell was modeled as either –5°C or –20°C, which corresponded to the highest and lowest reported values for the temperature at which ice nucleates within a cellular system [Toner, 1993]. Ice nucleation was modeled as a volumetric heat generation rate that was manifested when the cell reached the appropriate predefined nucleation temperature. Thermoelectric effects were neglected since these effects are proportional to current, which was negligible in the thermocouples.

The temperature response at the thermocouple junctions was observed as the parameters were varied. Mesh and time step refinement were performed to improve confidence in the selected methods and parameters. The mesh was refined locally at the cell volume because of the maximum temperature occurred at the cell due to the limitation on the number of elements 128,000 available.[1] The mesh (tetrahedral elements) sizes used for the cell were 10 μm for coarse, 5 μm for medium, and 2.5 μm for fine. The time steps were refined for the heat generation segment of the analysis only. The sizes used were two time steps for coarse, four steps for medium, and eight steps for fine.

Two alternative thermocouple geometries were modeled for the configuration of the device (Figure 4). Both designs had a wiring layer and a junction layer, but in the first case (Figure 4(a)) an intermediate "post" layer was inserted to reduce heat conduction to the wiring layer by allowing the use of thicker insulating material between the junctions and the wiring layer, and producing a larger voltage at the thermocouple junctions.

Heat conduction through the thermocouples was evaluated by considering a range of intermediate post diameters and heights to determine the best conditions. The lower bound on post diameter of ~3 μm was due to UV lithography resolution, and the upper bound on height of ~20 μm was limited by the thickness of the positive UV resist that could be spin coated without multiple coats, were as set by process limitations.

In order to measure the temperature in the neighborhood of a single nucleating cell, the center-to-center spacing of the thermocouples must be comparable to the cell diameter (~20 μm). However, as the spacing between the thermocouples is reduced, there is a strong possibility that a cell freezing

[1] This was determined by the University Advanced license of ANSYS (Version 9.0, ANSYS, Inc., Canonsburg, PA) for the software.

near one thermocouple might register strongly at adjacent thermocouples, making it difficult to distinguish individual freezing events as the device spacing is reduced. Intra-thermocouple spacing, center-to-center, of 45 μm, 50 μm and 55 μm was modeled to determine any trends and interference between adjacent thermocouples.

Thermal modeling helped determine appropriate dimensions for the prototype array and future systems. Posts of 3 μm and 5 μm in diameter were modeled, with heights of 20 μm and 50 μm. The control case was a temperature step input at the center of the cell without any thermocouples present. Due to lower thermal capacitance and less heat transfer away from the cell, a faster response time was expected with smaller thermocouples; this would lead to a measurement system that interfered less with the measured system. Results for both cooling rates 5°C/min and 500°C/min, with different geometries are shown in Figures 5 (a) and (b) respectively. The most favorable geometry, Case 2, 50 μm tall by 3 μm in diameter, had a 31% drop in temperature from the control case, and the least favorable geometry, Case 4, 50 μm tall by 5 μm in diameter, decreased 41%. Thinner and taller posts between the junction and wiring layers improved the simulated response temperature. It is interesting to note that the case without posts did not have the largest temperature drop. At a certain post diameter, between 3 μm and 4 μm, the heat conduction away from the cell was greater than that of the cell at close proximity to the wiring layer. For this reason, within the scope of the modeling it is observed that a larger temperature drop occurred with the largest diameter and tallest post size. The effect of post geometry on the response of an adjacent thermocouple to a cell freezing over a thermocouple was negligible (data not shown). Almost an identical thermocouple response, about 16% of the control case temperature rise, was detected at the thermocouple directly below the cell for all post geometries.

Two tests were performed to determine the effect of thermocouple spacing on response. The first observed the adjacent thermocouple response with a cell freezing directly above the center thermocouple for 45 μm, 50 μm and 55 μm center-to-center spacing (Table 2). The best results, with the smallest adjacent temperature rise, were shown at 55 μm spacing, with a 23% decrease in the temperature rise at the adjacent thermocouple compared to the 45 μm case. A comparison of the spacing of the thermocouples when a cell froze at different points between the center and an adjacent thermocouple was performed (Figure 6). Cells in close contact with the thermocouple junctions, minimized the temperature rise at the adjacent junctions. The tighter thermocouple spacing of 45 μm increased the likelihood that the cells would be near a thermocouple junction.

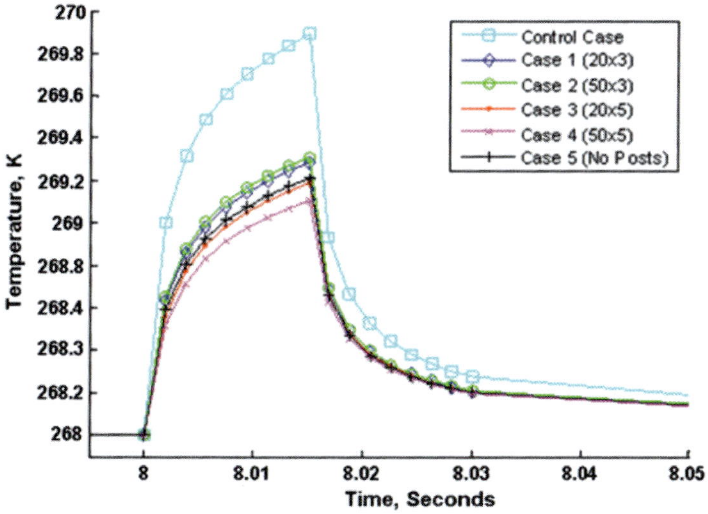

(a)

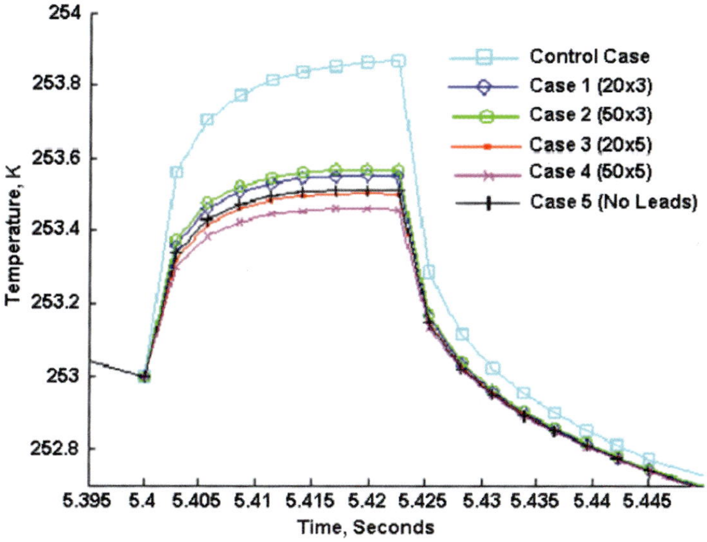

(b)

Figure 5. (a) Temperature (in K shown on y-axis) rise as a result of cell freezing for a 5°C per minute cooling rate with a –5°C nucleation temperature [Cygan, 2006]. Case 1 indicates posts having 20 μm height and 3 μm diameter. Case 2 indicates posts having 50 μm height and 3 μm diameter. Case 3 indicates posts having 20 μm height and 5 μm diameter. Case 4 indicates posts having 50 μm height and 5 μm diameter. Case 5 indicates only a wiring layer and junction without intermediate posts. (b) Temperature (in K shown on *y*-axis) rise as a result of cell freezing for a 5000°C per minute cooling rate with –200°C nucleation temperature for the cases described in (a).

Table 2. Maximum adjacent thermocouple temperature rise (°C) as a function of thermocouple center-to-center spacing [Cygan, 2006].

Configuration (height × diameter) (μm)	Maximum temperature rise (°C) in adjacent thermocouple		
	45 μm spacing	**50 μm spacing**	**55 μm spacing**
Case 1 (20 × 3)	0.35	0.30	0.27
Case 2 (20 × 5)	0.34	0.30	0.27
Case 5 (No posts)	0.35	0.30	0.27

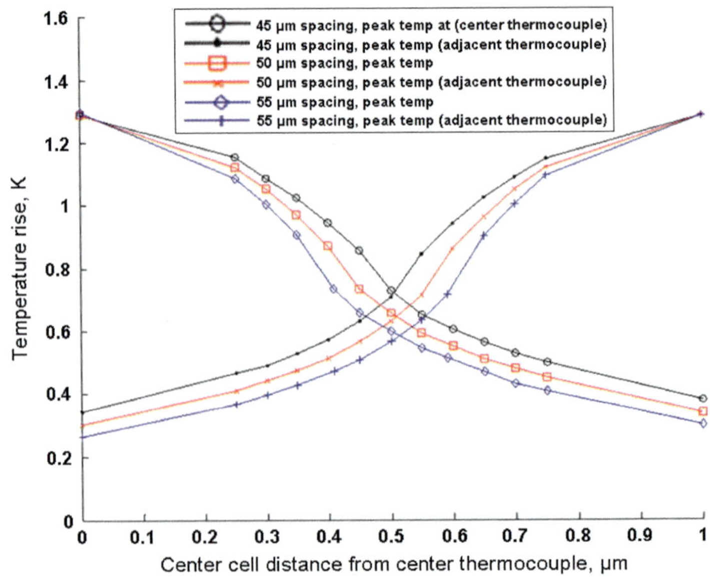

Figure 6. Thermocouple temperature increase (in °C) based on cell position [Cygan, 2006]. Temperature increase at the center and adjacent thermocouple is plotted versus non-dimensional cell position between two adjacent thermocouples. Plotted for Case 1 (20 μm × 3 μm) geometry.

4.2. *Fabrication*

The prototype arrays were fabricated using a multi-step, UV lithography-based microfabrication process. Five photo masks (chrome on glass) were ordered from Advance Reproductions Co. (Andover, MA), for use in the UV exposure step of the fabrication. The thermocouple junction area was limited by the tolerances (~±2 μm) arising from mask alignment [Cygan, 2006]. Figure 7 gives an overview of the steps in the microfabrication sequence. Positive photoresists, AZ4620 (AZ Electronic Materials, Somerville, NJ) and

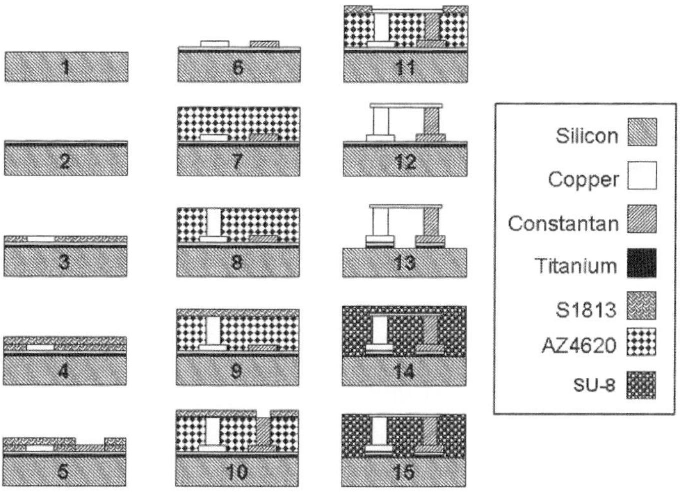

Figure 7. Fabrication outline. (1) Silicon wafer base. (2) E-beam deposit titanium and copper. (3) Spin S1813 resist, expose and deposit copper wiring layer. (4) Spin S1813 resist. (5) Expose and deposit constantan wiring layer. (6) Remove resist. (7) Spin AZ 4620 resist. (8) Expose and deposit copper post layer. (9) Spin S1813 resist. (10) Expose and deposit constantan post layer. (11) Expose and deposit copper junction layer. (12) Remove resist. (13) Etch copper and titanium base. (14) Cast SU-8. (15) Develop down to junction layer.

S1813 (Rohm and Haas, Philadelphia, PA), were used because of the need for multiple exposures of each layer. For the post layers, a thick film method of spin coating similar to an earlier study [Yoon *et al.*, 1998] was used. This yielded uniform resist thicknesses of approximately 20 μm for the post layer.

Type-T thermocouples have one copper and one constantan (Cu 55%, Ni 45%) electrode. Copper electrodeposition was done using the electrolyte composition containing copper sulphate and sulphuric acid. A suitable electrolyte composition for uniformly depositing the constantan alloy, based on the solution reported earlier [Panda and Podlaha, 2003] and plating parameters were determined using a rotating Hull cell (Brinkmann Instruments, Westbury, NY). This device allows us to test a range of electrodeposition currents in a single experiment, while accounting for mass transport effects from recess depth. For a rotating cylinder electrode (RCE) the boundary layer thickness (δ) for mass transport based on the speed of the rotation (S), was determined using the Eisenberg relation. (Equation 1)

$$\delta = 99.62 d^{-0.4} \nu^{0.344} D^{0.356} S^{-0.7} \qquad (1)$$

where d is the diameter of the RCE, ν is the viscosity of the solution, D is the diffusion coefficient of the solution. For micro-electrodeposition, Equation 1 relates the rotation rate of the RCE to the recess depth, simulating the conditions for recessed plating without a recessed substrate. Several rotating Hull cell experiments were run with different solution compositions and rotation rates. Rotation rates of 150, 400 and 600 rpm, which corresponded to recess depths of 100 μm, 50 μm, and 35 μm, were used in the experiments. An average current density was applied and the current density distribution along the electrode length was determined from the manufacturer's data. Using energy dispersive X-ray fluorescence (EDXRF, Omicron, Kevex, Atlanta, GA), the composition of the deposit at several points along the cylinder were analyzed, and plotted against the local current densities expected at those points.

The first step in the microfabrication process was to electron beam (e-beam) evaporate (Model BJD-1800 E-Beam Evaporator, Temescal, Wilmington, MA) a 0.01 μm thick titanium adhesion layer and a 0.05 μm thick copper plating base, to facilitate, electrodeposition in future steps (Figure 7, step 2). The e-beam deposited sample was cleaned using acetone, isopropyl alcohol (IPA), and rinsed in DI water to remove any impurities. A 2 μm thick layer of S1813 resist (Rohm and Haas, Philadelphia, PA) was spin coated (PWM101 Light Duty Photoresist Spinner, Headway Research, Garland, TX) on the wafer for the wiring layer. After spin coating, the sample was prebaked for 90 seconds at 95°C on a hot plate. A UV exposure was performed with vacuum contact on a Quintel UL7000-OBS Aligner and Exposure station (Quintel Corp., Morgan Hill, CA) at a wavelength of 365 nm with a dose of 75 mJ/cm^2. The resist was developed with MF-321 Developer (Rohm and Haas, Philadelphia, PA) for 120 seconds, and copper was deposited into the micro-patterned wiring layer recess (Figure 7, step 3).

The composition of the solution used for plating the copper wiring layer was Copper Sulphate (1.0M) and Sulfuric acid (100 mL/L). Plating was done galvano-statically, at a current density of –7 mA/cm^2 for 12 minutes. A second layer of 1 μm thick S1813 resist was spin coated (Figure 7, step 4) and prebaked over the first layer to prevent electrodeposition on the already deposited copper layer during constantan plating. An aligned exposure was performed for the constantan wiring, followed by development. Constantan was electrodeposited into the resulting set of recesses, using the optimized plating solution (1.0M Nickel Sulfate, 0.5M Copper Sulfate, 1.0M Sodium Citrate and 74 mL/L Ammonium Hydroxide) for the constantan-wiring layer. Plating was done galvano-statically, at a current density of 60mA/cm^2 for 15 seconds (Figure 7, step 5). The resist was completely removed after electrodeposition (Figure 7, step 6).

A 20 μm thick layer of AZ 4620 (AZ Electronic Materials, Somerville, NJ) resist was spin coated over the wiring layer (Figure 7, step 7). After spin coating, the sample was allowed to relax for 10 minutes before being soft baked in a convection oven (M326 mechanical convection oven, Blue M Electric, Watertown, WI) for 10 minutes at 85°C, and baked again on a hot plate at 115°C for 60 seconds. The sample was baked on a glass holder on the hotplate such that there was an air gap between the middle underside of the wafer and the hotplate, and the top was covered, as recommended by Yoon *et al.* [1998] for obtaining improved sidewalls on developed structures. After baking, the sample was allowed to cool for at least 20 minutes before exposure. An aligned exposure on the Quintel at a wavelength of 365 nm with an intensity of 600 mJ/cm^2 transferred the mask pattern to the resist and the recesses were developed using AZ developer (AZ 400).

The copper intermediate posts were electrodeposited (Figure 7, step 8). The composition of the copper post plating solution was the same as for the wiring layer, described earlier. A thin layer (~ 1 μm) of S1813 resist was spin coated (Figure 7, step 9) over the copper structures to prevent electrodeposition on the already deposited copper posts. The constantan post layer mask was aligned and exposed as for the copper post layer. Constantan was electroplated (Figure 7, step 10) in these recesses with the same nominal composition as that of the constantan wiring layer. When the same current density used for the constantan wiring layer was used to plate the posts, a heavily oxidized and uneven deposit formed over the posts. To alleviate this effect, the current density was reduced, and a pulse plating scheme [Panda and Podlaha, 2003] with a 0.25 duty cycle and 10 second on-time was used. A current density of 15 mA/cm^2 was applied for a total plating on-time of 20 minutes.

The junction layer was exposed without spinning another layer of resist. The sample was aligned and exposed with the final mask at a wavelength of 365 nm with a dose of 75 mJ/cm^2 on the Quintel exposure station, and developed for 2 minutes in MF 321 developer. The copper junctions were electrodeposited with the same plating solution as the copper post layer (Figure 7, step 11) with a current density of –7 mA/cm^2 for 45 minutes. The remaining resist was completely removed by soaking the sample in acetone (Figure 7, step 12). The sample was rinsed in IPA, DI water, and dried. The copper plating base was etched using APS copper etchant 100 (Transene Company, Inc, Danvers, MA) for 10 seconds, and then immediately placed in DI water and dried. The titanium adhesion layer was etched using a 1% hydrofluoric (HF) acid solution for 10 seconds, and immediately rinsed in DI water and dried (Figure 7, step 13).

In the final step, SU-8 2025 (MicroChem Corp., Newton, MA) was cast (Figure 7, step 14) on the substrate. First 10–15 ml of SU-8 was poured on the substrate and left on a flat level surface for 3 hrs to flatten out and form a uniform layer. The sample was placed on a level hotplate to soft bake the photoresist and baked at 65°C for 5 minutes followed by 15 minutes at 95°C. After baking the sample was allowed to cool to room temperature. The height of the resist was then measured and found to be 100 μm. The sample was milled on a fly cutting machine (Optimum 120, Precitech, Keene, NH) to 26 μm. The milled sample was aligned and exposed at a wavelength of 350 nm with a dose of 150 mJ/cm^2. A hot plate was used for the post exposure bake (PEB) at 65°C for 1 minute and then at 95°C for 5 minutes. The post baked sample was developed cast (Figure 7, step 15) in a beaker containing SU-8 developer (MicroChem Corp., Newton, MA), placed in an ultrasonic bath, for 5 minutes. The developed sample was sprayed with IPA and dried. The SU-8 layer provided rigidity and formed an insulated platform for the cells.

The results of the Rota-Hull experiments for the CuNi solution (consisting of 1.0M of Nickel Sulfate, 0.5M of Copper Sulfate, 1.0M of Sodium Citrate and 74 mL/L of Ammonium Hydroxide) are shown in Figure 8. The copper concentration is shown as a function of the current density at different rotation speeds. In an ideal case, the three curves corresponding to three different rotation rates mentioned above would be almost identical, indicating

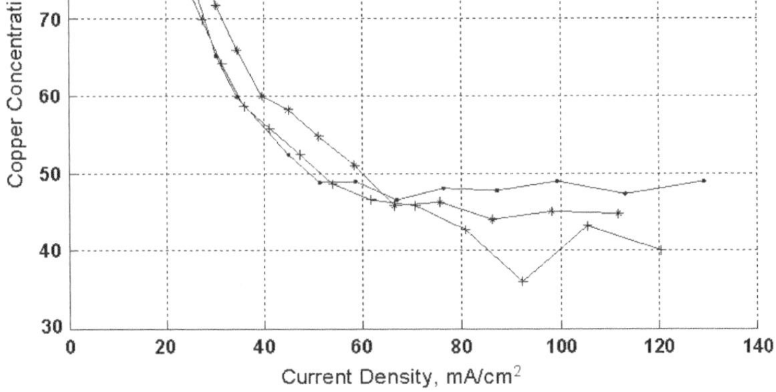

Figure 8. Copper composition curve for 3 different rotation rates of 650, 400 and 150 rpm of the working copper electrode with copper concentration on *y*-axis and current density on *x*-axis [Cygan, 2006].

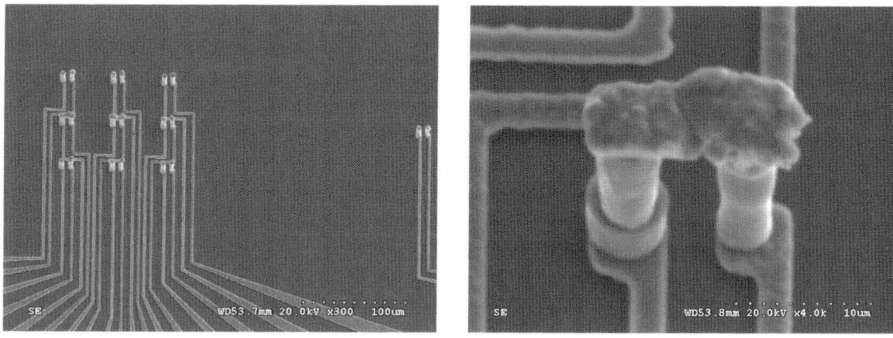

(a) (b)

Figure 9. SEM image of micro patterned device after removing the resist and etching of the seed and adhesion layers [Cygan, 2006]. (a) 3×3 array of 3 μm diameter and 20 μm high post thermocouples. (b) Successful thermocouple junctions.

recess depth, or mass transport effects do not influence composition. In practice similar trends were found for the CuNi solution composition, with less than 10% composition change between curves and some variation at higher current densities. In general, uniformity in composition should be best with small micro-patterned recess depths, but the curves show sufficient uniformity, even at 100 μm depths. This solution composition was chosen for the fabrication of the constantan posts and wiring layer.

Figure 9 shows SEM pictures of representative micropatterned devices after etching of the seed and adhesion layers. The actual alignment was better than ± 1 μm for all cases. A prototype, with an intermediate post layer, was fabricated on a wafer, with a yield of 16 of 20 junctions. A detailed tolerance analysis was performed to help determine the smallest size components that could be used in each successive layer [Cygan, 2006]. The tolerances that were determined for the system were adequate for the final fabrication of the device (± 2 μm).

4.3. *Testing and Calibration*

The electronics to process signals from nine Type T thermocouples was designed to detect small temperature differences in samples due to the latent heat released against a background temperature potentially 300 times greater. To accomplish this task, a tenth Type T thermocouple, fabricated on the same substrate, was used to measure the substrate temperature. Its output was amplified and differenced against the amplified signals of each of the sample thermocouples, which were multiplexed under software control. The differences were further amplified, and passed through a 60 Hz notch filter

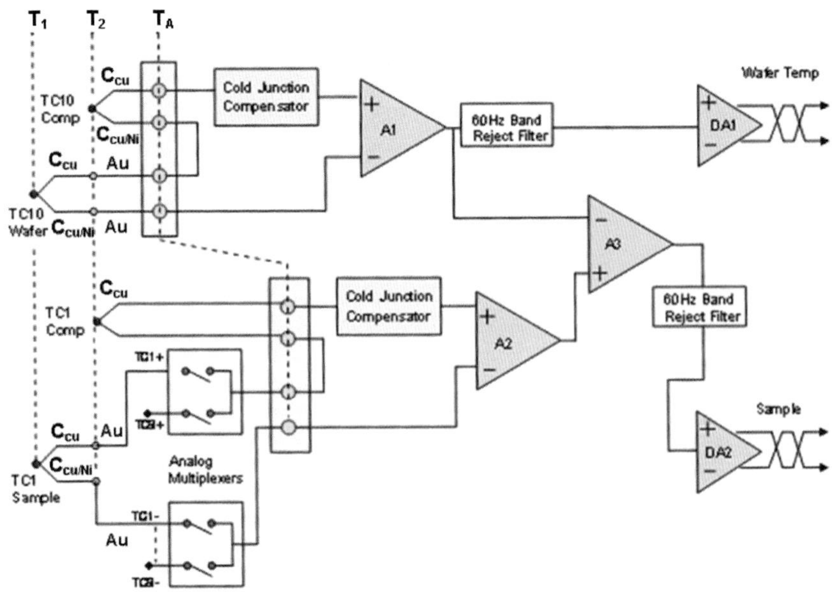

Figure 10. Schematic of the circuitry on the customized PC board showing the copper, copper-nickel and gold wire circuit (Cygan, 2006).

with 40 dB attenuation to reduce background AC line noise. The overall gain of the dedicated PCB was 3500. To minimize the errors in the multiplexer integrated circuit, terminal block connections, and the cold junction compensators, surface mount components were tightly placed. A National Instruments (Austin, TX) PCI-6229 data acquisition board was used to control the sample multiplexer and perform the 12 bit A/D conversion. A program was written in Labview (v. 7.1, National Instruments, Austin, TX) to acquire and plot the thermocouple array data. As shown in Figure 10, the wafer and sample thermocouples were both at an initial temperature T_1. The sample thermocouple would be exposed to the latent heat release of the cells while the wafer thermocouple would remain covered. In order to make external connections to the electronics, each thermocouple wiring layer was extended to the wafer circumference and contacts were made through spring-loaded gold connectors. These gold contacts create intermediate junctions at potentially different temperatures from the central test area and could produce measurement errors. To compensate for this, a second set of Type T thermocouples (Model 5TC-TT-T-40–36, Omega, Stamford, CT), labeled TC10 Comp and TC1 Comp, were located adjacent to the gold contacts in such a manner that the voltage output would be proportional to the difference between T_1 and T_A. This can be shown by summing the

Seebeck voltages around the loop, beginning with the sample and proceeding clockwise, canceling terms and re-arranging the resultant thermocouple output would give us Equation (2).

$$V_{out} = (\sigma c_u - \sigma cu/Ni)(T_1 - T_A) \quad (2)$$

A commercially available cooling stage (Model BCS-196, Linkam®, UK), with a recessed cooling element that uses liquid nitrogen to cool a 22 mm diameter circle at the center of the base and a heater to modulate the cooling rate, was used to provide controlled temperature inputs to the thermocouple array. The recessed cooling stage could not make direct contact with the bottom of the wafer containing the thermocouples, so a 22 mm diameter, 13 mm thick copper block (401 W/mK) was used to support the center of the micro-patterned wafer above the cooling element. The experiments were run over a temperature range from –20°C to 25°C. One thermocouple from the array could be calibrated at a time because of the design of the signal processing circuits. The experimental apparatus is displayed in Figure 11. The samples were cooled at approximately 5°C/min and the temperature at the reference and measuring junctions was recorded along with the voltage from the micro thermocouple. In order to distinguish between the voltage generated by the reference junction compensation thermocouple and the micro-thermocouple, a control case was also tested. The cooling rate was applied to the same system with the gold contact pads on the PCB offset so that they did not make contact with any of the thermocouple junctions i.e. no electrical connection. This provided data for the reference junction compensation thermocouple, for comparison with the micro-thermocouple array data.

A representative calibration curve showing a thermocouple sensitivity of 18.6 mV/°C after amplification was obtained (data not shown). The voltage varied linearly with respect to temperature, with voltage generated by the thermocouple on *y*-axis and the applied temperature on *x*-axis, $y = 0.0186x + 2.0127$ with $R^2 = 0.98$. The pre-amplification Seebeck coefficient estimated from the data by dividing the amplified thermocouple sensitivity by the overall gain was 5.3 μV/°C, which was significantly less than the nominal Seebeck coefficient for a Type T thermocouple of 38 μV/°C at 0°C[ASTM, 1970]. A possible reason for this discrepancy is the spatial variation in the thickness and the uneven composition along the wiring layer, which could be avoided by using a pulse plating scheme. Additionally, the presence of any titanium adhesion layer and copper seed layer on the silicon wafer will lower the overall Seebeck coefficient for the system.

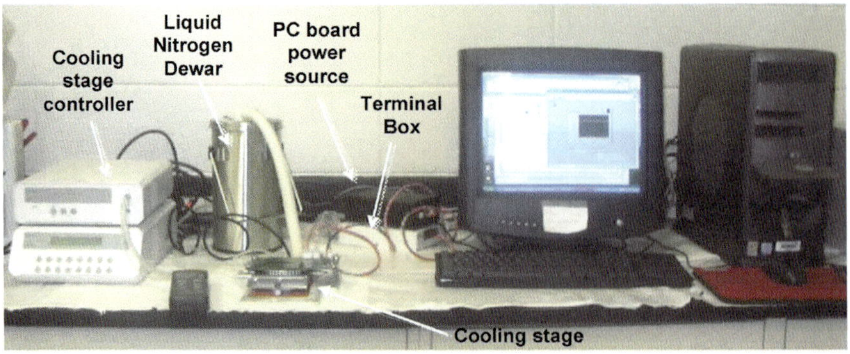

(a)

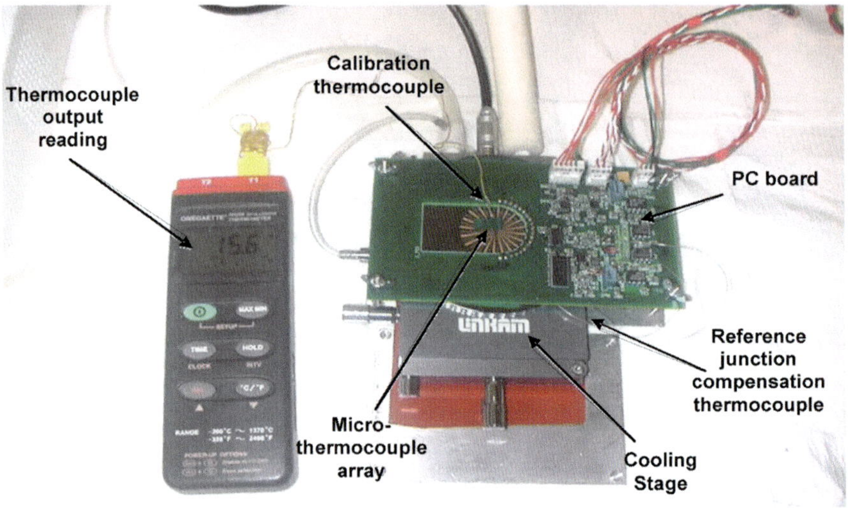

(b)

Figure 11. Experimental apparatus for calibration of the device [Cygan, 2006]. (a) Entire setup showing cooling stage and its controller, liquid nitrogen dewar, power source for PC board, computer. (b) Details of the cooling section of the experiment showing PC board, micro thermocouple array, calibration thermocouple and its thermal log used to record the output.

5. Flip Chip Bonded Device

An integrated device consisting of an array of thermal sensors (micro-thermocouples) and/or actuators (micro-thermoelectric coolers/heaters) would allow the measurement and control of temperature at the microscale. The fabrication of such a device would require a dense wiring layer below the devices (thermocouple and thermoelectric coolers/heaters) which will act as either a heat source and/or a sink and consequently, impede the device performance. To alleviate this problem, we have developed a technique to fabricate a series of microbump structures that will assist in the isolation of the dense wiring layer, within a flip-chip bonded device. Thermal modeling was used to determine the most suitable geometries for the device and assess the effect of the microbump size and structure on the performance of the thermoelectric cooler. For example, if the device is a micro-thermoelectric cooler, our analysis indicates that Bismuth-Telluride posts of 10 μm diameter and 20 μm height yield the optimal device performance with a microbump radius of 15 μm. Under these conditions, a maximum cooling of ~3.5K was achieved for an applied current of 23.7 mA. To avoid the complexities in the measurements due to the presence of a second junction, copper and constantan were chosen as the candidate microbump material. Electrodeposition along with a **L**ithographie, **G**alvanoformung, **A**bformung (LIGA) microfabrication technique was used to fabricate the microposts and microbumps. Briefly, copper and constantan microbumps, with mean diameters (heights) of 6.5 μm (7.8 μm) and 27.8 μm (12 μm) were fabricated with dimensional variation of ±0.5 μm with 95% confidence interval.

The primary advantages of flip-chip assembly are (i) small size due to the removal of often bulky and unwieldy wiring layer; (ii) high performance due to the reduced delay in inductance and capacitance; (iii) mechanically rugged connections; and (iv) low cost and actively used assembly technique in micro-electronics industry [Jittinorasett, 1999; Braun *et al.*, 2006]. Additional advantages of the use of micro-bumps within the flip-chip assembly are that they provide rigidity, act as a mechanical mounting of the device die to the substrate, prevent unwanted electrical contact between the die and substrate conductors and act as electrically conductive spacers [Jong-Kai *et al.*, 1996]. A variety of techniques, including vacuum deposition, stencil printing [Li and Thompson, 2000; Kloeser *et al.*, 2002], mechanical (wire) bumping [Neugebauer and Loughran, 1988], electro-deposition [Simon *et al.*, 1990; Karim and Schetty, 2000; Jordan, 2002] have been used to fabricate microbumps of different sizes and shapes with diameters ranging from 400 μm to 40 μm.

5.1. *Thermal Analysis of Micros-Thermo-Electric Cooler (μTEC)*

The cooling performance of a thermoelectric cooler (TEC) is typically determined by the dimensionless factor called the figure of merit, ZT and is defined as: $S^2\sigma T/K$ where T, S, σ, k are the absolute temperature (K), Seebeck coefficient (V K^{-1}), electrical conductivity ($\Omega^{-1}\,\mu m^{-1}$), , and thermal conductivity ($W\,\mu m^{-1}K^{-1}$), respectively [Rowe, 1996]. Materials like Bismuth-Telluride, with low thermal conductivity and high values of Seebeck coefficient and electrical conductivity, are utilized to obtain a high figure of merit or ZT. Prior attempts have been made to mathematically model the thermoelectric device performance, assuming the material properties being invariant with temperature [Rosi *et al.*, 1959; Sherman *et al.*, 1960; Rowe and Min, 1996]. Later, dimensionless entropy equations were derived that formed a close form expression for the TE system. These equations were later used to analyze the optimum design of thermoelectric cooler [Webb *et al.*, 1998; Huang and Duang, 2000; Xuan, 2002]. As the size of the TEC leg element scales down, the effect of contact resistance become increasingly important and is essential to accurately model the performance of μTEC. Models accounting for both the contact and electrical resistances [Da-Jeng *et al.*, 1999; Xuan *et al.*, 2002] have also been reported, but are not applicable when the temperature of the object to be controlled (passive load) is comparable to the size of μTEC. Prabhakar *et al.* [Prabhakar *et al.*, 2007] had modeled a single stage μTEC specifically for bio-applications. In all the above models it was assumed that the μTEC is connected to the external circuit by an inter-metallic planar wiring layer, which is an integrated part of the device die. However, in this section, a modular system of integrated sensors and actuators with an intermediate microbump layer is being modeled. Thus, it is essential to investigate the thermal effects of microbumps on the performance of a μTEC. Hence, a lumped parameter thermal analysis of the flip-chip bonded device (integrated cooling cell, a single stage μTC and microbump) was developed and analyzed, as described below.

Lumped parameter model development followed the earlier approach of Prabhakar *et al.* [2007] based on a set of rate equations describing one-dimensional thermal dynamics through the intermediate metal and ceramic end layers. Adiabatic boundary conditions were applied to four vertical side surfaces of the control volume. Symmetry within the device (Figure 12a) allows us to model a single TEC leg element sandwiched between two ceramic end layers (Figure 12b). The heat balance at the cold junction of the thermoelectric cooler is given by:

$$Q = S(T)IT_c - I^2\frac{R(T)}{2} - \lambda(T)(T_h - T_c),$$

where $S(T)IT_c$ represents the Peltier cooling effect, $I^2(R(T)/2)$ represents the Joule heating due to the resistance of the thermoelectric element and $\lambda(T)(T_h - T_c)$, represents the amount of heat flow from the hot to the cold junction. More specifically, $S(T)$ represents the temperature dependent Seebeck coefficient, I represents the current through the thermo-elements, T_c

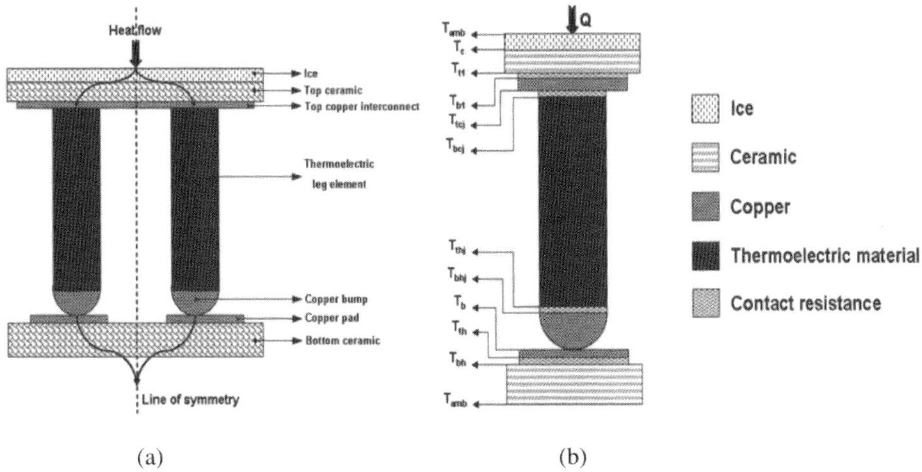

Figure 12. (a) Schematic of a single stage TEC showing only heat transfer along the line of symmetry which accounts for the cooling capacity. (b) Temperature distribution at each interface in a single μTEC element sandwiched between inter-metallic layer and ceramic plates. Redrawn from Prabhakar *et al.* [2007].

represents the cold junction temperature, R(T) represents the temperature dependent electrical resistance of the thermo-elements, $\lambda(T)$ represents the temperature dependent thermal conductance and T_h represents the hot junction temperature. The temperature dependence of S(T), $\kappa(T)$ and $\sigma(T)$ were obtained from literature as described elsewhere [Rowe, 1996; Prabhakar, 2006] and specifically are:

$S(T) = -0.001T^2 + 0.937T + 29.684$, $\mathrm{k}(T) = 0.0004T^2 - 0.231T + 53.21$, and $\sigma(T) = 1/(4.3516T - 279.9)$.

The thermal problem formulation involved a few assumptions as described earlier by Prabhakar *et al.* [2007], including (i) the presumption of a constant and zero heat load, $Q = 0$; (ii) negligible Joule heating due to interconnects when compared with that of a TEC element; (iii) treatment of the heat sink as a node at constant temperature; (iv) the cell to be subjected to temperature modulation by the μTEC being simulated as a heat generation input in its frozen state, or the uppermost layer in the model being represented by the properties of ice. Additional property data that is needed in the present analysis are the specific thermal and electrical contact resistances due to the microbumps. The electrical and thermal contact resistances are a result of impurities, variations in the stoichiometry of the thermoelectric compounds, variations in the crystal size and orientation, and defects at the interface, which are created in the fabrication process [da Silva and

Kaviany, 2004; 2005]. Both resistances cause reduction of the device performance. Thermal and electrical contact resistances are present at the metal and thermoelectric interfaces, and thermal contact resistances are present at the metal and insulator interfaces of the μTEC [da Silva and Kaviany, 2005]. Based on prior experimental investigations [Ilzycer *et al.*, 1983; Hmina and Scudeller, 1998; Orain *et al.*, 2001; Neidert *et al.*, 2004] a thermal contact resistance of 10^{-7} ($Km^2 W^{-1}$) was assumed. Similarly, utilizing published data the specific electrical contact resistance of the Bismuth Telluride-metal contact was assumed to be 2×10^{-11} Ωm^2.

A first law analysis of heat transfer from the cold junction to the hot junction of the μTEC, leads to a set of modified energy balanced equations given below:

$$
\begin{aligned}
&Q + \lambda_1(T_{amb} - T_c) - \lambda_2(T_{t1} - T_c) = 0\\
&\lambda_2(T_{t1} - T_c) - K_{R1}(T_{t1} - T_{b1}) = 0\\
&K_{R1}(T_{t1} - T_{b1}) - \lambda_3(T_{tcj} - T_{b1}) = 0\\
&\lambda_3(T_{tcj} - T_{b1}) - K_{R2}(T_{tcj} - T_{bcj}) = 0\\
&IS(T_{bcj})\, T_{bcj} - 1/2\, I^2(R + rce) - \lambda_4\,(T_{tcj} - T_{bcj}) - K_{R2}\,(T_{tcj} - T_{bcj}) = 0\\
&IS(T_{thj})\, T_{thj} - 1/2\, I^2(R + rce) - \lambda_4\,(T_{thj} - T_{bcj}) - K_{R3}\,(T_{thj} - T_{bhj}) = 0\\
&K_{R3}(T_{thj} - T_{bhj}) - \lambda_5(T_b - T_{bhj}) = 0\\
&\lambda_5(T_b - T_{bhj}) - \lambda_6(T_{th} - T_b) = 0\\
&\lambda_6(T_{th} - T_b) - K_{R4}(T_{th} - T_{bh}) = 0\\
&K_{R4}(T_{th} - T_{bh}) - \lambda_7(T_{bh} - T_{amb}) = 0
\end{aligned}
$$

where K_{R1}, K_{R2}, K_{R3}, and K_{R4} are given by $K_{Ri} = A_i/rct_i$ and rct_i represents thermal contact resistance at the interface of two different materials. A detailed nomenclature of the model variables is also provided in Table 3 and shown in Figure 12. Thermal contact resistance at the interface of copper microbump and copper pad was neglected and the initial device dimensions were chosen as previously optimized by Prabhakar *et al.* [2007], specifically the cellular space was modeled as (Length, L = 15 μm, Width, W = 15 μm and Height, H = 3 μm), the top ceramic plate (L = 15 μm, W = 14 μm and H = 5 μm), the top copper interconnect (L = 15 μm, W = 12 μm and H = to be determined, TBD) and bottom copper pad (L = 15 μm, W = 12 μm and H = TBD).

Figure 13(a), shows the influence of different μTEC leg element heights on the device performance (effective change in temperature, ΔT ($T_{bh} - T_c$)). It is found that an increase in the height of the μTEC leg element increases the heat load at the cold junction, leading to a small junction temperature

Table 3: Nomenclature for first law analysis of μTEC.

Symbol	Definition	Units
ZT	Dimensionless figure of merit	
$S(T)$	Temperature dependent Seebeck coefficient	VK^{-1}
K	Temperature dependent thermal conductivity	$W\ \mu m^{-1}K^{-1}$
$\sigma(T)$	Temperature dependent electrical conductivity	$\Omega^{-1}\mu m^{-1}$
T	Absolute temperature of the system	K
Q	Heat flow rate at the cold junction	W
T_c	Cold junction temperature	K
T_h	Hot junction temperature	K
T_{amb}	Ambient temperature	K
I	Current through the thermo-elements	A
R	Electrical resistance of the thermo-elements	Ω
V	Voltage across the thermo-elements	V
Q_h	Heat rejection of the μTEC	W
Q_c	Cooling capacity of the μTEC	W
A	Cross-sectional area of the thermo-elements	μm^2
H	Height of the thermo-elements	μm
L	Length of the thermo-elements	μm
W	Width of the thermo-elements	μm
λ	Thermal conductance	WK^{-1}

decrease. This implies that there is an inverse proportionality between the height of the leg element and effective change in temperature (K). Hence in order to achieve better temperature drop the μTEC leg element height should be as small as possible. However, the smallest achievable leg height is limited by the ability to plate bismuth telluride uniformly in recess, previously shown to be ~20 μm [Panda and Podlaha, 2003]. Hence the height of the μTEC leg element in the subsequent analysis was assumed to be 20 μm.

Figure 13b, shows the influence of different diameters of μTEC leg elements on the device performance. Increasing the diameter of the μTEC element increases the conductance, so the junction temperature increases. Hence, to achieve a larger effective temperature drop the μTEC element diameter should be as large as possible. However, the proposed end application of modulating cell temperature imposes a design constraint on the maximum diameter of the μTEC element, viz the space between two cells should be ~30 μm. Hence, the diameter of each element was restricted to a maximum of 10 μm, with the space between leg elements being ~10 μm.

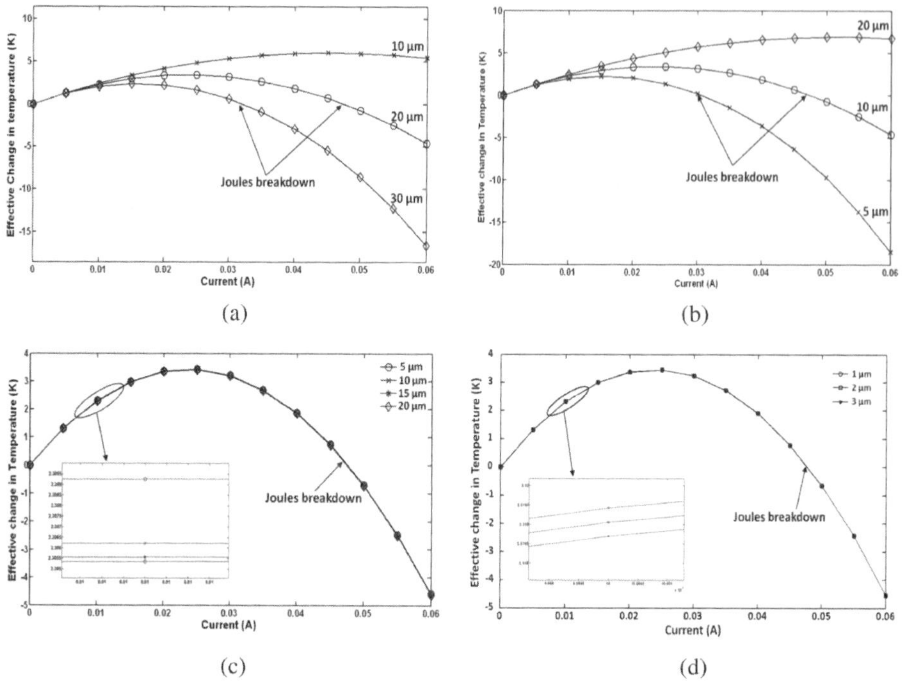

Figure 13. (a) Influence of the μTEC leg element height at zero external heat load. (b) Influence of the μTEC leg element diameter at zero external heat load. (c) Influence of the copper bump radius at zero external heat load. (d) Influence of the height of the top copper interconnect at zero external heat load. Redrawn from Prabhakar *et al.* [2007].

After fixing the dimensions of μTEC leg element (height 20 μm, diameter 10 μm) it was also essential to explore the effect of the metal microbump radius and interconnect height on the μTEC system performance (Figure 13(c)). As seen in Figure 13(c), the temperature drop across the device increased as the radius of the microbump decreased. However, the smallest achievable microbump dimension is limited by the fabrication processes as described in Section 4.2 and hence, the radius of the microbump was fixed at 15 μm. Figure 13d, shows that the temperature drop increases as the height of the copper interconnect is decreased. To satisfy the requirement of appropriate area for current passage through the μTEC leg elements, the height of the copper interconnect is presumed to be ~1/10 of the μTEC leg element [80]. Hence, the height of the copper interconnect was selected to 2–3 μm. Finally we found that the temperature drop decreased as the height of the bottom copper pad is decreased (data not shown). However, to generate a copper film with a good surface finish, the height of the bottom copper pad was selected to be ~0.5 μm.

Figure 14(a) represents the overall temperature profile along the length of the μTEC for different applied currents. The regions in yellow represent

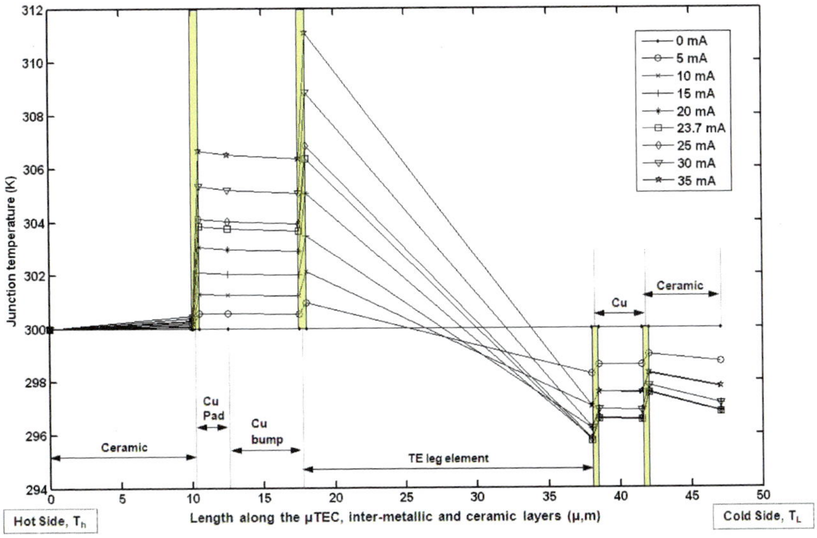

(a)

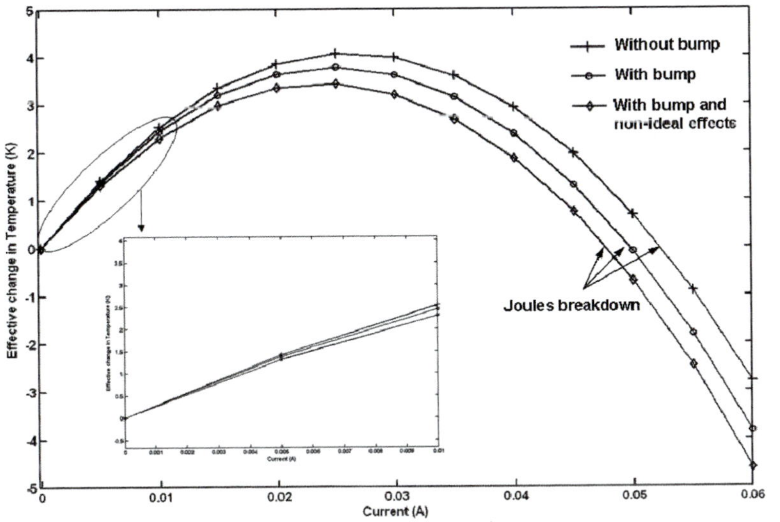

(b)

Figure 14. (a) Temperature profile along the length of the μTEC for different applied current (0 – 35 mA). Regions highlighted represent contact resistance present at the interface. (b). Influence of bump and non-ideal effects like electrical and thermal contact resistance on the effective temperature change ($T_{bh} - T_c$) at zero external heat load ($Q = 0$). Redrawn from Prabhakar *et al.* [2007].

the presence of contact resistances at the interface of different materials. These contact resistances act as heat generators at the interface as can be inferred from the steep temperature rise at the interfaces. Figure 14(a) also shows that the effective change in temperature ($\Delta T_{max} = T_{bh} - T_c = 3.44$K at

23.7 mA) was always less than the temperature drop across the μTEC leg element ($\Delta T_{te} = T_{bcj} - T_{thj} = 10.6$K at 23.7 mA), due to the presence of contact resistances at different interfaces. Figure 14(b) gives the effective change in temperature, ΔT ($T_{\mathrm{bh}} - T_c$), as a function of applied current at zero external heat load ($Q = 0$), for three different conditions: without a microbump, with a microbump, and with a microbump coupled with non-ideal electrical and thermal contact resistances. As shown in the inset of the Figure 14(b), there are no differences between the three scenarios at low currents (0 to 5 mA). The simulations also suggest that a maximum effective change in temperature of ~3.5K occurs at an applied current of 23.7 mA and the corresponding Joules breakdown current was found to occur at 47.7 mA. Future work includes the development of a more detailed Finite element analysis of the μTEC and models to account for the possibility of Joule heating within the interconnect as well as the heat loss to the surrounding polymer.

5.2. *Microfabrication of Microbumps for Flip Chip Bonded Device*

As described earlier we found that a microbump of approximately 15 μm in radius is optimal for the proposed FCB device. Hence we developed a technique to fabricate microbumps using an UV lithography-based microfabrication technique and electroplating. This method gives high flexibility of adjusting the geometry, material composition and scalability. It is important to note that Ueno *et al.* [1998] had electroplated copper column bumps on printed wire board with diameter of 75 μm, height of 50 μm with a pitch of 200 μm. Tomita *et al.* [2000] had successfully fabricated tin capped copper column bumps with a height of 5 μm that were spaced 100 μm apart for 3D LSI chip integration. Yamada and Togasaki [2003] fabricated copper column based solder bumps of various sizes by electroplating, with diameters of 120 μm, 85 μm, 65 μm, a height of 70 μm and a pitch of 300 μm for CCD micro-camera visual inspection system. Ho-Young *et al.* [2007] electroplated Cu (60 μm)/Sn/Ag (20 μm) double bump with a diameter of 60 μm and a pitch of 100 μm. However, there are no published reports of fabricating Constantan (copper-nickel) microbumps. Additionally, the fabrication of hemi-spherical shaped copper bumps is unique to the present study and differs from the commonly reported cylindrical shaped microbumps.

Figure 15 gives an overview of the steps in the UV lithography-based microfabrication technique performed on a silicon wafer (4" diameter single side polished, 1–20 Ω–cm). A 0.01 μm titanium adhesion layer and a 0.05 μm copper plating base is then e-beam (Model BJD-1800 E-Beam Evaporator, Temescal, Wilmington, MA) evaporated onto the surface of the silicon wafer.

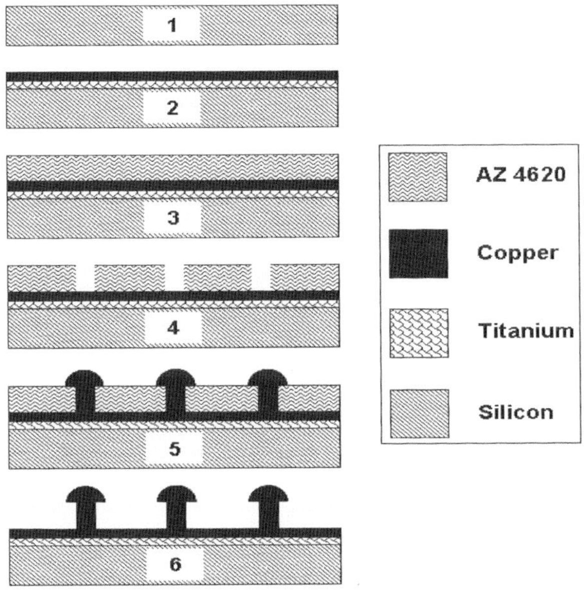

Figure 15. Overview of the microfabrication steps. (1) Silicon wafer base. (2) E-beam deposit titanium and copper. (3) Spin AZ 4620 resist. (4) Expose and develop the resist. (5) Deposit copper in the recess to form the bump. (6) Remove the resist.

The wafer is cleaned using acetone, isopropyl alcohol (IPA), and rinsed in DI water to remove any oxides or organic residue. The wafer is then dehydration baked at 95°C on a hot plate for 5 minutes to promote adhesion of the resist with the substrate. AZ 4620 resist is spin coated at 2000 rpm for 5 seconds to get a nominal thickness of 20 μm. The sample is then given a 10 minute relaxation time before being soft baked in a convention oven (M326 mechanical convection oven, Blue M Electric, Watertown, Wisconsin) for 10 minutes at 85°C. The sample is then re-baked on a hot plate at 115°C for 60 seconds with an air gap between the middle underside of the wafer and the hotplate, with the top covered [Paydenkar *et al.*, 2001]. The air gap prevents the sudden heating at the base of the resist, which can cause harmful thermal stresses, while the top cover helps to maintain a uniform temperature over the resist surface [Yong-Min *et al.*, 2005]. The high bake temperature was experimentally found to improve aspect ratio, but also reduced exposure sensitivity and hence, was kept rather short at 60 seconds [Cygan, 2006]. After baking, the sample is allowed to cool for at least 20 minutes before further UV exposure. The UV exposure is performed under high vacuum contact on a Quintel UL 7000-OBS Aligner and Exposure station (Quintel Corp., Morgan Hill, CA) with a dose of 600 mJ cm^{-2} at 365 nm wavelength.

The sample is developed for 5 minutes in AZ 400K developer (AZ Electronic Materials, Somerville, NJ), diluted 4:1 with DI water, rinsed with DI water, and dried by blowing nitrogen gas. Developing was done in Bransonic ultrasonic cleaner bath (Bransonic 2510R-DTH, Branson Ultrasonics Corporation, Danbury, CT 06813).

Galvanostatic Electrodeposition was performed using a plating jig, as described elsewhere [Cygan, 2006; Kardak, 2008]. Prior to deposition, the patterned sample was placed in a vacuum chamber and ~5–10 mL of deionized (DI) water was placed over the patterns. The whole assembly was then immersed in the plating solution for electrodeposition (Figure 16). The initial electrodeposition was done in kinetic region where the ions tend to follow the electric field through the least resistant path giving rise to a concave profile at the top of the deposit, Figures 16(a) and 4(b). When the deposition of the metal reaches the top of the resist the current is ramped up to near limiting value. At short time scales, when the diffusion layer thickness is small compared to the size of the active region, each spot or region generates its

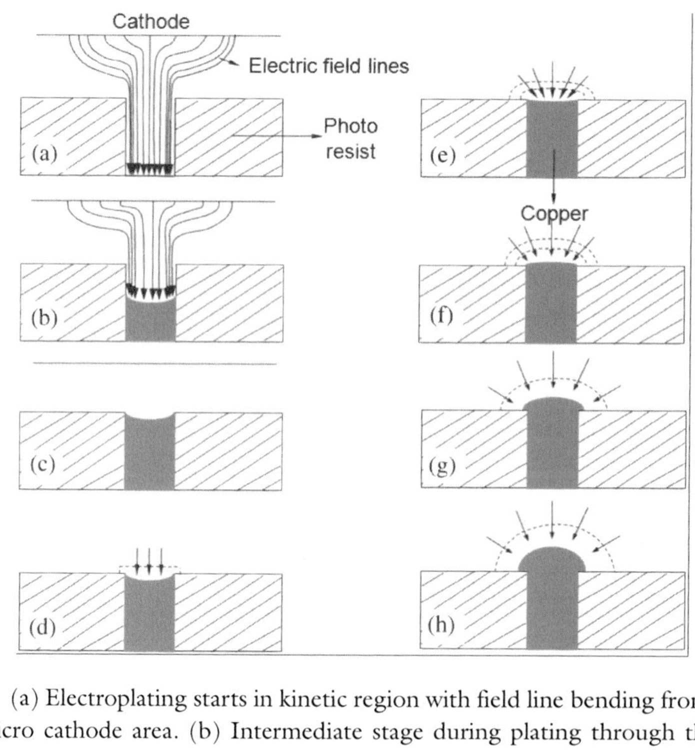

Figure 16. (a) Electroplating starts in kinetic region with field line bending from the faraway anode to micro cathode area. (b) Intermediate stage during plating through the recess. (c) Shows a profile when the top of the resist is reached. (d) Potential at the cathode or working electrode is stepped up leading to the evolution of planar diffusion field. (e) Plating at the center of the profile. (f) At an intermediate stage where bump formation begins and the transition of the shape of the diffusion field to become spherical. (g), (h) Formation of uniform spherical diffusion field with more uniform plating on all sides. Redrawn from Kardak [2008].

own planar diffusion field, as shown in Figure 16(d). This allows for more deposition at the center of the recess (Figure 16(e)). At longer time scales, the individual linear diffusion is augmented by a radial component or a spherical diffusion field (Figures 16(f) and 16(g)). Thus, facilitating the formation of hemispherical microbump (Figure 16(h)). A platinum coated titanium mesh counter electrode was used along with a mechanical stirrer to provide agitation. A horizontal substrate orientation was utilized to allow the escape of any hydrogen gas formed during electrodeposition.

The copper plating solution (Chomnawang 2002) consisted of 1.0M Copper Sulphate and 100 mL/L of Sufuric Acid while the Constantan plating solution consisted of 1.0M Nickel Sulfate, 0.5M Copper Sulfate, 1.0M Sodium Citrate and 74 mL/L Ammonium Hydroxide. Copper was electrodeposited in the patterned recess until the top surface of the resist, under kinetic control, at 7 mA cm^{-2} for 75 minutes. Over-plating of copper was done at 300 mA cm^{-2} so as to have diffusion controlled electroplating for the bump formation. Over-plating for copper microbumps was done for 3 different plating times of 5, 10 and 15 minutes. Constantan was electrodeposited in the patterned recess until it reached to the top surface of the resist, under kinetic control, at 15 mA cm^{-2} with a 0.25 duty cycle and 10 seconds on time was used for a total plating on-time of 20 minutes. Constantan over-plating was done under diffusion control at 0.75 A cm^{-2} with 0.25 duty cycle and 5 seconds on-time for a total on-time of 10 minutes. The final structure was rinsed in DI water and blown dry with nitrogen gas. The diameter and the height of the microposts was then measured using a Measurescope (MM-22, Nikon Corp., Kawasaki, Japan) and non-contact profilometer (Manovea ST400, Micro Photonics Inc., Allentown, PA) respectively. The measured values are represented by the mean value ± dimensional values representing the 95% confidence interval in Table 4.

Table 4. Measured mean diameter of the copper microposts for 3 different plating times on two base structures of 5 and 3 μm diameter. The values represent mean values ± dimensional variation with 95% confidence interval (CI).

		3 μm Diameter recess		5 μm Diameter recess	
Micropost Material	**Over-plating time (mins)**	**Diameter (μm ± CI)**	**Height (μm ± CI)**	**Diameter (μm ± CI)**	**Height (μm ± CI)**
	5	6.55 ± 0.45	7.81 ± 0.22	10.23 ± 0.52	8.49 ± 0.24
Copper	10	18.01 ± 0.47	9.78 ± 0.21	20.32 ± 0.52	10.26 ± 0.22
	15	29.56 ± 0.25	12.31 ± 0.24	30.16 ± 0.26	12.70 ± 0.24
Constantan	10	27.8 ± 0.37	12.0 ± 0.25	28.5 ± 0.51	12.4 ± 0.26

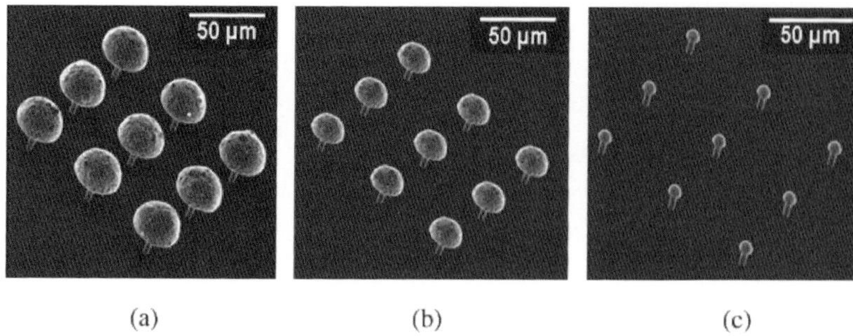

Figure 17. SEM micrograph of copper bump electroplated on 3 μm diameter base structure for different over-plating times (Kardak, 2008). (a) 15 minutes (b) 10 minutes (c) 5 minutes.

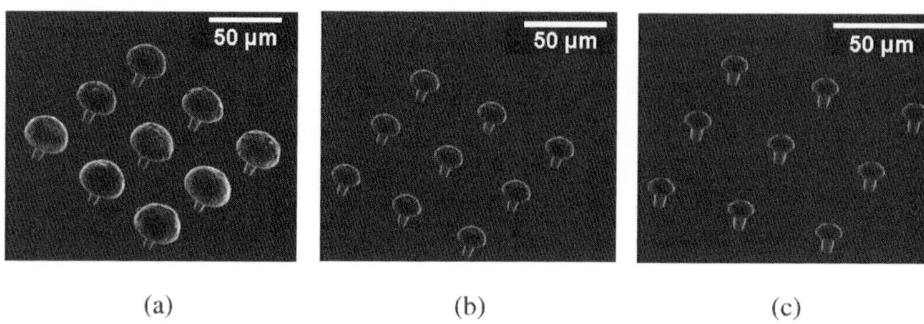

Figure 18. SEM micrograph of copper bump electroplated on 5 μm diameter base structure for different over-plating times [Kardak, 2008]. (a) 15 minutes (b) 10 minutes (c) 5 minutes.

Figures 17 and 18 shows SEM micrographs of copper microbumps electroplated on 3 and 5 μm diameter base structures for over-plating times of 5, 10 and 15 minutes. The data in Table 4 suggests that as the over-plating time is increased, the difference between the mean diameter and height of the copper microposts decreased between both the 3 and 5 μm base structures. This suggests that at the initial stage of over-plating, the area exposed to the plating solution does indeed have an effect on the final micropost diameter and height. However, this difference can be minimized by increasing the over-plating time. The corresponding data for Constantan microbumps with an overplating time of 10 min is also shown in Table 4 while the corresponding SEM images, and energy dispersive spectrometric analysis performed to confirm the presence of copper and nickel in the microbump, are not shown in the interest of brevity [Kardak, 2008]. Although galvanostatic plating without pulse plating scheme was used to fabricate copper bumps, further studies are needed to employ reverse pulse plating to improve the surface morphology and form compact electrodeposit [Ueno *et al.*, 1998]. Further studies with reduction of total over-plating on time to form smaller constantan bumps of the order few micrometers are also suggested.

6. Nanostructures for Micro-ThermoElectric Devices

The fabrication, modeling and testing of prototype devices for temperature measurement (by microsized thermocouples) and temperature modulation (by microsized thermoelectric coolers) suggested that there is a clear need to improve the thermoelectric efficiency of thermoelectric coolers (and consequently, the cooling capacity or the temperature drop in these devices). Previous theoretical predictions and experimental studies by Dresselhaus and her colleagues [Hicks and Dresselhaus, 1993; Dresselhaus *et al.*, 1999; Sun *et al.*, 1999; Lin *et al.*, 2000; Chen *et al.*, 2003; Lin and Dresselhaus, 2003; Yang *et al.*, 2005; Dresselhaus *et al.*, 2007] identified that low dimensional materials (thin films, nanowires or nanotubes) increase the thermoelectric efficiency significantly when compared to their bulk counterparts. These modeling studies have further suggested that nanotubes might be more effective in increasing the thermoelectric efficiency (*ZT*) due to their reduced lattice thermal conductivity and strong phonon blocking effects. Electrodeposition of nanotubes although similar to electrodeposition of nanowires require a different set of conditions to facilitate hollow tubular formation. Various approaches have been used for nanotube formation including pore wall modification through the use of templates with anionic functional group pore walls, selective etching, and the use of low current efficiency electrolytes [Davis *et al.*, 2010; Gupta *et al.*, 2010; Gupta and Podlaha, 2010; Pinisetty *et al.*, 2011a, Pinisetty *et al.*, 2011b]. Thus, prototype thermoelectric coolers (TECs) composed of nanostructured (nanowire or nanotube) composites were fabricated, characterized and tested for their thermoelectric efficiency (*ZT*). Bismuth-telluride (Bi_2Te_3) and antimony-telluride (Sb_2Te_3) that received the most attention for near room temperature thermoelectric applications [Dresselhaus *et al.*, 1999; Chen *et al.*, 2003; Dresselhaus *et al.*, 2007] were chosen to be the n-type and p-type elements for our thermoelectric cooler devices, respectively.

6.1. *Fabrication of Nanostructures*

Electrodeposition is a technique particularly well suited for depositing into nano-scaled recesses or pores to create low-dimensional Bi_2Te_3 and Sb_2Te_3 nanowires or nanotubes using a template approach [Pinisetty and Devireddy, 2010; Pinisetty *et al.*, 2011a; Pinisetty *et al.*, 2011b]. These nanostructures can then be integrated into a TEC. Since a single one-dimensional

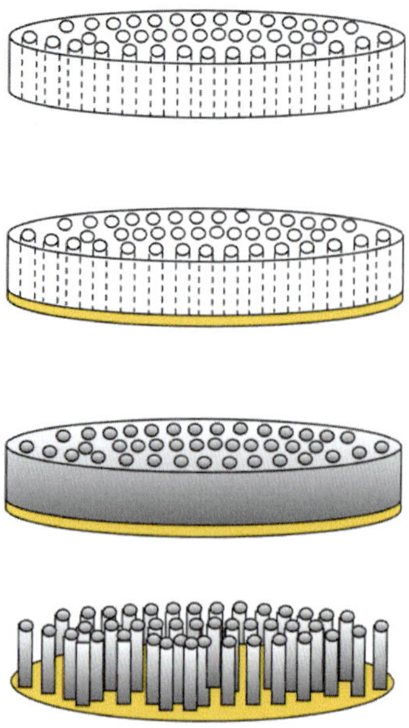

Figure 19. Schematic of nanowire or nanotube fabrication process using template based deposition: Figures from top to bottom represent: Isometric view of porous template, Sectional side view of the template with sputtered gold, Template after performing electro-deposition and Free standing nanowires or nanotubes after the template is dissolved. Reproduced with permission from Pinisetty *et al.* [2011b].

nanostructure cannot carry sufficiently high enough current, synthesis of dense, conductive nanostructure arrays is critical to the development of realistic TECs. A schematic of nanowires and/or nanotubes that can be fabricated using electrodeposition technique in conjunction with a template is shown in Figure 19. Briefly, the process involves the following steps. A thin layer of gold (Au) is sputtered onto one side of either alumina or polycarbonate membrane to render them conductive. The template with the conductive gold substrate is given a negative charge and is used as the cathode in the electrochemical cell. The unsputtered face of the template is exposed to the electrolyte for electrodeposition. As the only conductive surface is the gold base at the bottom of the pores, a bottom up fill inside the pores is achieved. Within short intervals double layer charging causes a large current response, which decays very quickly to a steady Faradaic current. Finally, as the wires start to grow out of the pores a mushroom shaped cap is formed resulting in an increase in electrode area and hence the current density. This increase in current is used as an end point marker during the electrodeposition process. After the electrodeposition, the templates are dissolved in suitable solvents. For example, alumina or polycarbonate templates are dissolved in potassium hydroxide

(KOH) or methylene chloride (CH_2Cl_2), respectively. Once the templates material is dissolved, free standing nanowires or nanotubes are obtained (as shown in Figure 19). And finally, electrodeposition employing mild deposition conditions such as room temperatures and atmospheric pressue conditions, faciliatates fabrication of large numbers of free standing nanostructures.

The electrodeposition of Bi_2Te_3 and Sb_2Te_3 nanowires/nanotubes was carried out at room temperature in a typical three-electrode cell setup. Track-etched polycarbonate (PC) (Poretics PCTE, GE Osmonics, Minnetonka, MN) sputtered with gold on one side was used as the working electrode. PC templates with average pore diameters of 400 nm and 100 nm were used for the fabrication of nanotubes and nanowires, respectively. The gold surface was kept in contact with a copper plate held inside a polyetheretherketone (PEEK) stationary holder and a circular area of 2 cm^2 was exposed to the electrolyte for electrodeposition. A square platinum mesh was used as the counter electrode positioned horizontally above the working electrode (Figure 20). An Acumet saturated calomel reference electrode (SCE) was used to record the potential at the working electrode. The reaction cell was a

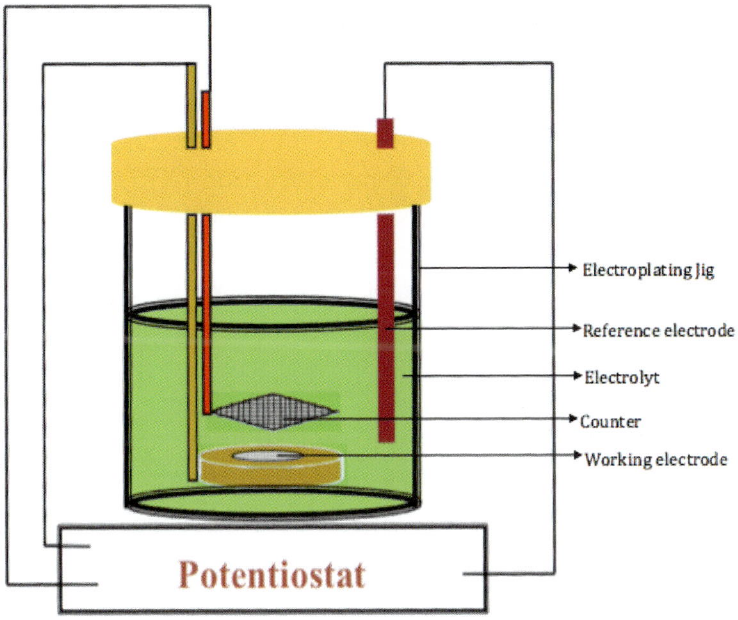

Figure 20. Experimental apparatus used to deposit Sb_2Te_3 or Bi_2Te_3 nanowires or nanotubes. Porous membrane is used as the working electrode, a platinum mesh was used as the counter electrode and a saturated calomel electrode (SCE) was used as the reference electrode.

1000 ml beaker, filled with electrolyte, optimized by rota-hull cell experiments (2.3 mM Sb_2O_3, 4.2 mM TeO_2, 330 mM of tartaric acid, and 3M of HNO_3), for Sb_2Te_3 deposition, and (10 mM Bi_2O_3, 10 mM TeO_2 and 2.5 M HNO_3) for Bi_2Te_3 deposition. The electrolyte was magnetically stirred during experiments to ensure proper mixing of ions and to prevent depletion of ions near the electrode surface. All experiments were carried out with a Princeton Applied Research (PAR) potentiostat/function generator. The Sb_2Te_3 nanowires/nanotubes were galvanostatically deposited while the Bi_2Te_3 nanowires/nanotubes were potentiostatically deposited.

6.2. *Characterization of Nanostructures*

The theoretical advantages of the thermoelectric materials chosen (Bi_2Te_3 and Sb_2Te_3) can only be extracted if the stoichiometry of the materials is maintained in the fabricated nanostructures. The stoichiometry ratio (39% Bi and 61% Te for Bi_2Te_3 and 39% Sb and 61% Te for Sb_2Te_3) is also critical in maintaining their specific carrier type, i.e. Bi_2Te_3 to be n-type and Sb_2Te_3 to be p-type. Therefore, it is necessary to characterize the nanostructures to study their morphology, composition, and crystal structure. Morphological analysis was performed using a Scanning electron microscope (SEM), specifically a JEOL 840A SEM operated at an accelerating voltage of 20 kV (Figure 21). The composition analysis was performed with wavelength dispersive spectrometry (WDS), using a JEOL 733 super probe operated at an accelerating voltage of 15 kV and a probe current of 10 μA (Figure 22). The crystal structure analysis was performed using X-ray diffraction (XRD), with a Bruker/Siemens D5000 machine with, CuKα radiation (data not shown).

Figure 21 shows that the Sb_2Te_3 nanowires are arranged regularly, parallel to one another, implying that most of the pores are filled. Further more all the Sb_2Te_3 nanowires have the same height, indicating that the nanowires are deposited within the pores at the same rate, which is important for future fabrication of a nanostructured device. The average diameter and length of the deposited nanowires were found to be ~100 nm and 7 μm, respectively. As shown in Figure 21 (bottom figure), the use of larger pore diameter PC template of ~400 nm resulted in tubular deposits. The thickness of the nanotube walls was found to be ~60 ± 3 nm.

The formation of a nanotube, rather than a nanowire, is attributed to the partial filling of the gold on the bottom sidewalls of the PC template membrane that promotes deposition in that region. Subsequent gas evolution, a byproduct of the electrodeposition process performed at current densities much higher to the reduction potentials of the materials, possibly restricts

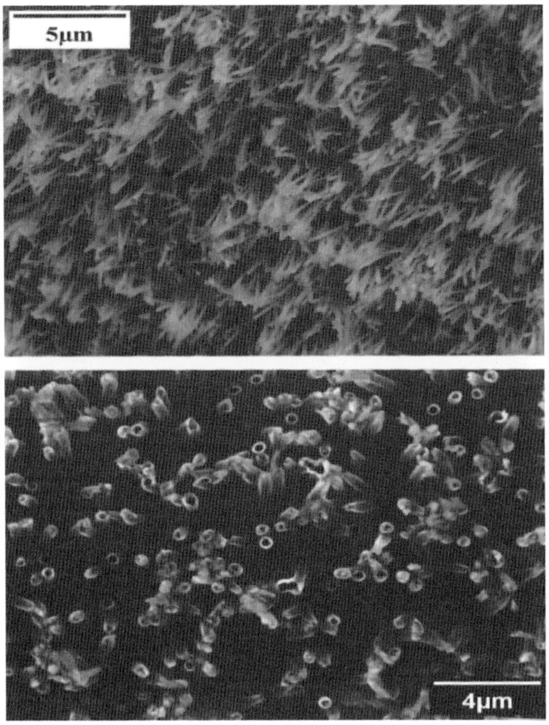

Figure 21. SEM images of Sb_2Te_3 nanowire arrays electrodeposited at ~10 mA cm^{-2} using a polycarbonate (PC) template with an average pore diameter of ~100 nm (top) and Sb_2Te_3 nanotube arrays electrodeposited at ~5.5 mA cm^{-2} using a PC template with an average pore diameter of ~400 nm (bottom). Similar images were obtained for Bi_2Te_3 nanowire arrays electrodeposited at ~–61 mV using a PC template with an average pore diameter of ~100 nm and Bi_2Te_3 nanotube arrays electrodeposited at ~–320 mV using a PC template with an average pore diameter of ~400 nm (data not shown). Reproduced with permission from Pinisetty *et al.* [2011b].

further deposition in the central regions of the pore, thus promoting a tubular shape [Prabhakar, 2006]; although, this may not be the only mechanism responsible for tube formation. It is also observed that the growing rate of Sb_2Te_3 nanotubes is not as uniform as that of the Sb_2Te_3 nanowires (for reasons as yet, unknown).

Figure 22 shows the dependence of the Sb_2Te_3 nanowires (top figure, Figure 22) and nanotubes (bottom figure, Figure 22) composition (wt%) on the applied current density. The cathodic current density range was selected from 2.75 mA cm^{-2} to 25 mA cm^{-2}. The dotted line in Figure 22 shows the desired composition of Sb (~39%). The composition (wt%) of nanowires deposited at a cathodic current density of ~10 mA cm^{-2} was found to be

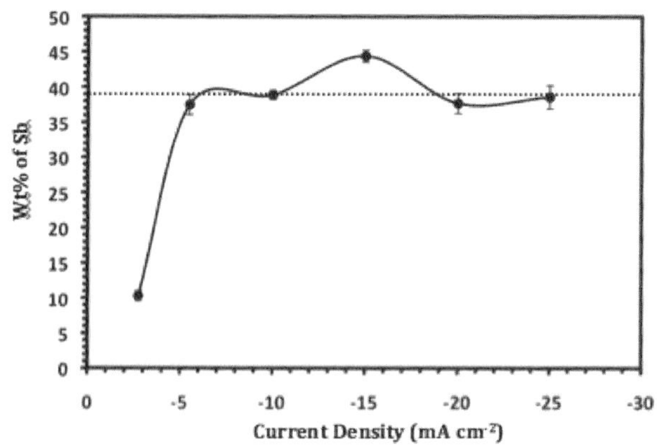

(a)

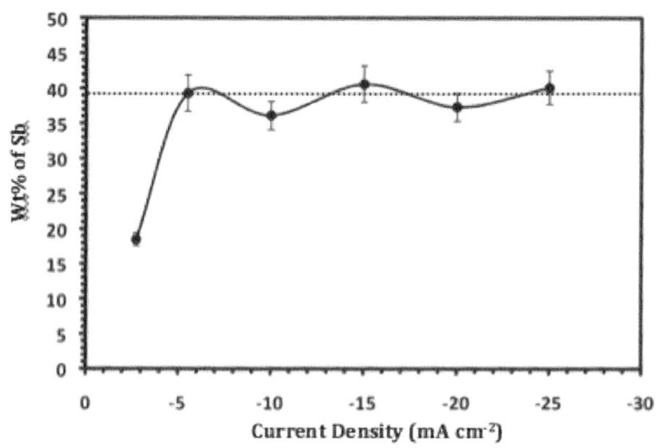

(b)

Figure 22. Composition (wt%) of (a) Sb-Te deposit inside pores of the polycarbonate (PC) template with an average diameter of ~100 nm and (b) Sb-Te deposit inside pores of the polycarbonate (PC) template with an average diameter of ~400 nm, as a function of applied current density (mA cm^{-2}). The dotted line depict the desired wt% of Sb. Reproduced with permission from Pinisetty *et al.* [2011b].

~38.9 ± 0.6% (top figure, Figure 22) whereas in the case of nanotubes, deposition at a cathodic current density of ~5.5 mA cm^{-2} resulted in a composition (wt%) of ~39.2 ± 2.6% (bottom figure, Figure 22). Although depositions at an higher current density of ~25 mA cm^{-2} resulted in a composition (wt%) closer to the desired composition, both for nanowires and nanotubes, XRD measurements showed that the deposits were amorphous. The results indicate that the nanowires and nanotubes deposited at current densities of 10 mA cm^{-2} and 5.5 mA cm^{-2} respectively, are p-type. Similarly, the composition (wt%) of Bi within Bi_2Te_3 nanowires deposited at a cathodic potential of ~61 mV was found to be ~38.8 ± 1.4% whereas in the case of Bi_2Te_3 nanotubes, deposition at a cathodic potential of ~320 mV resulted in a composition (wt%) of ~38.7 ± 1.9%.

6.3. *Seebeck Coefficient and Electrical Resistance of Nanostructures*

Seebeck coefficient measurements were performed on Physical Property Measurement System using a custom built device. Figure 23 shows the room temperature Seebeck coefficient of Sb_2Te_3 nanowire or nanotubes as a function of applied current density. Sb_2Te_3 nanowires deposited at a current density of ~10 mA cm^{-2} yielded the highest Seebeck coefficient value of 359 ± 11 $\mu V\ K^{-1}$ (Figure 23, Top) while the Sb_2Te_3 nanotubes deposited at a current

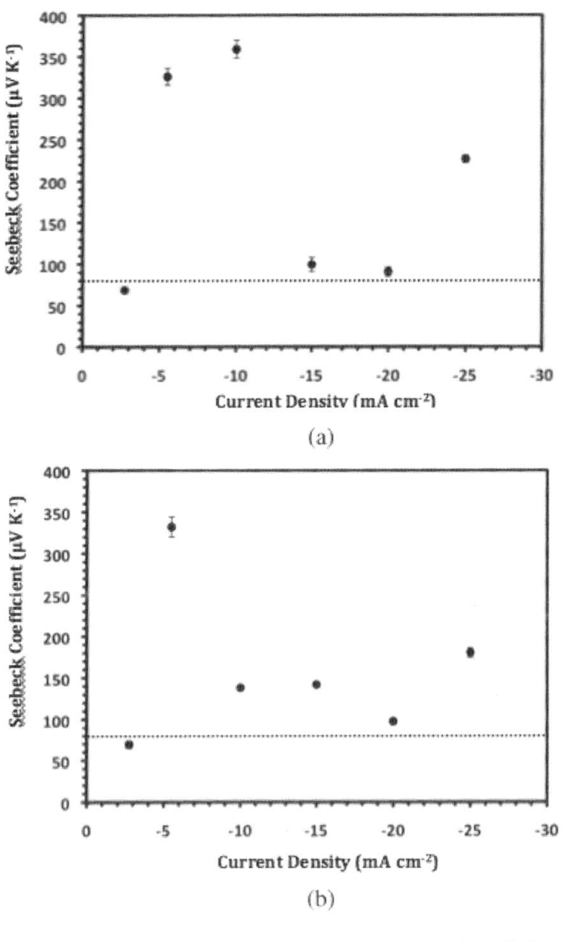

Figure 23. Seebeck coefficient of Sb_2Te_3 nanowire arrays (top) and Sb_2Te_3 nanotube arrays (bottom), as a function of applied current density (mA cm^{-2}). The dotted line shows the bulk Seebeck coefficient value of Sb_2Te_3.

density ~5.5 mA cm^{-2} yield the highest Seebeck coefficient value of 332 ± 12 $\mu V\ K^{-1}$ (Figure 23, bottom). The positive Seebeck coefficient values indicate that the nanowires and nanotubes are p-type. The Seebeck coefficient values reported here are much higher than those commonly measured for bulk Sb_2Te_3 (~79 $\mu V\ K^{-1}$) [53, 58]. The enhancement of the Seebeck coefficient values for the nanowires and nanotubes may be due to the single-carrier effect in nanostructures, as previously postulated.

Similar data was obtained for the room temperature Seebeck coefficient value of Bi_2Te_3 nanowire or nanotubes as a function of applied deposition potential. Specifically, Bi_2Te_3 nanowires deposited at a current density of ~–61 mV yielded the highest Seebeck coefficient value of –118 ± 3.6 $\mu V\ K^{-1}$ (data not shown) while the Bi_2Te_3 nanotubes deposited at a current density ~–320 mV yielded the highest Seebeck coefficient value of –143 ± 4.3 $\mu V\ K^{-1}$ (Figure 25(b)). The negative Seebeck coefficient values indicate that the nanowires and nanotubes are n-type. These Seebeck values are lower than those commonly measured for bulk Bi_2Te_3 (~–181 $\mu V\ K^{-1}$). Intriguingly, the Seebeck coefficient values are lower than bulk bismuth telluride values and may be due to the adverse contribution from the polycarbonate template, which has a very high surface resistivity.

Further studies of the electrical resistance of Bi_2Te_3 and Sb_2Te_3 nanowires or nanotubes as a function of temperature showed that electrical resistance decreases with increasing temperature and exhibits a negative temperature coefficient of resistance (TCR), i.e. a typical semiconductor-like behavior. The TCR of the nanotubes is found to be smaller than that of nanowires, indicating that the nanotubes, for which electrical transport is expected to happen only through its wall thickness, have lower resistance sensitivity to temperature.

6.4. *Measurement of Thermoelectric Efficiency (ZT) of µTEC*

The thermoelectric efficiency of the device composed of nanostructures (nanowires or nanotubes) is measured based on Harmans technique [Rowe, 1996; Rowe and Min, 1996]. A custom built device (Figure 24) is used to mount the sample and perform the measurements. The nanostructured samples (n-type and p-type) are loaded by manganin wires with 0.02 mm in diameter and are put in vacuum in order to replicate sufficiently an adiabatic condition. The attachment of the wires to the samples is shown in Figure 24. The wires were connected to the sample by using silver epoxy (Epotek H20E). In order to maintain a sufficiently isolated condition to have precise *ZT* measurement, a thermal radiation shield was also set up. A custom

Figure 24. Snapshot of the device used to measure *ZT* (based on 'Harmans Technique').

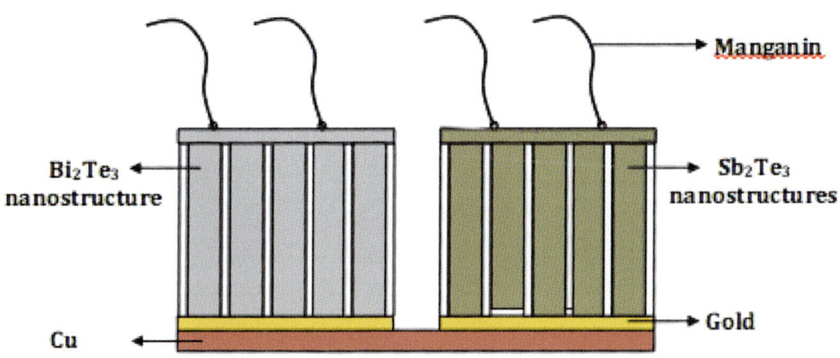

Figure 25. Schematic of a bulk nanostructured TE device with connections to measure *ZT* by using 'Harmans Technique'.

designed temperature control and measurement program, created with LabView software operating a Dell PC, run was used to the experiment. Heater current was supplied by a Keithley 220 DC current source, and the thermo-emf and differential thermocouple voltage were measured by Keithly 182 nanovoltmeters and recorded by the PC. All the measurements were performed at room temperature.

The schematic of the bulk nanostructured device is shown in Figure 25. When the nanowires or nanotubes are deposited within a gold sputtered PC template they are over-deposited such that they form a thin film on the top of the substrate. Individually n-type nanowires or nanotubes and p-type nanowires or nanotubes are attached to a copper tape and assembled into a

Table 5. ZT measurement of thermoelectric device composed of nanowires or nanotubes.

TE Device	Bi_2Te_3 (n-type)	Sb_2Te_3 (p-type)	ZT
1	Nanowires	Nanowires	0.11
2	Nanotubes	Nanotubes	0.31
3	Nanowires	Nanotubes	0.4
4	Nanotubes	Nanowires	0.22

bulk TE device. The copper tape serves the electrical junction required between n-type and p-type structures. When the current is passed through the manganin wires, it is expected to pass through the nanowires or nanotubes via the film and a thermoelectric response should be generated. The *ZT* calculated for all the combinations in the bulk nanostructured device is shown in Table 5. It is observed that a combination of Bi_2Te_3 nanowires as n-type and Sb_2Te_3 nanotubes as p-type yielded the best *ZT* value of 0.4 at 300K.

7. Conclusion

This chapter described our efforts to model, fabricate, and characterize microscale arrays of thermoelectric devices based on the Peltier effect and the Seebeck effect. A set of thermodynamic models were developed to determine device geometry and dimensions before micro-fabrication techniques were utilized to fabricate thermocouple arrays as well as a microscale thermoelectric cooler with nanostructures. The development of the latter included the fabrication, characterization of nanowires and nanotubes of commonly used thermoelectric materials like Bi_2Te_3 and Sb_2Te_3. Our efforts to fabricate microscale bumps for use in a integrated device to limit the issues with wiring and heat transfer are also detailed. Future work includes the development of multi-module multi-stage TEC devices to provide for larger temperature differences.

Acknowledgments

The work described in this chapter would not have been possible without the guidance, intellectual support and technical know-how of Prof. Murphy in the Department of Mechanical Engineering at the Louisiana State University (LSU), Prof. Podlaha-Murphy in the Department of Chemical Engineering at Northeastern University, Prof. Young in the Department of Physics at LSU and Prof. Spivey in the Department of Chemical Engineering at LSU. Several graduate students in the department of Mechanical Engineering performed the work described in this chapter, as part of their graduate thesis work, including A. Cygan (MS), A. Prabhakar (MS), A. Kardak (MS), D. Kandra (MS) and D. Pinisetty (PhD). Funding is acknowledged from the NSF Graduate Fellowship (for A. Cygan), LSU Economic Development Assistantship Fellowship (for D. Pinisetty), the Department of Mechanical Engineering (A. Kardak and D. Kandra) and the LSU system office (A. Prabhakar). And finally, acknowledgements are also due to S. Shaik for help with formatting the document.

References

ASTM (1970). E-20 on Temperature Measurement, Manual on the use of Thermocouples in temperature measurement.

Bernard, A. and B. J. Fuller (1996). Cryopreservation of human oocytes: a review of current problems and perspectives. *Human Reproduction Update*, 2(3), pp. 193–207.

Bischof, J. C. (2000). Quantitative measurement and prediction of biophysical response during freezing in tissues. *Annual Review of Biomedical Engineering*, 2, pp. 257–288.

Bischof, J., K. Christov and B. Rubinsky (1993). A morphological study of cooling rate response in normal and neoplastic human liver tissue: cryosurgical implications. *Cryobiology*, 30(5), pp. 482–492.

Braun, T., K. F. Becker, M. Kohl, B. V. R. Aschenbrenner and H. Reichl (2006). High temperature potential of flip chip assemblies for automotive applications. *Microelectronics and Reliability*, 46, p. 144.

Chen, G., M. S. Dresselhaus, G. Dresselhaus, J. P. Fleurial and T. Caillat (2003). Recent developments in thermoelectric materials. *International Materials Reviews*, 48(1), pp. 45–66.

Chomnawang, N (2002). *Three-Dimensional Micromachined On-Chip Inductors for High Frequency Applications.* Master of Science in Electrical and Computer Engineering, Louisiana State University.

Cygan, A. (2006). *Modeling and Fabrication of a Microthermocouple Array.* Master of Science Thesis in Mechanical Engineering, Louisiana State University.

da Silva, L. W. and M. Kaviany (2004). Micro-thermoelectric cooler: interfacial effects on thermal and electrical transport. *International Journal of Heat and Mass Transfer*, 47(10–11), pp. 2417–2435.

da Silva, L. W. and M. Kaviany (2005). Fabrication and measured performance of a first-generation microthermoelectric cooler. *Journal of Microelectromechanical Systems*, 14(5), pp. 1110–1117.

Da-Jeng, Y., K. Chang-Jin, G. Chen, J. P. Fleurial and H. B. Lyon (1999). Spot cooling using thermoelectric microcoolers. *Eighteenth International Conference on Thermoelectrics, 1999.*

Davis, D., M. Zamanpour, M. Moldovan, D. Young and E. J. Podlaha (2010). Electrodeposited, GMR CoNiFeCu Nanowires and Nanotubes from Electrolytes Maintained at Different Temperatures. *Journal of the Electrochemical Society*, 157(6), pp. D317–D322.

Day, S. H., D. A. Nicoll-Griffith and J. M. Silva (1999). Cryopreservation of rat and human liver slices by rapid freezing. *Cryobiology*, 38(2), pp. 154–159.

Devireddy, R. V., P. R. Barratt, K. B. Storey and J. C. Bischof (1999a). Liver freezing response of the freeze-tolerant wood frog, *Rana sylvatica*, in the presence and absence of glucose. I. Experimental measures. *Cryobiology*, 38(4), 310–326.

Devireddy, R. V., P. R. Barratt, K. B. Storey and J. C. Bischof (1999b). Liver freezing response of the freeze-tolerant wood frog, *Rana sylvatica*, in the presence and absence of glucose. II. Mathematical modeling. *Cryobiology*, 38(4), pp. 327–338.

Devireddy, R. V. and J. C. Bischof (1998). Measurement of water transport during freezing in mammalian liver tissue: Part II — The use of differential scanning calorimetry. *Journal of Biomechanical Engineering*, 120(5), pp. 559–569.

Devireddy, R. V. and J. C. Bischof (2003). Recent advances in cryobiology using calorimetry 99. In S. Kakac, H. Smirnov and M. R. Mila (eds.) *Low Temperature and Cryogenic Refrigeration* (Kluwer Academic Publishers, Dordrecht, The Netherlands), pp. 265–294.

Devireddy, R. V., J. E. Coad and J. C. Bischof (2001). Microscopic and calorimetric assessment of freezing processes in uterine fibroid tumor tissue. *Cryobiology*, 42(4), pp. 225–243.

Devireddy, R. V., M. R. Neidert, J. C. Bischof and R. T. Tranquillo (2003). Cryopreservation of collagen-based tissue equivalents. I. Effect of freezing in the absence of cryoprotective agents. *Tissue Engineering*, 9(6), pp. 1089–1100.

Devireddy, R. V., D. Raha and J. C. Bischof (1998). Measurement of water transport during freezing in cell suspensions using a differential scanning calorimeter. *Cryobiology*, 36(2), pp. 124–155.

Devireddy, R. V., D. J. Smith and J. C. Bischof (1999c). Mass transfer during freezing in rat prostate tumor tissue. *AIChE Journal*, 45(3), pp. 639–654.

Diller, K. R. (1982). Quantitative low temperature optical microscopy of biological systems. *Journal of Microscopy*, 126(Pt 1), pp. 9–28.

Dresselhaus, M. S., G. Chen, M. Y. Tang, R. G. Yang, H. Lee, D. Z. Wang, Z. F. Ren, J. P. Fleurial and P. Gogna (2007). New directions for low-dimensional thermoelectric materials. *Advanced Materials*, 19(8), pp. 1043–1053.

Dresselhaus, M. S., G. Dresselhaus, X. Sun, Z. Zhang, S. B. Cronin and T. Koga (1999). Low-dimensional thermoelectric materials. *Physics of the Solid State*, 41(5), pp. 679–682.

Fennema, O. R., W. D. Powrie and E. H. Marth (1973). *Low Temperature Preservation of Foods and Living Matter* (Marcel Dekker, Inc., New York).

Gupta, M., D. Pinisetty, J. C. Flake and J. J. Spivey (2010). Pulse electrodeposition of Cu-ZnO and Mn-Cu-ZnO nanowires. *Journal of the Electrochemical Society*, 157(9), pp. D473–D478.

Gupta, M. and E. J. Podlaha (2010). Electrodeposition of CuNiW alloys: thin films, nanostructured multilayers and nanowires. *Journal of Applied Electrochemistry*, 40(7), pp. 1429–1439.

Hicks, L. D. and M. S. Dresselhaus (1993). Thermoelectric figure of merit of a one-dimensional conductor. *Physical Review B*, 47(24), pp. 16631–16634.

Hmina, N. and Y. Scudeller (1998). Thermal interface resistance and subsurface effusivity of submicron metallic films on dielectric substrates: an experimental method for simultaneous determination. *International Journal of Heat and Mass Transfer*, 41(18), pp. 2781–2798.

Ho-Young, S., J. Gi-Jo, L. Jun-Kyu, C. Joon-Young and P. Kyung-Wook (2007). Cu/SnAg double bump flip chip assembly as an alternative of solder flip chip on organic substrates for fine pitch applications. *Proceedings of the 57th Electronic Components and Technology Conference, 2007 (ECTC '07)*.

Huang, B. J. and C. L. Duang (2000). System dynamic model and temperature control of a thermoelectric cooler. *International Journal of Refrigeration*, 23(3), pp. 197–207.

Hubel, A. (1997). Parameters of cell freezing: implications for the cryopreservation of stem cells. *Transfusion Medicine Reviews*, 11(3), pp. 224–233.

Ilzycer, D., A. Sher and M. Shiloh (1983). Measurements of thermoelectric properties of a thermoelectric cooler. *Energy Conversion and Management*, 23(4), pp. 241–244.

Jittinorasett, S. (1999). *UBM Formation on Single Die/Dice for Flip Chip Applications*. Master of Science Thesis, Virginia Polytechnic Institute and State University.

Jong-Kai, L., J. Drye, W. Lytle, T. Scharr, R. Subrahmanyan and R. Sharma (1996). Conductive polymer bump interconnects. *Proceedings of the 46th Electronic Components and Technology Conference, 1996.*

Jordan, J. (2002). Gold stud bump in flip-chip applications. *27th Annual IEEE/SEMI International Electronics Manufacturing Technology Symposium, 2002 (IEMT 2002).*

Kardak, A. (2008). *Fabrication of Micro Bumps for Micro Scale Thermal Management*. Master of Science Thesis in Mechanical Engineering, Louisiana State University.

Karim, Z. S. and R. Schetty (2000). Lead-free bump interconnections for flip-chip applications. *26th IEEE/CPMT International Electronics Manufacturing Technology Symposium, 2000.*

Kim, H., M. C. Murphy and E. J. Podlaha (2012). Electrodeposition of high aspect ratio super invar microstructures. *Journal of the Electrochemical Society*, 159(9), pp. D549–D554.

Kloeser, J., P. Coskina, R. Aschenbrenner and H. Reichl (2002). Bump formation for flip chip and CSP by solder paste printing. *Microelectronics Reliability*, 42(3), pp. 391–398.

Landecker, K. (1970). Improvement of the performance of Peltier junctions for thermoelectric cooling. *Journal of Physics C: Solid State Physics*, 3(10), p. 2146.

Levin, R. L., E. G. Cravalho and C. E. Huggins (1976). A membrane model describing the effect of temperature on the water conductivity of erythrocyte membranes at subzero temperatures. *Cryobiology*, 13(4), pp. 415–429.

Li, L. and P. Thompson (2000). Stencil printing process development for flip chip interconnect. *IEEE Transactions on Electronics Packaging Manufacturing*, 23(3), pp. 165–170.

Lin, Y. M. and M. S. Dresselhaus (2003). Thermoelectric properties of superlattice nanowires. *Physical Review B*, 68(7), p. 14.

Lin, Y. M., X. Z. Sun and M. S. Dresselhaus (2000). Theoretical investigation of thermoelectric transport properties of cylindrical Bi nanowires. *Physical Review B*, 62(7), pp. 4610–4623.

Love, R. M. (1966). The freezing of animal tissue. In H. T. Meryman (ed.) *Cryobiology* (Academic Press, London), Chap. 7, pp. 317–405.

Lovelock, J. E. (1953). The haemolysis of human red blood-cells by freezing and thawing. *Biochimica et Biophysica Acta*, 10(3), pp. 414–426.

Lozano-Morales, A. and E. J. Podlaha (2008). Electrodeposition of NiCu-matrix nanocomposites using a rotating cylinder Hull cell. *Journal of Applied Electrochemistry*, 38(12), pp. 1707–1714.

Magri, P., C. Boulanger and J. M. Lecuire (1996a). Synthesis, properties and performances of electrodeposited bismuth telluride films. *Journal of Materials Chemistry*, 6(5), pp. 773–779.

Magri, P., C. Boulanger and J. M. Lecuire (1996b). Volumetric analysis for the control of bismuth and tellurium solutions and alloys. *Analysis*, 24(5), pp. 186–189.

Majumdar, A., J. P. Carrejo and J. Lai (1993). Thermal imaging using the atomic force microscope. *Applied Physics Letters*, 62(20), pp. 2501–2503.

Mansoori, G. A. (1975). Kinetics of water loss from cells at subzero centigrade temperatures. *Cryobiology*, 12(1), pp. 34–45.

Mazur, P. (1963). Kinetics of water loss from cells at subzero temperatures and the likelihood of intracellular freezing. *The Journal of General Physiology*, 47, pp. 347–369.

Mazur, P. (1970). Cryobiology: the freezing of biological systems. *Science*, 168(3934), pp. 939–949.

Mazur, P. (1984). Freezing of living cells: mechanisms and implications. *The American Journal of physiology*, 247(3 Pt 1), pp. C125–142.

Mazur, P., S. P. Leibo and E. H. Chu (1972). A two-factor hypothesis of freezing injury. Evidence from Chinese hamster tissue-culture cells. *Experimental Cell Research*, 71(2), pp. 345–355.

McGrath, J. J. (1988). Membrane transport properties. In J. J. McGrath and K. R. Diller (eds.) *Low Temperature Biotechnology: Emerging Applications and Engineering Contributions* (ASME Press, New York), pp. 273–330.

Nagata, H., T. Takahashi, N. Chikushi and T. Takenaka (2000). A series of bismuth layer-structured ferroelectrics. *Ferroelectrics*, 241(1–4), pp. 309–316.

Neidert, M. R., R. V. Devireddy, R. T. Tranquillo and J. C. Bischof (2004). Cryopreservation of collagen-based tissue equivalents. II. Improved freezing in the presence of cryoprotective agents. *Tissue Engineering*, 10(1–2), pp. 23–32.

Neugebauer, C. A. and J. A. Loughran (1988). Method of fabricating gold bumps on IC's and power chips, Google Patents.

Nuwayhid, R. Y. (2002). The essential physics of thermoelectricity for mechanical engineers. *International Journal of Mechanical Engineering Education*, 30(4), p. 351.

Oegema, T. R., Jr., L. B. Deloria, M. M. Fedewa, J. C. Bischof and J. L. Lewis (2000). A simple cryopreservation method for the maintenance of cell viability and mechanical integrity of a cultured cartilage analog. *Cryobiology*, 40(4), pp. 370–375.

Orain, S., Y. Scudeller, S. Garcia and T. Brousse (2001). Use of genetic algorithms for the simultaneous estimation of thin films thermal conductivity and contact resistances. *International Journal of Heat and Mass Transfer*, 44(20), pp. 3973–3984.

Panda, A. and E. J. Podlaha (2003). Nanoparticles to improve mass transport inside deep recesses. *Electrochemical and Solid State Letters*, 6(11), pp. C149–C152.

Paydenkar, C. S., F. G. Jefferson and D. F. Baldwin (2001). Flip chip assembly process development, reliability assessment and process characterization for polymer stud grid array-chip scale package. *Proceedings of the 51st Electronic Components and Technology Conference, 2001.*

Pazhayannur, P. V. and J. C. Bischof (1997). Measurement and simulation of water transport during freezing in mammalian liver tissue. *Journal of Biomechanical Engineering*, 119(3), pp. 269–277.

Pinisetty, D. and R. V. Devireddy (2010). Thermal conductivity of semiconductor (bismuth-telluride)-semimetal (antimony) superlattice nanostructures. *Acta Materialia*, 58(2), pp. 570–576.

Pinisetty, D., D. Davis, E. J. Podlaha-Murphy, M. C. Murphy, A. B. Karki, D. P. Young and R. V. Devireddy (2011a). Characterization of electrodeposited bismuth-tellurium nanowires and nanotubes. *Acta Materialia*, 59(6), pp. 2455–2461.

Pinisetty, D., M. Gupta, A. B. Karki, D. P. Young and R. V. Devireddy (2011b). Fabrication and characterization of electrodeposited antimony telluride crystalline nanowires and nanotubes. *Journal of Materials Chemistry*, 21(12), pp. 4098–4107.

Poleo, G. A., R. S. Denniston, B. C. Reggio, R. A. Godke and T. R. Tiersch (2001). Fertilization of eggs of zebrafish, Danio rerio, by intracytoplasmic sperm injection. *Biology of Reproduction*, 65(3), pp. 961–966.

Poleo, G. A., R. R. Godke and T. R. Tiersch (2005). Intracytoplasmic sperm injection using cryopreserved, fixed, and freeze-dried sperm in eggs of Nile tilapia. *Marine Biotechnology*, 7(2), pp. 104–111.

Prabhakar, A. (2006). *Modeling, Fabrication and Characterization of a Bio-Micro-Thermoelectric Device for Highly Localized Temperature Control.* Master of Science Thesis in Mechanical Engineering, Louisiana State University.

Prabhakar, A., M. C. Murphy and R. V. Devireddy (2007). Modeling of a Bio-Thermo-Electric Micro-Cooler. *The Open Biotechnology Journal*, 1, pp. 1–8.

Rall, W. F. (1992). Cryopreservation of oocytes and embryos: methods and applications. *Animal Reproduction Science*, 28(1–4), pp. 237–245.

Rosi, F. D., B. Abeles and R. V. Jensen (1959). Materials for thermoelectric refrigeration. *Journal of Physics and Chemistry of Solids*, 10(2–3), pp. 191–200.

Rowe, D. M. (1996). *CRC Handbook of Thermoelectrics.* (CRC Press, London).

Rowe, D. M. and G. Min (1996). Design theory of thermoelectric modules for electrical power generation. *Science, Measurement and Technology, IEE Proceedings*, 143(6), pp. 351–356.

Rowley, S. D. (1992). Hematopoietic stem cell cryopreservation: a review of current techniques. *Journal of Hematotherapy*, 1(3), pp. 233–250.

Rubinsky, B. and M. Ikeda (1985). A cryomicroscope using directional solidification for the controlled freezing of biological material. *Cryobiology*, 22(1), pp. 55–68.

Rubinsky, B., C. Y. Lee, J. Bastacky and T. L. Hayes (1987). The mechanism of freezing process in biological tissue. *CryoLetters*, 8, pp. 370–381.

Schrijnemakers, E. W. and F. Van Iren (1995). A two-step or equilibrium freezing procedure for the cryopreservation of plant cell suspensions. *Methods in Molecular Biology*, 38, pp. 103–111.

Sherman, B., R. R. Heikes and R. W. Ure (1960). Calculation of efficiency of thermoelectric devices. *Journal of Applied Physics*, 31(1), pp. 1–16.

Simon, J., E. Zakel and H. Reichl (1990). Electroless deposition of bumps for TAB technology. *40th Electronic Components and Technology Conference, 1990.*

Sun, X., Z. Zhang and M. S. Dresselhaus (1999). Theoretical modeling of thermoelectricity in Bi nanowires. *Applied Physics Letters*, 74(26), pp. 4005–4007.

Takahashi, M., M. Kojima, S. Sato, N. Ohnisi, A. Nishiwaki, K. Wakita, T. Miyuki, S. Ikeda and Y. Muramatsu (2004). Electric and thermoelectric properties of electrodeposited bismuth telluride (Bi2Te3) films. *Journal of Applied Physics*, 96(10), pp. 5582–5587.

Tomita, Y., M. Tago, Y. Nemoto and K. Takahashi (2000). Copper bump bonding with electroless metal cap on 3 dimensional stacked structure. *Proceedings of the 3rd Electronics Packaging Technology Conference, 2000 (EPTC 2000).*

Toner, M. (1993). Nucleation of ice crystals in biological cells. In P. L. Steponkus (ed.) *Advances in Low-Temperature Biology* (JAI Press, London), pp. 1–52.

Ueno, F., T. Motomura, H. Hirai, O. Shimada, Y. Sonoda and Y. Fukuoka (1998). New flip chip attach technology for fine pitch interconnections using electroplated copper bumps formed on a substrate. *2nd IEMT/IMC Symposium, 1998.*

Walsh, J. R. (1998). *Modeling and Design of Algal Cryopreservation Protocols.* Ph.D. Thesis, University of Texas at Austin, Austin, Texas.

Webb, R. L., M. D. Gilley and V. Zarnescu (1998). Advanced heat exchange technology for thermoelectric cooling devices. *Journal of Electronic Packaging*, 120(1), pp. 98–105.

Xuan, X. C. (2002). Optimum design of a thermoelectric device. *Semiconductor Science and Technology*, 17(2), p. 114.

Xuan, X. C., K. C. Ng, C. Yap and H. T. Chua (2002). Optimization and thermodynamic understanding of conduction-cooled Peltier current leads. *Cryogenics*, 42(2), pp. 141–145.

Yamada, H. and T. Togasaki (2003). Resin flow characteristics of underfill encapsulant for flip-chip interconnection. *IEEE Transactions on Components and Packaging Technologies*, 26(1), pp. 268–274.

Yang, R. G., G. Chen and M. S. Dresselhaus (2005). Thermal conductivity of simple and tubular nanowire composites in the longitudinal direction. *Physical Review B*, 72(12), p. 7.

Yong-Min, K., J. Young-Doo, P. Kyung-Wook, K. Jung-Do and L. Jin-Woo (2005). Assembly yield enhancement of electroless Ni-P UBM/Pb-free solder joints. *Proceedings of the 7th Electronic Packaging Technology Conference, 2005 (EPTC 2005)*.

Yoon, J.-B., C.-H. Han, E. Yoon and C.-K. Kim (1998). Novel two-step baking process for high-aspect-ratio photolithography with conventional positive thick photoresist. *Conference on Materials and Device Characterization in Micromachining*.

Chapter 7

Microdevices for Measurement of Cell Membrane Biophysical Properties

Gang Zhao, Zhongping Huang and Dayong Gao

1. Introduction

Cryopreservative agents (CPAs) have been widely used in cryopreservation of cells, tissues and organs along with their discovery in 1950 [Smith and Polge, 1950]. Although CPAs are usually indispensable for preventing cells from being damaged during freezing/thawing, the addition and removal of them before freezing and after thawing may cause osmotic injuries if inappropriate steps are applied (Figure 1). In order to avoid the osmotic injuries that may be introduced by CPA loading/unloading, theoretical optimization based on mathematical models is necessary. The Kedem–Katchalsky and the modern two-parameter (2-*p*) models are the most commonly used ones [Kedem and Katchalsky, 1958]. For solving of either of the models, the following biophysical parameters of cell membrane are necessary: (a) the cell membrane hydraulic conductivity (L_p); (b) the cell membrane permeability of CPAs (P_s); and (c) the activation energies of the two parameters (E_{Lp}, E_{Ps}). Knowing the cell membrane permeability properties will make it possible to develop optimal protocols for the addition and removal of CPAs [McGrath, 1997; Kleinhans, 1998; Woods *et al.*, 1999].

To quantify the cell membrane transport properties, various methods have been developed by means of nuclear magnetic resonance [Paganelli and Solomon, 1957], "stop flow" [Macey and Farmer, 1970], electron paramagnetic resonance [Kleinhans *et al.*, 1992], electronic sizing [Mcgann *et al.*, 1982; Gilmore *et al.*, 1995; Liu *et al.*, 1995], and light microscopy [Leibo, 1980; Mcgrath, 1985; Gao *et al.*, 1994; Gao *et al.*, 1996; Chen *et al.*, 2007;

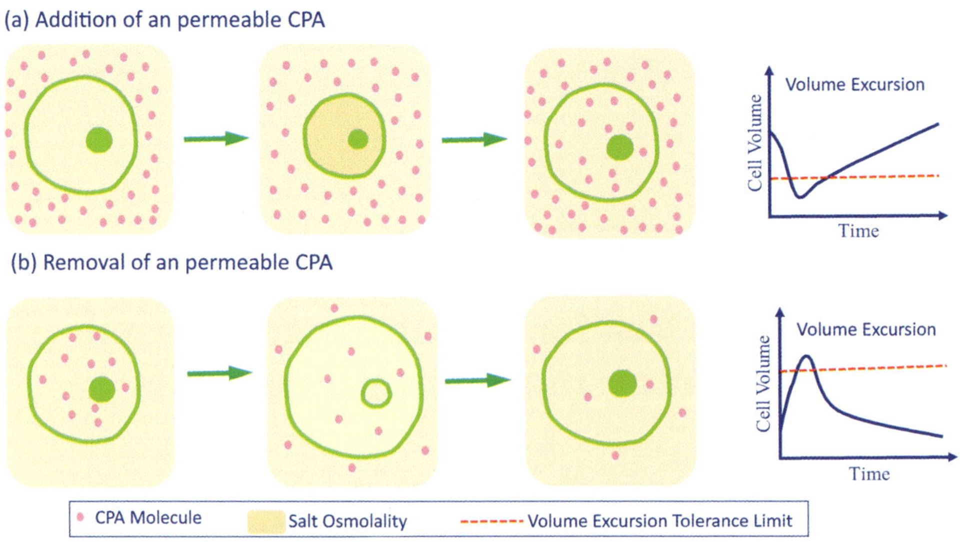

Figure 1. Cell volume responses during addition and removal of a permeable CPA.

Liu *et al.*, 2015]. Among all these methods, light microscopy is the only method that is applicable to determine membrane transport properties of both cases of a large population of cells and a single cell. This technique can be used to directly observe, trace, and analyze the volume changes of single cells in transient anisotonic environments.

In 1985, McGrath developed a microdiffusion system [McGrath, 1985] for investigation of cell osmotic behaviors which has been proved to be a very powerful tool in determining the membrane permeability of many different types of cells. This system belongs to the typical light microscopy method: it comprises a main channel for the bulk flow of desired solutions, a chamber which serves as a sample region, and a dialysis membrane, between the chamber and the channel, to immobilize the cells. Nevertheless, two potential problems may occur during using this system: (a) ripping the dialysis membrane unintentionally when the microdiffusion chamber is manufactured and (b) introducing additional mathematical complexity due to the presence of the dialysis membrane [McgGath, 1985; McGrath, 1997].

Gao *et al.* [1994] developed a micropipette technique to determine the membrane transport properties of mouse oocytes. In their experiment, individual mouse oocytes are held using a micropipette by applying a small negative pressure to the zona pellucida. The immobilized mouse oocyte is then directly perfused using another micropipette with an anisotonic solution. Using this technique, the permeability coefficients of water, dimethylsulfoxide (Me_2SO) and propylene glycol (PG) of mouse oocytes were determined. Although the micropipette technique simplifies the mathematical formulation and overcomes the problems related with the microdiffusion technique, the glass micropipette can only be used with cell types having "shells" (e.g. zona pellucida for oocytes). Moreover, each glass micropipette can hold only one cell during perfusion, and as a result very limited numbers of cells could be measured for each run. Combining the main elements of the microdiffusion chamber and the micropipette perfusion method, Gao *et al.* [1996] later presented a microperfusion chamber being capable of manipulating multiple or single cells, and different cell types (e.g. mouse oocytes, golden hamster pancreatic islet cells, etc.). Rather than being held by the micropipette, cells are immobilized on a transparent porous membrane (average pore size: 1–2 μm) at the bottom of the cavity and are then perfused instantaneously by the second solution, whose flow rate is controlled by a Hamilton Syringe. The chamber cavity and the porous membrane overcome the single cell limitation of micropipette perfusion and avoid the mathematical complexity of the microdiffusion method. However, in order to ensure the effectiveness of this device, the membrane pores have to be large enough and quantity sufficient to produce a fast extracellular environment change, and also small enough to prevent cells of interest from being squeezed out. That means customized porous membranes may be needed to immobilize different types of cells and meanwhile provide the ability to instantaneously replace the medium. On the other hand, washing the chamber and reassembling all parts together after the experiments makes this method labor intensive.

Takamatsu *et al.* [2004] developed a sandwich structured perfusion chamber to study the osmotic response of human prostatic adenocarcinoma cells (PC-3) being immobilized on a glass surface or a thin mesh cloth, which has the following advantages: (a) low cost and easy to make, and (b) precisely experimentally predetermined concentration of extracellular solution. It has only been successfully used to measure the hydraulic conductivity of cell membrane of several different types of cells [Takamatsu *et al.*, 2004], while no applications on cotransport of water and CPA across cell membrane were reported.

Microfluidic systems fabricated in poly(dimethylsiloxane) (PDMS) using soft lithography techniques have many advantages over conventional systems [Duffy *et al.*, 1998; McDonald *et al.*, 2000; Applegate *et al.*, 2004], and the microfluidic chips have been applied to the study of the osmotic behaviors of cells in recent years [Ateya *et al.*, 2005; Huh *et al.*, 2005; Chao *et al.*, 2006]. Chen *et al.* [Chen *et al.*, 2007; Chen *et al.*, 2008] developed a microfluidic perfusion system which allows for studies of cell osmotic behaviors, and by using it, they systematically studied mouse dendritic cell membrane transport properties of water and CPAs. The system is comprised of an inlet and an outlet (both made of silicone tube), a microfluidic channel (made from PDMS), and a slide class, which makes it cost effective and disposable for multiple usage.

Recently, according to the achievements of the former researchers, following the sandwich structure design of Takamatsu, Zhao designed a low-cost and high-precision microperfusion chamber for quantitative analysis of biophysical properties of cell membranes [Zhao, 2012]. Liu *et al.* [2015] manufactured a novel microperfusion chamber based on Zhao's design, and the measurements of the cell membrane permeabilities of spodoptera-frugiperda ovarian cells (Sf21 cells) and the pig iliac endothelial cells (PIECs) were successfully accomplished. In this system, the thermocouple was embedded in the microchannel of the chamber during manufacturing, and the concentration profile for the change of the extracellular solution during an osmotic shift process was carefully determined according to both experimental examination and computational finite element analysis. Due to these two evident improvements, the precision of this system has been greatly enhanced compared to the above mentioned traditional ones.

This chapter gives an introduction to the commonly used microfluidics based perfusion systems [Chen *et al.*, 2007; Chen *et al.*, 2008; Liu *et al.*, 2015] and the outline for their developments. It covers both the instruments and their principles, followed by a discussion of the data processing approaches as well as examples of typical applications on cell osmotic behaviors.

2. Membrane Transport Theory

2.1. *Three Parameter Equations (Kedem–Katchalsky Equations)*

The three parameter formalism is developed by Kedem and Katchalsky [1958]. This formalism assumes the solute and water transport across the cell membrane through a cotransport channel. A reflection coefficient σ is used to describe the degree of interaction between the solute and solvent. Therefore, in the Kedem–Katchalsky equations, three parameters are utilized to characterize the membrane permeability [Elmoazzen *et al.*, 2009]: the water permeability (hydraulic conductivity, L_p); the solute mobility (ω); and the reflection coefficient (σ). The solute mobility is expressed as a solute permeability $P_s = \omega RT$, where R is the universal gas constant and T is absolute temperature. The rate of total volume change of the cell was described in Elmoazzen *et al.* [2009] as follows:

$$\frac{dV_{w+s}}{dt} = -L_p ART\left\{\left(C_n^e - C_n^i\right) + \sigma\left(C_s^e - C_s^i\right)\right\} \quad (1)$$

where A is the surface area of the cell, w refers to water, s refers to solute, the value C is the molarity with the superscripts denoting the internal cell solution (i) and the solution external to the cell (e) and the subscripts denoting the nonpermeating solutions (n) and the permeating solutes (s). The change in intracellular, permeating solute is given as [Elmoazzen *et al.*, 2009]:

$$\frac{dN_s}{dt} = (1-\sigma)\left(\frac{1}{2}\right)\left(C_s^e + C_s^i\right)\frac{dV_{w+s}}{dt} P_s A\left(C_s^e - C_s^i\right) \quad (2)$$

$$0 \leq \sigma \leq 1 - \frac{P_s v_s}{RTL_p} \quad (3)$$

where N_s is the number of moles of solute in the cell, v_s is the partial molar volume of solute.

2.2. *Modern Two-Parameter Equations*

Kleinhans [1998] pointed out the reflection coefficient σ in the K–K formalism was often unnecessary as the two parameter (2-P) formalism worked just as well as the K–K formalism for a number of different transport situations including (a) bilayer transport in which the solute and water diffuse across the bilayer; (b) transport in which water only moves through a selective channel and the solute diffuses across the bilayer; and (c) transport in which the water and solute use separate channels.

The 2-P formalism is developed by Jacobs and Stewart [Jacobs and Stewart, 1932; Jacobs, 1933]:

$$\frac{dV_w}{dt} L_p RT\left(\pi^e - \pi^i\right) \quad (4)$$

where L_p ($m^3N^{-1}s^{-1}$) is the membrane hydraulic conductivity and is a measure of the rate of water movement across the cell membrane, R is the universal gas constant, T is the absolute temperature, π^e is the extracellular solution osmolarity, and π^i is the intracellular solution osmolarity. The solute flux is described as:

$$\frac{dN_s}{dt} = P_s A\left(C_s^e - C_s^i\right) \tag{5}$$

where P_s (cm/s) is the solute permeability and is a measure of the rate of solute permeability across the cell membrane, C_s is the solute molarity, superscript e indicates the extracellular condition and i indicates the intracellular condition.

3. Microperfusion Systems

3.1. *Traditional Measurement Systems*

To quantify the membrane transport properties, many measurement systems are developed. Figure 2(a) is the macroscopic stopped-flow system made by Solomon and his colleagues [Sidel and Solomon, 1957]. A volume of cells is mixed quickly with a second solution to make the final solution concentration of interest. After mixing, the flow is stopped and the cells enter an optical

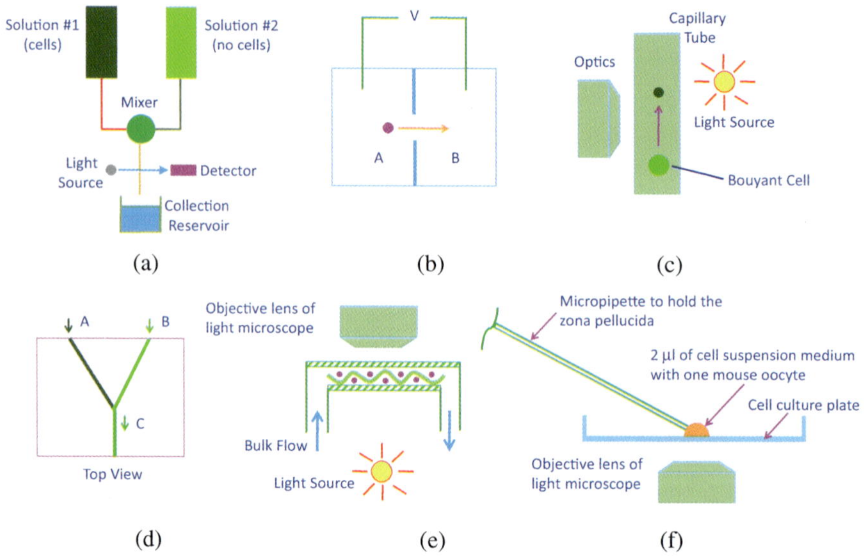

Figure 2. (a) A macroscopic stopped-flow system [Sidel and Solomon, 1957]. (b) an electronic particle size counter [Kubitschek, 1958]. (c) photomicroscopic method [Leibo, 1980]. (d) microscopic stopped-flow method [Diller and Bradley, 1984]. (e) Perfusion and freezing cryomicroscope system [Walcerz and Diller, 1991]. (f) micropipette perfusion method [Gao *et al.*, 1996]. The figures were redrawn from the references.

interrogation area. Cell volume is correlated with light scattering or absorption so that the dynamics of cell volume changes are detected by a light-sensitive device. This method has the advantage that if it is set up properly, the change in extracellular concentrations is fast enough so as to be assumed instantaneous relative to the cell volume response. However, the membrane transport properties determined with this method are population-averaged values because the transient light signal measured is a consequence of the response of many cells interacting with the incident light beam. That is to say, determining the properties of individual cells with this method is not possible. Figure 2(b) shows the electronic particle size counter [Kubitschek, 1958]. The orifice separates two liquid solutions which are electrically conducting. The orifice itself has a characteristic dimension of the order of the cell characteristic dimension so that as a cell passes through the orifice it distorts the electric field. The relationship between cell volume and electronic signal is defined by calibration. This method has many features in common with the method in Figure 2(a). If it is set up properly, the change in extracellular concentrations may be assumed instantaneous relative to the cell volume response. However, the membrane transport properties determined with this method are population-averaged values because the transient electronic signal measured is a consequence of the response of many cells passing through the orifice. In contrast to systems in Figures 2(a) and (b) which require relatively large volumes and large numbers of cells, systems in Figures 2(c)–(f) are developed to determine the membrane transport properties of individual cells. Figure 2(c) is the photomicrosopic method developed by Leibo [1980] to measure the hydraulic conductivity of the individual mouse oocyte. A capillary tube filled with hypertonic sodium chloride solution is aligned with the gravitational field. A light microscope is rotated to allow imaging of the individual oocytes in isotonic NaCl solution. Mouse oocytes in isotonic NaCl solution are less dense than the hypertonic solution in the capillary tube so that when an individual oocyte is introduced into the bottom of the capillary tube it rises while it dehydrates and shrinks. The cell volume history is quantified by photographing the oocyte and calculating volume assuming sphericity. This method is also a one-shot method since altering the extracellular solution is impractical. Figure 2(d) is the microscopic stopped-flow method by Diller and Bradley [Diller and Bradley, 1984], which is capable of producing data where cell numbers are limited. This device is designed in a thin sandwich arrangement. Two fluid streams are mixed to produce an instantaneous step change of extracellular concentration and then the flow is stopped. In contrast to system in Figure 2(a), this system allows imaging of single cells. This allows data for individual cells to be obtained, making it feasible to produce information about distribution of cell membrane transport properties in a population. However, mixing at the interface (near location c) was not adequate so that only cells very close to the interface between the two streams could be assumed to experience an "instantaneous" step change. Figure 2(e) shows the perfusion and freezing cryomicroscope system made by Walcerz and Diller [1991]. This system immobilizes the cells by a mesh between two transparent sheets which allows virtually instantaneous step changes of extracellular concentrations. This approach has made it possible to perfuse, freeze, and thaw islets in a controlled manner. However, the manufacture and control of this unique device is complex. Figure 2(f) is the micropipette perfusion method introduced by Gao and his colleagues [Gao *et al.*, 1996]. Individual oocytes are immobilized using a micropipette to apply a small suction pressure to the zona pellucida of an oocyte. The advantages of this method include the fact that it is of simple construction and that it is applicable to individual cells. Disadvantages include that it requires a holding pipette and that the cell must have a zona pellucida or other structure that provides the means of "grabbing" the cell while at the same time not distorting the cell or otherwise introducing artifacts in the osmotic behavior or its interpretation.

3.2. *Membrane Based Perfusion System*

As shown in Figure 3(a), McGrath developed a microdiffusion chamber [McGrath, 1985] with a desire to measure membrane transport properties of scarce materials. Cells in an initial solution are placed in the sample region, which is formed between a cover glass and a dialysis membrane. A bulk flow of this initial solution is driven by a gravity feed or pump beneath the dialysis membrane. The dialysis membrane serves to immobilize the cells by absorbing the shear stress associated with the bulk flow beneath the dialysis membrane. Constituents of new bulk flow solution begin to equilibrate with the initial solution in the sample region by a diffusive process. This causes cells to respond osmotically but they remain essentially stationary and excellent optical quality is possible for recording the dynamic response. This method makes it possible to quantify the dynamic osmotic responses of small numbers of immobilized cells contained in small volumes before, during, and after osmotic perturbation. The primary disadvantages of this method are associated with the dialysis membrane. One class of problems occurs during the time the sample region is formed just prior to running the experiment. The other issue is associated with mathematical complexity due to the presence of the dialysis membrane. Care must be taken in handling the dialysis membrane

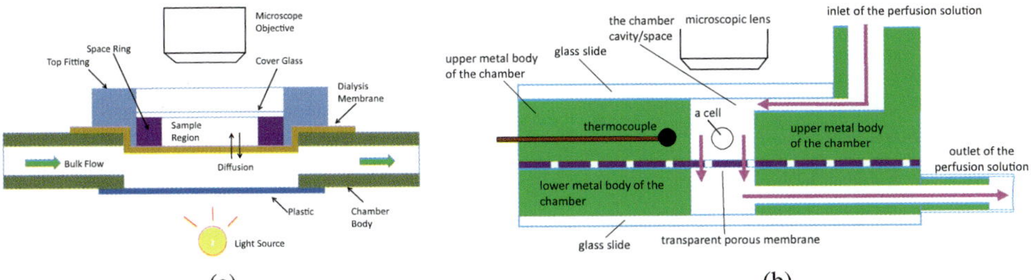

Figure 3. (a) Microdiffusion chamber [McGrath, 1985]. (b) microperfusion chamber [Gao *et al.*, 1996]. The figures were redrawn from the references.

when the cells are placed between the top cover glass and the dialysis membrane to form the sample region. Problems that may arise include ripping the dialysis membrane, introducing air bubbles into the sample region or losing the small number of cells from the field of view by lateral movement of cells when the dialysis membrane sample region is formed.

To measure cell volume changes as a function of time, four technical requirements must be satisfied: (a) The change of the extracellular condition must be feasible, rapid and controllable; (b) the cell must be held at a fixed location under the microscope during the change of the extracellular medium. (c) the cell response to the change of extracellular condition must be measurable and resolvable; (d) the temperature of the cell suspension must be controllable. The microperfusion chamber developed by Gao and his colleagues [Gao *et al.*, 1996] meets all these requirements. The design is shown in Figure 3(b). In experiments, a cell suspension is first loaded at the solution inlet and then aspirated into the chamber cavity (height: 2 mm, diameter: 1 mm, volume: 1.6 μl) using the Hamilton Syringe. With continued aspiration, part of the cell suspension solution moves out of the chamber, but the cells remain on the transparent porous membrane at the bottom of the cavity. A perfusion medium is then loaded at the solution inlet and made to pass the cell or cells by aspiration. The perfusion flow rate is controlled using the Hamilton Syringe and by adjusting the rpm of a DC motor connected with the Hamilton Syringe. The total volume of the medium used to perfuse the cells can be measured by reading the scales on the Hamilton Syringe. Based on the Hamilton Syringe reading, the original isotonic solution in the perfusion chamber cavity is quickly replaced by the new perfusion medium within 0.5 s. The small size of the chamber cavity and downward flow of the perfusion medium immobilizes the cells in the chamber during the perfusion process. The cells are continuously perfused for another 30 s. Over 90 μl of the perfusion medium runs through the cells during the perfusion. Before, during, and after the perfusion process, the history of cell volume changes is recorded by a video camera until osmotic equilibrium is achieved.

The chamber cavity and porous membrane overcome the single cell limitation of micropipette perfusion and avoid the mathematical complexity of the microdiffusion method. However, in order to assure the effectiveness of this device, the membrane pores have to be large enough and or numerous enough to produce a fast environment change, yet also small enough to prevent cells of interest from squeezing out. That means customized porous membranes may be needed to immobilize different cell types and still provide the ability to instantaneously replace the medium. On the other hand, washing the chamber after use and reassembling all parts together for each use makes this process labor intensive.

3.3. *Microfluidics-based Direct-Perfusion Systems*

3.3.1. *Sandwich structured microperfusion chamber*

The setup of the sandwich structured microperfusion chamber [Takamatsu *et al.*, 2004] is shown in Figure 4(a). The chamber is designed in a thin sandwich arrangement between two slide glasses (76 mm × 52 mm × 1.3 mm, S9213, MATSUNAMI). A 50-mm thick silicone rubber sheet with a cutout 40 mm long and 4 mm wide is sandwiched, providing a chamber of the cutout size. The bottom glass plate has two holes for the inlet and outlet of the solution at both ends of the chamber. A temperature-controlled stage that is 12.5 mm in thickness, which includes a thin film heater (Minco Products, Inc.), a flow channel for the coolant and a thermocouple for temperature measurement, is attached on top of the glass plate. The temperature-controlled stage has a 10-mm-diameter hole at the center for observation by transmitted light. These components are mounted on the stage of an inverted microscope (TE2000-U, Nikon). In the experiment using cells, the top slide glass is covered after putting cell suspension in the chamber, and the rest of the components are then assembled on the microscope stage. Within about ten minutes of setting up the devices, isolated cultured cells, including relatively less adhesive cells such as cancer cells, attached lightly on the glass surface. Therefore, many cells did not move against the solution flow at a low velocity. To avoid washing out of cells from the chamber, a thin mesh cloth (350S, Asahi Sakai Corp.), which is woven using a nylon fiber 30 mm in diameter with about 40-mm square meshes, is optionally used and set in the chamber to immobilize cells.

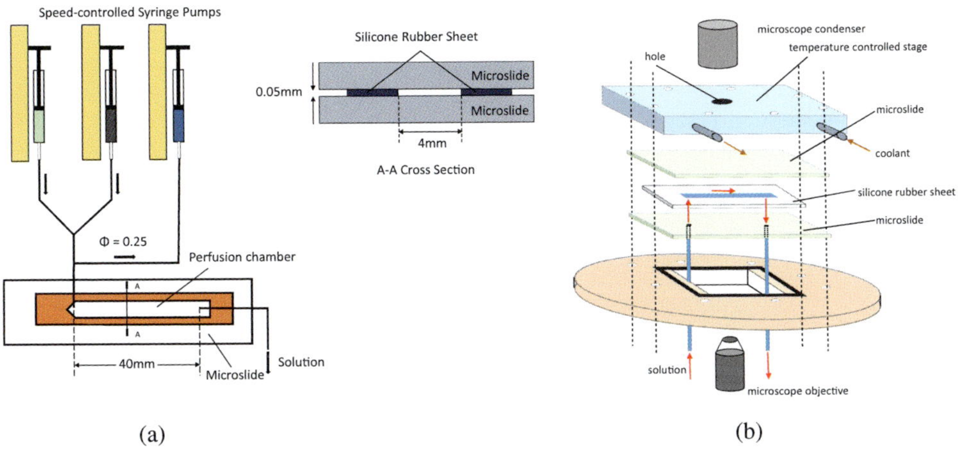

Figure 4. (a) Schematic of the perfusion microscope system. (b) Schematic showing the structure of the perfusion chamber mounted on the microscope stage. The figures were redrawn from [Takamatsu *et al.*, 2014].

Figure 4(b) shows the schematic of the experimental setup. The solution is supplied to the perfusion chamber using two variable-speed syringe pumps (IC3210, KD Scientific). One syringe (5 ml, Hamilton Co.) is filled with an isotonic solution and the other with a hypertonic solution. If we intend to increase the tonicity in the chamber abruptly, we switch the syringe pump from that of the isotonic solution to that of the hypertonic solution. Returning to the isotonic solution is achieved in a similar manner. If we intend to increase the concentration gradually at a constant rate, we supply the isotonic and the hypertonic solutions at the same time with different flow rates that change with time; we decrease the flow rate of the isotonic solution and increase that of the hypertonic solution linearly with time, keeping the total flow rate constant. The syringes and the chamber are connected by capillary tubes (TYGON Tube, Saint-Gobain PPL Corp.) with inner diameter of 0.25 mm. The flow rate of the solution, which is set at a velocity of about 1 mm/s in the perfusion chamber, is 7 ml/min. Due to a considerably low flow rate of the solution, it takes about 70 s for the solution to reach the center of the chamber from the junction of tubes from the two syringe pumps.

The actual change in concentration was examined using a laser interferometer. The perfusion chamber was removed from the microscope stage and set up in a Michelson–Morley interferometer, as shown in Figure 5. In this case, the perfusion chamber was composed of 1.1-mm-thick Pyrex glass

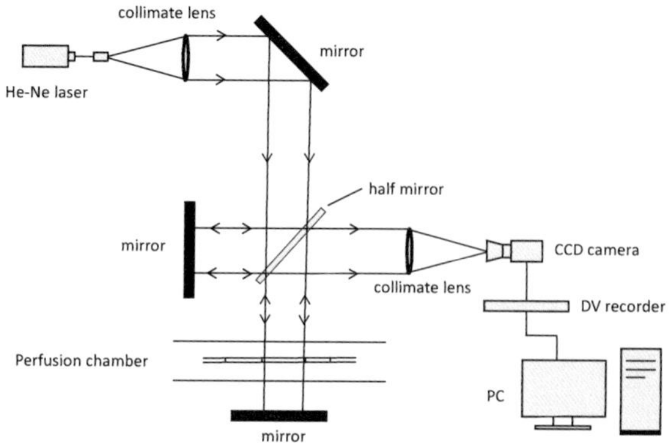

Figure 5. Schematic of the laser interferometer for the measurement of NaCl concentration change in the perfusion chamber. The figures was redrawn from [Takamatsu *et al.*, 2004].

plates. A He-Ne laser beam (wave length: 632.8 nm) is split in two parts: one proceeds through the perfusion chamber and returns through the same path, the other proceeds through a different optical path, and recombined. The interference pattern is recorded using a CCD camera (CS5110C, Tokyo Electronic Industry Co.) and a digital video recorder (TRV17K, Sony). Having the chamber perfused with the isotonic solution, the half mirror was adjusted to obtain the fringe pattern that is nearly parallel to the flow. After the experiment, the recorded motion video was captured on PC at 0.5-s or 5-s intervals, and analyzed using image processing software (SWALLOW 2001DV, Digimo). The concentration of NaCl solution measured using the laser interferometer confirmed that the concentration was acceptably uniform in the cross direction irrespective of the manner of change in concentration, and that the concentration changes in replicate experiments agreed well. The actual change in concentration agreed fairly well with the intended change produced by controlling two syringe pumps, when the concentration was changed linearly with time. The highest rate of increase or decrease in NaCl concentration produced in the present setup, which is achieved during the abrupt change, was also determined by the measurements.

The advantages of Takamatsu *et al.* [2004] microperfusion chamber are: (a) cost-effectiveness; (b) easier fabrication; (c) precise concentration collection. However, it also has several disadvantages, including (a) limited observation area; (b) incovenient sample loading process; (c) imprecise temperature collection as the measurement is made outside the microperfusion chamber; (d) inconvenience brought by concentration setting using laser interferometer before experiments.

3.3.2. *PDMS-based microperfusion chamber*

Using soft lithography, Chen *et al.* [2007, 2008] developed a microfluidic perfusion system. As shown in Figure 6, the microfluidic perfusion system immobilizes single cells while preventing them from forming multiple layers by restricting the volume available above them. In region I, the area encompassed by dotted lines represents the microfluidic perfusion chamber: cells flow from inlet (left in this diagram) to outlet (right) and are trapped by a sieving structure, while the fluid keeps flowing downward through the bottom of the microchannel, h_2 confines the movement of cells. To allow cells to flow freely from the inlet to the chamber, h_2 should be larger than the diameter of cells (Φ) but less than $2 \times \Phi$ (to maintain a monolayer of cells in the channel). The height of the slit (h_1), between the bottom of the channel and the block, should be smaller than Φ. It is optimal to make $h_1 < \Phi/2$ in order to prevent cells from squeezing into the slot. Since Φ ranges from 10 μm to 15 μm, the dimensions for h_1 and h_2 were chosen to be 5 μm and 20 μm, respectively, and the channel width was chosen to be 150 μm wide. In region II, a chamber is formed to allow bulky liquid to flow through (2 liters per minute), in order to locally control the temperature.

A three-dimensional schematic overview of the experimental setup is shown in Figure 7. The microfluidic perfusion chamber device, made in PDMS and irreversibly sealed on a glass slide, lying on top of the temperature-controlled chamber is mounted on the stage of an inverted microscope

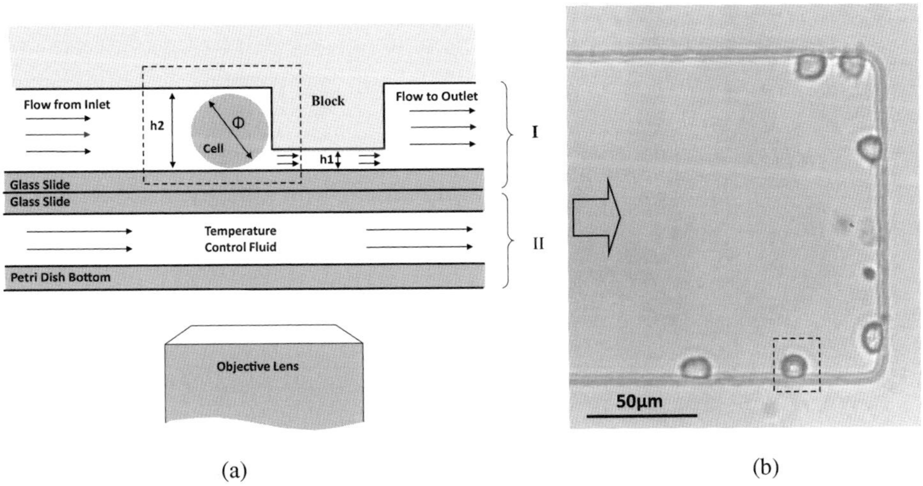

Figure 6. (a) Schematic diagram of the microfluidic perfusion chamber. (b) mDCs trapped within the perfusion chamber. Reproduced from [Chen *et al.*, 2008] with permission.

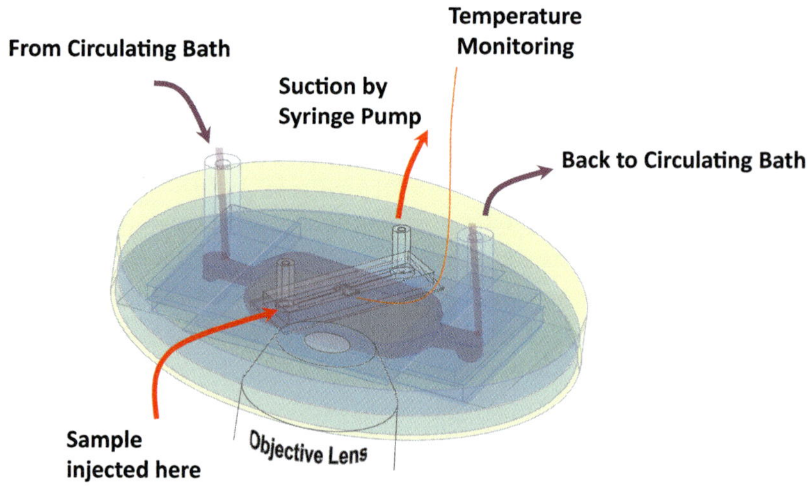

Figure 7. 3D schematic overview of the experimental setup. Cell suspension and prepared medium with different concentrations are injected into inlet while fluid is sucked continually and automatically from outlet using Hamilton Syringe, controlled by a DC motor. Cell volume response is observed using an inverted microscope throughout the experiment. The local temperature in the neighboring area of the microfluidic perfusion chamber is dominated by the bulky fluid underneath and monitored by a thermocouple. Reproduced from [Chen *et al.*, 2008] with permission.

(TE2000-S, Nikon, and Tokyo, Japan). In order to perform continuous aspiration, a Hamilton Syringe, controlled by a DC motor (out of range and not shown here), was connected through a silicone tube to the outlet of the device.

This microfluidic perfusion system made by Chen *et al.* [2008] is cost-effective and disposable for multiple usages, and it is easy to customize for the needed size and geometry of the chamber with just a change of the photomask. However, the fabrication is difficult, and the temperature and concentration collection is not precise. Therefore, the following sandwich structured microperfusion chamber is developed.

3.3.3. *Improved sandwich structured perfusion chamber*

Based on the microperfusion chamber of Takamatsu *et al.* [2004], Zhao [2012] designed an improved sandwich structured perfusion microscope, and Zhao's group [Liu *et al.*, 2015] manufactured the improved microperfusion chamber. The structure of the microperfusion system is shown in Figure 8. The chamber is designed as a sandwich unit (Figure 8(a)), where the silica gel thin film (100 μm in thickness) with a microchannel is sandwiched between

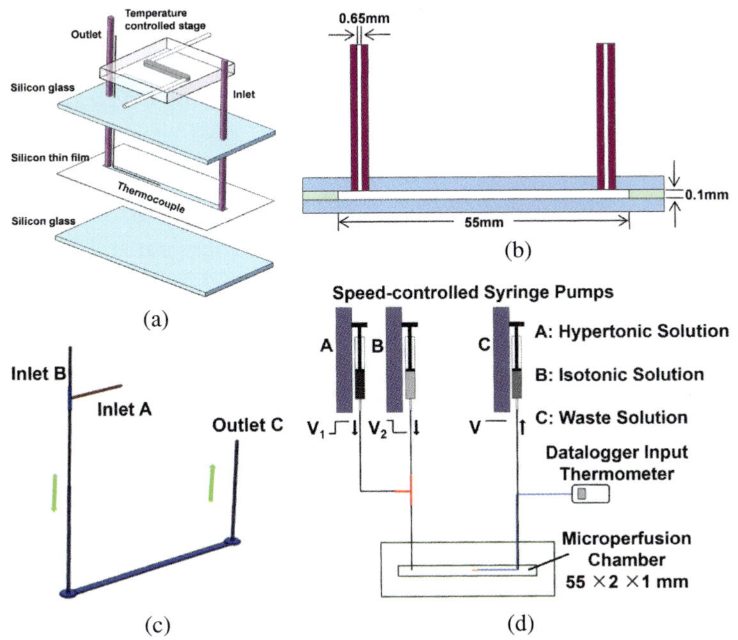

Figure 8. The microperfusion chamber. (a) Schematic showing the arrangement of the major components. (b) Side elevation of microperfusion chamber. (c) The calculation domain extracted from both the capillary tubes and the microchannel. (d) Schematic of the microperfusion system. Reproduced from [Liu *et al.*, 2015] with permission.

two pieces of slide glass. The microchannel is ~55 mm long and ~2 mm wide (Figure 8(b)). Two small holes with diameter of ~1 mm are drilled in the upper slide glass, corresponding to the two ends of the microchannel, which are designed for insertion of the inlet and outlet of the solution.

In this design, there was a narrow groove on the lower slide glass, which was used for locating the thermocouple (TT-K-36, Omega, USA). Moreover, all the capillary tube of the microperfusion system was replaced by Teflon tube with a much thicker wall (1.2 mm), so that the pressure fluctuation of the flowing liquid was avoided. Lastly, being different from all the previous studies, the concentration profile for the extracellular solution during an osmotic shift was carefully calibrated according to both experimental examination and computational fluid dynamics-based finite element analysis on the fluid flow in both the capillary tube and the microchannel. Between the inlet and outlet was a temperature-controlled stage connected to a water bath with constant temperature. A 40 × 5 × 10 mm rectangular chamber was located on the stage for coolant flow and observation (Figure 8(a)), covered by a thin glass. These components were tightly clamped using a file clip and then the assembled structure as a whole was placed on an inverted microscope (ECLIPSE Ti-U, Nikon, Japan).

There were three programmable syringe pumps (Legato 110, KD Scientific, USA) in the microperfusion system, two of which were used to inject the perfusion solution, and the third one was used to drain the waste solution and to close the system for a stable flow. An isotonic solution was injected as follows: 15 μL/min for 4 min, and then paused for 12 min. In contrast, a hypertonic solution was injected as follows: paused for the first 4 min, and then injected at speed of 15 μL/min for 12 min. A polytetrafluoroethylene (PTFE) inlet tube was connected to two speed-controlled syringe pumps through a "T" type three-way tube. The whole pipeline was airtight and the PTFE tube had a thick, rigid wall to ensure that the fluid flow was stable throughout the entire system (Figure 8(d)).

The improved sandwich structured perfusion chamber [Liu *et al.*, 2015] is cost-effective. The cell loading and removing are conducted by specified channels. Temperature collection is precise as the thermocouple is just embedded in the microchannel during manufacturing. Concentration collection precision is assured by finite element analysis (FEA) simulation with a preliminary calibration.

4. Theory for Solution Mixing in Microperfusion Chamber

4.1. *Finite Element Analysis of Transient Concentration Field in the Microchannel*

A theoretical model was developed to study the mixing process of different concentrations of solutions in the microchannel. The typical convection-diffusion theory under the conditions of dilute aqueous solution was applied. The control-volume of the model includes the continuous fluid in the "T" type three-way tube, the microperfusion chamber, the inlet and the outlet of chamber. The convection-diffusion model is a multiphysical field weak coupling problem, including two physical processes: mass migration and fluid flow. The mass migration physical process was described by the convection-diffusion equation [Li *et al.*, 2012]:

$$\frac{\partial c_s}{\partial t} + \nabla \cdot \left(-D_s \nabla c_s\right) + \mathbf{u} \cdot \nabla c_s = R_s \tag{6}$$

where c_s is solute concentration, t is the time, D_s is the solute diffusion coefficient [Philibert, 2005], $\mathbf{u}$ is the distribution of the velocity vector in the velocity field and R_s is the source term, which could be described by the change of solute concentration in the field. In this model, the Peclet number is very small, which means diffusion dominates the mass migration; therefore the equation could be solved well by FEA. Here, the flow in the microperfusion chamber is a single-phase laminar flow, so we can use the incompressible Navier–Stokes (NS) equation to describe it as follows:

$$\rho\frac{\partial \mathbf{u}}{\partial t}+\rho(\mathbf{u}\cdot\nabla)\mathbf{u}=\nabla\cdot\left[-p\mathbf{I}+\mu\left(\nabla\mathbf{u}+(\nabla\mathbf{u})^{T}\right)\right]+\mathbf{F} \tag{7}$$

$$\rho\,\nabla\cdot\mathbf{u}=0 \tag{8}$$

Equation (7) is the momentum conservation equation for the laminar flow and Eq. (8) is the mass conservation equation, where ρ and μ are the density and viscosity of the fluid, ρI means the pressure in X, Y or Z axis, and F is the body force, T is transposition operator. The second term on the left hand side of Eq. (7) is the inertial term and it cannot be omitted because the flow has a particular inertia under the microscope.

A dilute aqueous solution is assumed when solving the equations. A dilute aqueous solution means the concentration is so dilute that no matter how unevenly distributed, the properties of the solution (such as the fluid's density and viscosity) will not vary. In this case, the calculation of mass migration and fluid flow are independent of one another and the fluid field has a one-way effect on mass migration. The NS equation can be solved first, and then the results can be substituted to the mass migration equation to obtain the concentration distribution. Considering the boundary conditions for the Navier–Stokes equation, at the inlets and outlets:

$$\mathbf{u}=-u_i\,\mathbf{n} \tag{9}$$

where u_i follows the real injection speed of the syringe pumps respectively. At the walls, no slip conditions state that the velocity is zero:

$$\mathbf{u}(u,v,w)=0. \tag{10}$$

For mass migration, the assumption that solute molecules interact only with water molecules implies that it is sufficient to consider a single solute of concentration c_s. At the inlets, the concentration boundary conditions were assumed as the values of isotonic (or hypotonic) solution and hypertonic solution, respectively. At the outlets, the outflow boundary condition was set as convective flux condition, stating that the diffusive transport normal to the boundary is negligible:

$$-n\cdot D_i\nabla c_i=0. \tag{11}$$

which means the concentration gradients in the flow direction are eliminated.

As the typical properties of the fluid, the dynamic viscosity and the diffusion coefficient should be predetermined before solving Eq. (6). The electrical conductivity method [Xiao *et al.*, 2009] based on the finite diffusion theory was used to measure the diffusion coefficient. The diffusion coefficient was reliable, according to Gu and Wu [Wu *et al.*, 1990; Gu *et al.*, 2002]. A single cylinder rotary viscometer (SNB-2, Shanghai Jingtian Electronic Instrument Co., Ltd., China) was used to determine the dynamic viscosity.

To ensure the accuracy of the simulation used by the software and obtain a general understanding of the solution concentration change, a color solution ($CuCl_2$) was chosen to simulate the actual experiment, and the cell-free experimental results obtained by optical image processing compared with simulation results. The fundamental parameters of $CuCl_2$ solution used in the verification model were set as follows: diffusion coefficient $D = 1.02 \times 10^{-5}$ cm^2/s, dynamic viscosity $\mu_h = 1.81$ mPa·s and $\mu_l = 1.22$ mPa·s with the concentrations 2.963 and 0.593 mol/L, respectively, at the temperature of 25°C. Before doing the experiment with $CuCl_2$, the pictures of $CuCl_2$ solutions with different concentrations in microperfusion chamber were recorded. The grayscale of the picture was found to have a linear relationship with concentration (Figure 9(c)). Therefore, concentration during the $CuCl_2$ experiment could be inferred from the grayscale. During the cell-free experiments, the $CuCl_2$ solution with low concentration, $C_l = 0.5926$ mol/L, was used to mimic the isotonic solution, while the $CuCl_2$ solution with high concentration, $C_h = 2.9630$ mol/L was used to mimic the hypertonic solution. Then the concentration profile was measured upon a typical osmotic shift and used as the experimental result.

Simulation results were obtained using a model created to match the size of the actual microperfusion system and the boundary conditions were close to reality. The concentration profiles of the $CuCl_2$ solution are shown in Figures 9(a) and 9(b). The concentration change at the mid-position 1 of the channel was compared between the two methods (Figure 9(e)). According to this comparison, the simulation results produced by the FEA model were valid, although there were slight disagreements between two sets of data. It is likely that many details such as the junction of flexible pipes, the elasticity of the rubber hose, the frictional forces at the tube wall, the junction of the inlet and chamber, and other factors potentially influencing the results were omitted in the theoretical model.

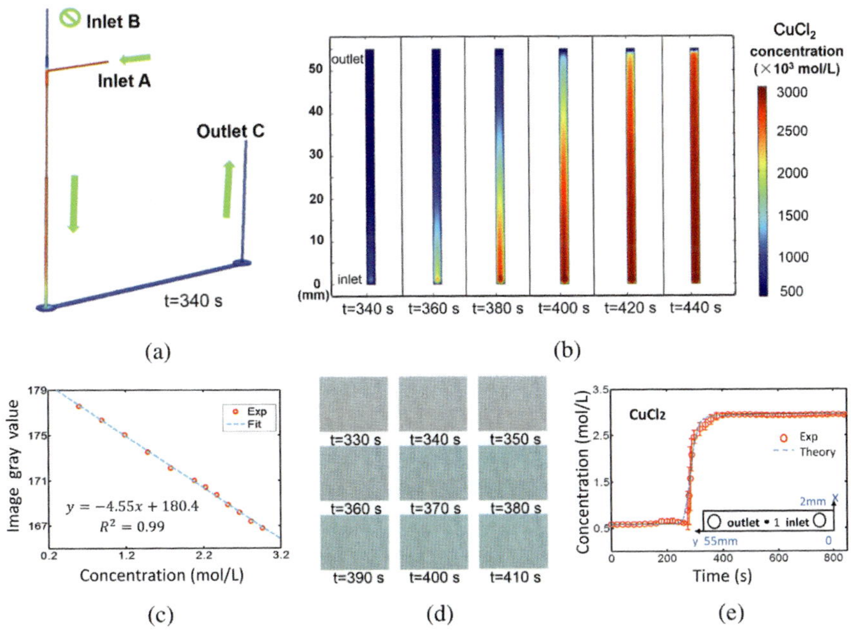

Figure 9. Determination of the concentration profile in the microchannel. (a) A three-dimensional plot of concentration distribution at t = 340 s. (b) A sectional view of concentration change in $CuCl_2$ solution in the microchannel. (c) The relationship between the image grey value and the concentration. (d) The sequential photos of the $CuCl_2$ solution. (e) Simulated versus measured concentration profiles for the $CuCl_2$ solution at the typical point (point 1) ($R^2 = 0.997$). Reproduced from [Liu *et al.*, 2015] with permission.

4.2. *Concentration Profile for Extracellular Solution during Osmotic Shift*

4.2.1. *Simulated concentration profile*

The microperfusion chamber model can be used to simulate the solution concentration change according to the model validation. The simulation of extracellular solution concentration was divided into two kinds of solution simulation: NaCl solution and DMSO solution. The parameters of the model can also be altered because of the diffusion coefficient ($D_{\text{NaCl}} = 1.25 \times 10^{-5}$ cm²/s and $D_{\text{DMSO}} = 1.03 \times 10^{-5}$ cm²/s) and dynamic viscosity of the solution. The dynamic viscosity values are shown in Figure 10 (the viscosity of 0.154 mol/L NaCl is used to approximate the 1× PBS).

The concentration (C_e and $C_{e,CPA}$) changes at different temperatures are shown in Figures 10(a)–(b). There is almost no distinction between the effects

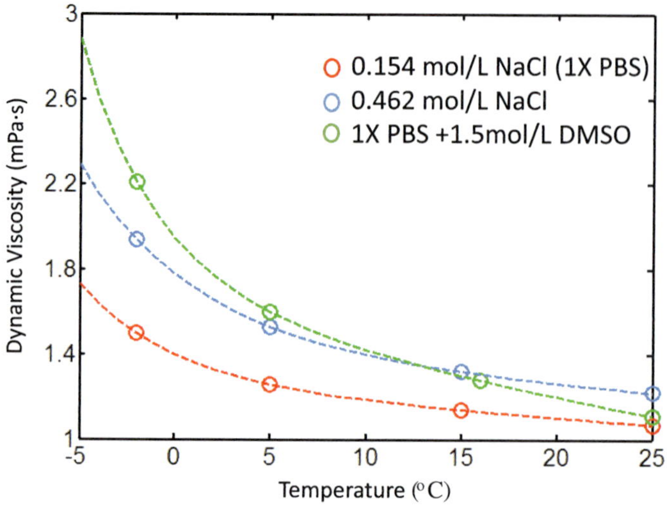

Figure 10. Dynamic viscosity of solutions measured at different temperatures. Reproduced from [Liu *et al.*, 2015] with permission.

of different temperatures. Thus, there is no need to distinguish the conditions with regard to temperature when fitting the cell parameters.

The change of the NaCl and DMSO concentration field at 25°C from 340 to 440 sec is shown in Figure 11(c)–(d), where five positions in the channel were chosen. The comparison of the NaCl concentration changes between the five positions is made in Figure 12. From Figure 11 and Figure 12, it can be known that the concentration change in the whole channel displays an axially symmetric distribution. Besides, there is a sharp, high concentration band in the middle of the channel when hypertonic solution infuses into the isotonic solution, meaning that the concentration changes are different at the same position on the Y axis. This is a potential factor affecting the precision of experimentally determined cell membrane permeability. Thus, it is necessary to choose a position near the middle of the channel to observe cell volume changes, as has never been discussed or even referred to in the previous studies [Adams *et al.*, 1967; McGrath, 1997; Takamatsu *et al.*, 2004; Chen *et al.*, 2008]. Finally, it can be judged that the total step time is about 100 sec when the isotonic solution shifts to the hypertonic solution.

4.2.2. *Self-adaptive time-varying function for extracellular concentration*

For almost all the currently existing studies on measurement of L_p (and/or P_s), it has been assumed that the concentration of extracellular solution during osmotic shift follows an ideal step function (especially for CPA(s) and

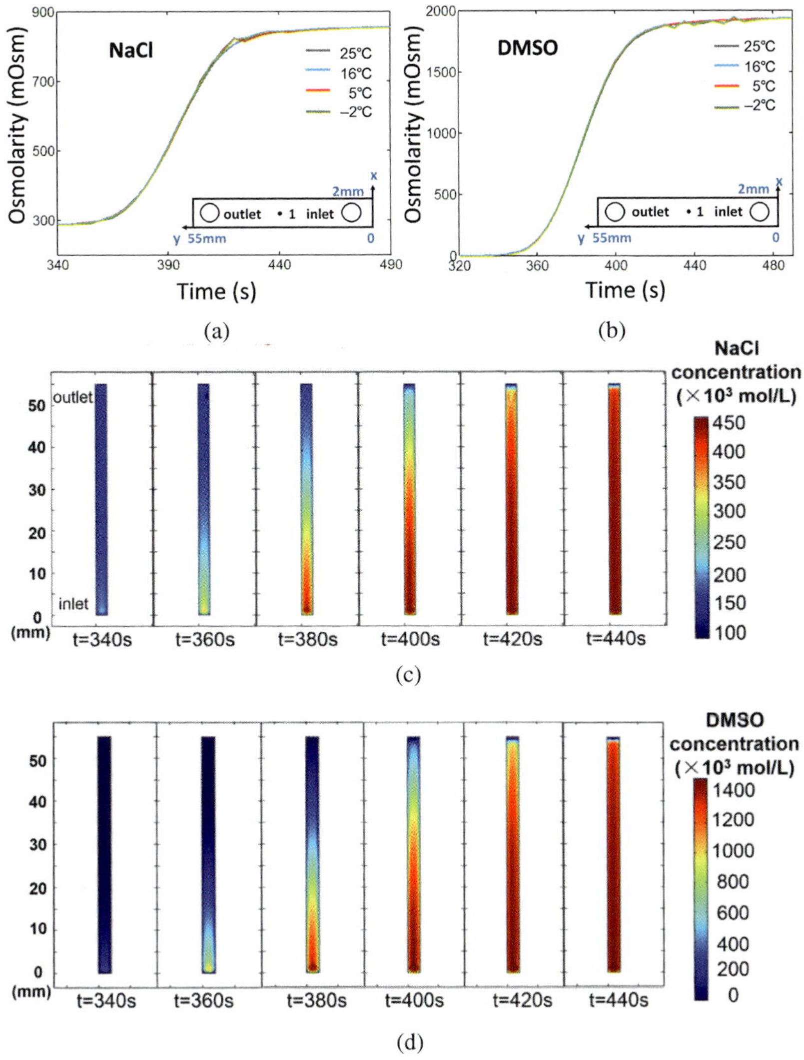

Figure 11. The comparison of solution concentration profiles and sectional view of solution concentration distributions in the chamber. (a) The comparison of NaCl solution concentration profiles at different temperatures. (b) The comparison of DMSO solution concentration profiles at different temperatures. (c) A sectional view of concentration distributions of NaCl solution. (d) A sectional view of concentration distributions of DMSO solution. Redrawn from [Liu *et al.*, 2015] with permission.

salt(s)). Zhao [2010] studied the influence of non-ideal step change for extracellular concentration on the curve-fitting values of L_p during osmotic experiments. It was found that the assumed self-adaptive time-varying function for extracellular concentration could always produce the very satisfactory fitting results, and there were no significant differences between the hydraulic conductivity values fitted using the laser interferometer measured extracellular

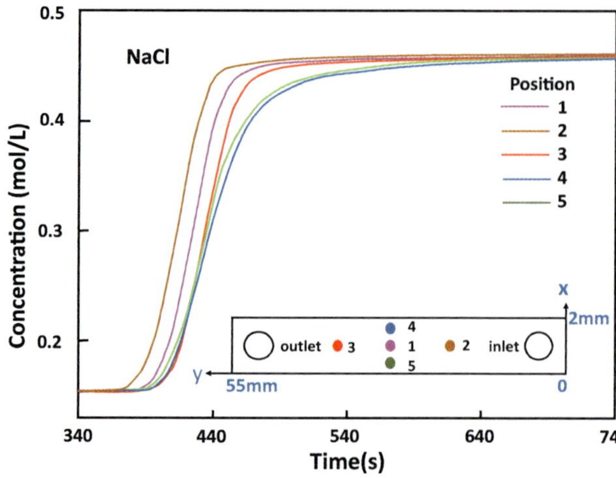

Figure 12. A comparison of NaCl concentration change at five positions. Reproduced from [Liu *et al.*, 2015] with permission.

concentration profile and the predicted one using the assumed self-adaptive time-varying function (student's t test, $p > 0.05$). Due to the adaptive feature of the assumed function for the concentration of extracellular solution, the function was suggested to be used for all the similar studies for measurement of cell membrane permeability by osmotic shift.

Concentration profiles for extracellular solution

The volume changes of the bovine carotid artery endothelial cells during osmotic shift (being transferred from 0.15 to 0.5 mol/L NaCl solution) obtained from the perfusion microscope [Takamatsu *et al.*, 2004; Zhao, 2010; Zhao *et al.*, 2012] will be used for all the following analysis. The following functions for $C_e(t)$ will be investigated.

(i) The fitted function for the laser interferometer measured extracellular concentration changes

A real $C_e(t)$ fitted according to the experimental data, marked as $C_{e,r}(\mathrm{t})$, firstly used in Takamatsu *et al*'s studies as follows [Takamatsu *et al.*, 2004]:

$$C_{e,r}(t) = \begin{cases} 0.15 & t < 0 \\ 0.15 + 0.35\left[-8\left(\frac{t}{22}\right)^6 + 30\left(\frac{t}{22}\right)^5 - 39\left(\frac{t}{22}\right)^4 + 18\left(\frac{t}{22}\right)^3\right] & 0 \le t \le 22. \\ 0.5 & t > 22 \end{cases} \quad (12)$$

(ii) The approximate piecewise function

An approximate piecewise function for $C_{e,r}(t)$ with a fixed time interval from the initial to the final concentration (the ramp time, $\Delta t = \Delta t_0 = 22$s, is determined according to $C_{e,r}(t)$), marked as $C_{e,p}(t)$:

$$C_{e,p}(t) = \begin{cases} 0.15 & t < 0 \\ 0.15 + \left[(0.5 - 0.15)/\Delta t_0)\right] \cdot t & 0 \le t \le \Delta t_0. \\ 0.5 & t > \Delta t_0 \end{cases} \quad (13)$$

(iii) The ideal stepwise function

An ideal step function, marked as $C_{e,s}$(t)

$$C_{e,s}(t) = \begin{cases} 0.15 & t < 0 \\ 0.5 & t \ge 0 \end{cases}. \quad (14)$$

(iv) The self-adaptive time-varying function

An approximate piecewise function for $C_{e,r}(t)$ with an adjustable Δt (Δt need to be fitted together with L_p and the range for Δt may be estimated according to experiment data), that is $C'_{e,p}$(t):

$$C'_{e,p}(t) = \begin{cases} 0.15 & t < 0 \\ 0.15 + \left[(0.5 - 0.15)/\Delta t)\right] \cdot t & 0 \le t \le \Delta t. \\ 0.5 & t > \Delta t \end{cases} \quad (15)$$

In order to be intuitionistic, the above mentioned four kinds of functions for $C_e(t)$ are shown in Figure 13. And it should be pointed out that the concentration change may not be triggered just from $t = 0$ during the real experiments, in this case, right translation of these functions should be necessary.

Fitting results by different concentration profiles

Two typical cells during the osmotic shift (one at 11°C and the other at 23°C), the corresponding curve-fitting processes and results are shown in Figure 14, and all the fitted parameters are listed in Tables 1 and 2. The fitted results can further be divided into two groups, that is, V_b is predetermined or fitted. For brevity, only the results associated with fitted V_b are shown in Figure 14, while all the fitted parameters are list in Tables 1 and 2. If V_b is not fitted, it will be calculated before curve-fitting according to the equilibrium volumes of the cell (V_0 and V_f) in the solutions of the initial and the final concentrations ($C_{e,0}$ and $C_{e,f}$). In one sentence, if V_b is fitted, the fitted parameters are (L_p & V_b) or (L_p, Δt & V_b), and if V_b is predetermined and fixed, the fitted parameters are (L_p) or (L_p & Δt).

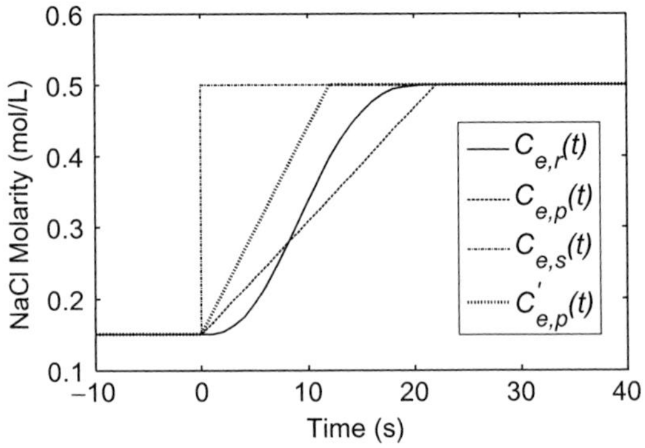

Figure 13. Typical functions for extracellular NaCl molarity. Reproduced from [Zhao, 2010] with permission.

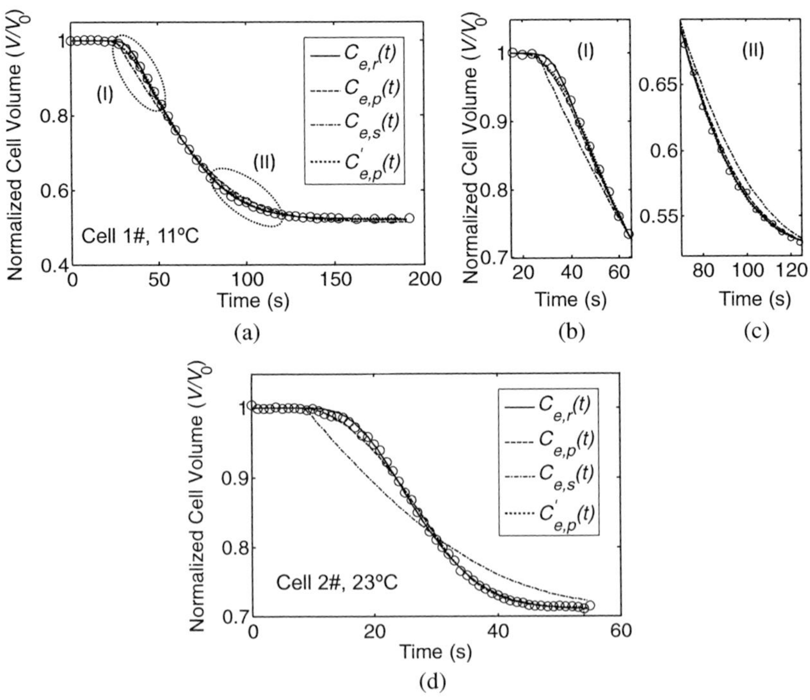

Figure 14. Effect of different functions for extracellular concentration $C_e(t)$ on curve-fitting process of membrane hydraulic conductivity L_p (two typical cells, separately shifted from 0.15 to 0.5 mol/L NaCl solution at 11°C and 23°C). Reproduced from [Zhao, 2010] with permission. *Note*: hollow circles represent experimental data; subfigures (b) and (c) are partial magnification of (a); fitted parameters are (L_p, V_b) or $(L_p, V_b, \Delta t)$.

Table 1. Fitting results of L_p with different $C_e(t)$ (cell 1#, 11°C) [Zhao, 2010].

Extracellular NaCl molarity	Fitted parameters	V_b/V_0	Δt(s)	L_p (μm/min/atm)	R^2	Relative error of L_p (%)*
$C_{e,r}(t)$	L_p & V_b	0.32	—	0.1242	0.9998	—
$C_{e,p}(t)$	L_p & V_b	0.31	22	0.1188	0.9994	–4.4
$C_{e,s}(t)$	L_p & V_b	0.30	—	0.1030	0.9946	–17.0
$C'_{e,p}(t)$	L_p, V_b & Δt	0.32	26.6	0.1272	0.9999	2.4
$C_{e,r}(t)$	L_p		—	0.1247	0.9998	—
$C_{e,p}(t)$	L_p	0.32**	22	0.1194	0.9994	–4.3
$C_{e,s}(t)$	L_p		—	0.1046	0.9939	–16.2
$C'_{e,p}(t)$	L_p & Δt		26.7	0.1274	0.9999	2.2

*, L_p value with $C_{e,r}(t)$ is assumed to be the real value.
**, V_b is predetermined using V_0 and V_f if not fitted.

Table 2. Fitting results of L_p with different $C_e(t)$ (cell 2#, 23°C) [Zhao, 2010].

Extracellular NaCl molarity	Fitted parameters	V_b/V_0	Δt(s)	L_p (μm/min/atm)	R^2	Relative error of L_p (%)*
$C_{e,r}(t)$	L_p & V_b	0.59	—	0.2202	0.9994	—
$C_{e,p}(t)$	L_p & V_b	0.59	22	0.2772	0.9992	25.9
$C_{e,s}(t)$	L_p & V_b	0.47	—	0.1083	0.9574	–50.8
$C'_{e,p}(t)$	L_p, V_b & Δt	0.59	16.7	0.2324	0.9996	5.5
$C_{e,r}(t)$	L_p		—	0.2264	0.9992	—
$C_{e,p}(t)$	L_p	0.59**	22	0.2762	0.9992	22.0
$C_{e,s}(t)$	L_p		—	0.1210	0.9425	–46.6
$C'_{e,p}(t)$	L_p & Δt		17.3	0.2406	0.9996	6.3

*, L_p value under $C_{e,r}(t)$ is assumed to be the real value.
**, V_b is predetermined using i and V_f if not fitted.

From Figure 14 and Tables 1 and 2, it can be seen that:

(i) Although all the four assumed extracellular concentration functions can provide reasonable fitted L_p values, $C_{e,p}(t)$ and $C'_{e,p}(t)$ give out the better fitting results ($R^2 > 0.999$) than $C_{e,s}(t)$ ($R^2 < 0.995$), and $C'_{e,p}(t)$ can always provide the best coefficient of determination no matter V_b is regarded to be a predetermined constant or to be a parameter for fitting. For each group, $C'_{e,p}(t)$ provides the best fitting results while $C_{e,s}(t)$ provides the worst fitting results.

(ii) During the initial stage of the cell shrinkage (typically from 25 to 50 s), only $C_{e,p}(t)$ and $C'_{e,p}(t)$ can provide satisfying agreement for the volume changes between the fitted and the experimental data.

(iii) In each group, the piecewise function with an adjustable slope for concentration change ($C'_{e,p}(t)$) provides the better fitting results than those of the piecewise function with a fixed slope for concentration change ($C_{e,p}(t)$) according to R^2 value, while both of the two functions provide very good predictions for cell volume change, as can been seen from Figure 14.

(iv) Each function for $C_e(t)$ could provide the better fitting results if V_b is regarded as a parameter to be fitted. This means if V_b is predetermined according to the several initial and final equilibrium volumes (V_0 at $C_{e,0}$ = 0.15 mol/L, and V_f at $C_{e,f}$ = 0.5 mol/L), there may be some calculated error for V_b, although it is not so big. Perhaps, the error is caused by the neglect of the other cell volume values during cell shrink. In our opinion, V_b affects the approach to equilibrium of the cell volume in the new concentration, as will be subsequently discussed.

Whether V_b could be and should be Fitted?

In most of the currently existing studies, the average V_b is calculated according to Boyle-van't Hoff (BVH) plot before L_p is fitted. For the electronic particle size counter method, the L_p measured is the average value of many cells of the same type, and so it is all right to predetermine the average V_b based on BVH relationship for the cells of the same type before curve-fitting. However, it may not be appropriate for the single-cell-focused methods, e.g. the microfluidic device method, the micropipette perfusion method, and the microperfusion chamber method etc., since the L_p value measured is the L_p of the single cell at each run for these methods. So, V_b of the cell should be incorporated into the model to fit out the correct L_p value for itself instead of using the average V_b value of many cells. The influences of V_b on the cell volume response lie in not only the final cell volumes but also the cell volumes during the recovery stage, even if the cells are not in a new equilibrium state now. And so, in order to use full information provided by the volumes, V_b should be fitted instead of being predetermined using only V_0, V_f, $C_{e,0}$ and $C_{e,f}$. Only by doing so, can the experimental errors be effectively decreased during determination of V_b.

5. Applications of the Microperfusion Chambers

5.1. *Membrane Transport Properties for Mouse Dendritic Cell*

The PDMS-based microperfusion chamber developed by Chen *et al.* [2008] was used to investigate the osmotic behaviors of mouse dendritic cells (mDCs) toward water and CPAs at different temperatures, and the ideal step function, $C_{e,s}(t)$, was used for the curve-fittings.

On the day of the experiment, mDCs of diameters in the range of 10–15 μm were collected, counted by a hemacytometer, centrifuged, and resuspended in 1 × phosphate buffered saline (PBS) (300 mOsm) at a cell density of about 5×10^4 /mL, and used within 3 h. In experiments, 1× PBS is first injected into the inlet to rinse the channel (Figure 15). Due to the oxygen plasma treatment, the surfaces of glass and PDMS inside the channel are highly hydrophilic, and the medium fill the channel quickly. Bubbles form inside the chamber, but quickly disappear. The cell suspension is later injected into the same inlet. With continual suction by the Hamilton Syringe, cells flow through the channel and are trapped in the perfusion chamber. At the same time, part of the cell suspension solution continue moving downstream through the bottom slit. Once the number of cells is enough for the experiment (usually 10–15), the original cell suspension solution is replaced by the new PBS solution with different concentrations of electrolytes (2× and 3×) or 1.5 M Me_2SO solution which is isotonic with respect to NaCl (0.9% NaCl). The new medium is injected in less than half a second in a sufficient volume (0.5 mL) until the silicone tubing (length ~10 mm) or the cubical hole

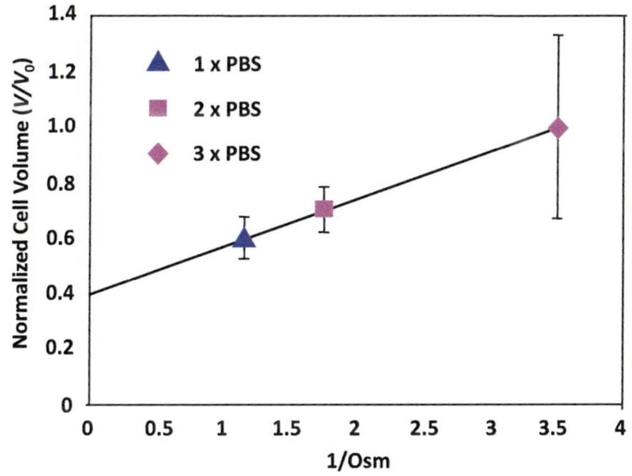

Figure 15. Boyle Van't-Hoff relationship for mDCs at room temperature (22°C) by diffusion method. Cells were randomly collected for measurements. Inactive cell volume, V_b, was determined to be 40% of original/isotonic volume. Reproduced from [Chen *et al.*, 2008] with permission.

(4 mm × 4 mm × 2 mm) fills so that residual cell suspension (~1 mL) will not affect the desired concentration. The liquid (~1.3 mL) inside the chamber of the temperature control device is made from a 1:1, in volume, mixture of distilled water and propylene glycol (CQ Concept Inc., McHENRY, IL), circulated by a circulating bath (Poly-Science, Niles, IL). The temperature control device allows the local temperature around the perfusion chamber to be controlled between −20°C and 40°C (±0.5°C). A T-type thermocouple (SA-1 T, Omega, Stamford, CT) is embedded inside the PDMS near the perfusion chamber for local temperature monitoring. Before, during, and after the perfusion process, the cell volume changes are recorded by a CCD camera (Fusion, Logitech Inc., Fremont, CA), which is connected to the computer, until osmotic equilibrium between the intracellular and extracellular environments is achieved.

A Boyle Van't Hoff plot is shown in Figure 15. The inactive volume V_b of mDCs was determined to be 40% [$n = 15$ for each concentration, correlation coefficient (R^2) of 0.99] of the original/isotonic volume, obtained by the diffusion methods [Lucke and McCutcheon, 1932; Leibo, 1980].

Figure 16 shows photographs of mDCs perfused with hypertonic 3× PBS at 22°C. The arrow depicts the origin of the flow. Cells were trapped within the perfusion chamber while the medium flowed downstream by suction as shown in Figure 16(a). A magnified view, Figure 16(b), shows a series of photographs where a cell was exposed to the hypertonic solution at $t = -10$, 0, 10, 20, 35, and 60 s, respectively. The cell decreased in size until it reached the osmotic balance between intracellular and extracellular solutions. The history of volume change of mDC was then processed using Matlab, shown in Figure 16(c). The water permeability coefficient, L_p, was determined by curve-fitting, as shown in Figure 17. Other individual cells, perfused at 12, 25, and 32°C, were processed by the identical method aforementioned. Values of L_p determined at different temperatures are listed in the first major row I of Table 3.

5.2. *Membrane Transport Properties of Bovine Carotid Artery Endothelial Cells*

The osmotic properties of bovine carotid artery endothelial cells (BCAECs) associated with cryobiology were determined [Zhao *et al.*, 2012] using the traditional perfusion microscope developed by Takamatsu *et al.* [2004]. The self-adaptive time-varying function for extracellular concentration, $C'_{e,p}(t)$, was incorporated into the water transport equation during the curve-fitting processes.

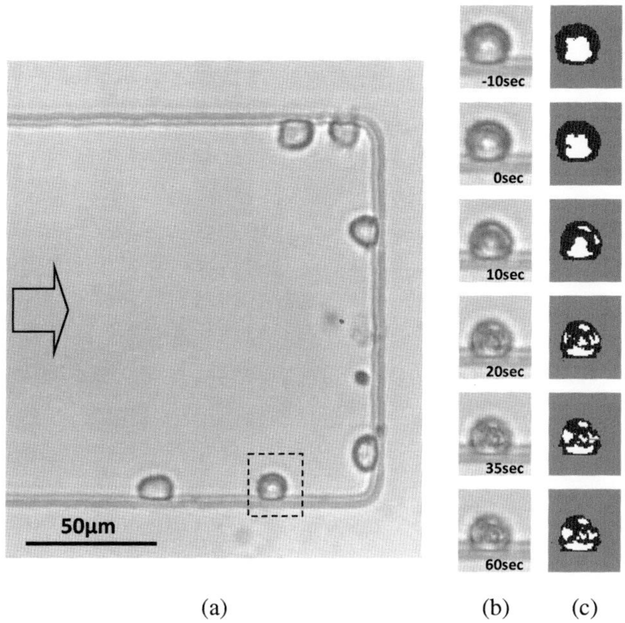

Figure 16. (a) mDCs trapped within the perfusion chamber. (b) A magnified view inside dashed line when cells were perfused by 3× PBS at the moments of $t = -10$, 0, 10, 20, 35, and 60 s, respectively. (c) Post-processed images corresponding laterally to (b). Reproduced from [Chen *et al.*, 2008] with permission.

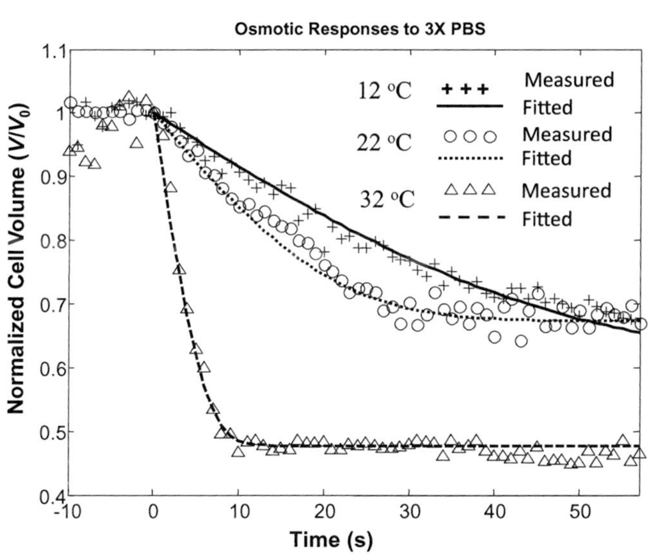

Figure 17. Volume change of mDCs in response to 3× PBS. Reproduced from [Chen *et al.*, 2008] with permission.

Table 3. Water permeability (L_p) at four different temperatures and Me_2SO permeability (P_s) at room temperature [Chen *et al.*, 2008].

	Temperature (°C)	L_p (μm/min/atm)	P_s (10^{-3} cm/min)
		Mean ± SD (n, R2)	Mean ± SD (n, R2)
I. (3× PBS)	12	0.11 ± 0.02 (8, > 0.87)	N/A
	22	0.16 ± 0.03 (4, > 0.83)	N/A
	25	0.95 ± 0.29 (6, > 0.93)	N/A
	32	1.54 ± 0.66 (6, > 0.90)	N/A
II. (1.5M DMSO + 0.9% NaCl)	22	0.17 ± 0.04 (5, > 0.66)	0.63 ± 0.29 (5, > 0.66)

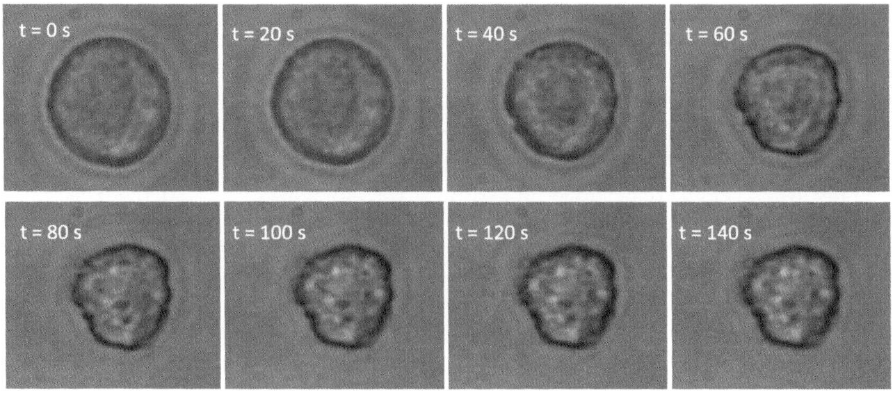

Figure 18. Bovine carotid artery endothelial cells (BCAECs) responses during osmotic change from 0.15M to 0.5M NaCl solution at 11°C. Reproduced from [Zhao *et al.*, 2012] with permission.

Figure 18 shows the photographs reproduced from the video that show cell shrinkage after exposure to hypertonic (0.5M) NaCl solution from the isotonic condition (0.15M) at 11°C. To calculate the volume of a cell, the projected area of the cell is measured after image processing procedures (Figure 19): enhancing the contrast of the original image, removing the background, encompassing the boundary and filling the enclosed area. Figure 20 shows the volume of a cell as a function of time. The decrease of cell volume obtained by the experiment agreed well with that calculated for the increase of the osmolarity in the extracellular solution that was estimated from the measured concentration change. In the example shown in Figure 20 at 23°C, the increase in the osmolarity of the intracellular solution was 10–20 s behind the increase in the extracellular solution because of the cell membrane as the

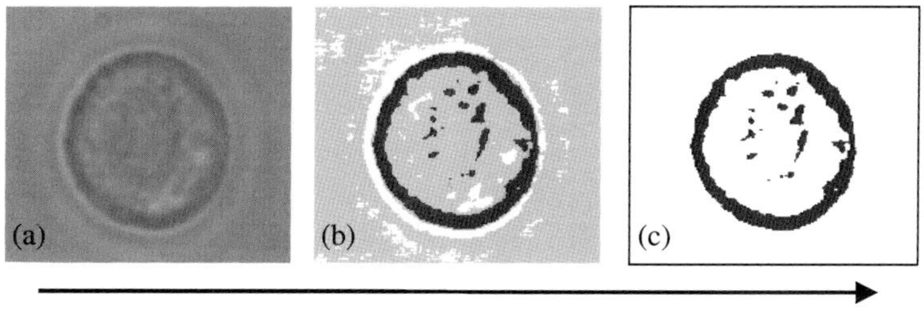

Figure 19. Image processing procedure for measurement of the projected area of a cell. (a) Original micrograph. (b) Enhancement of contrast. (c) Removal of the background. Reproduced from [Zhao *et al.*, 2012] with permission.

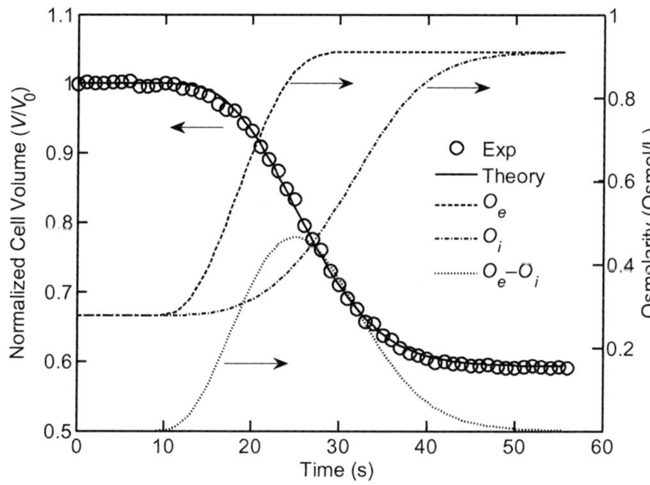

Figure 20. Volume of a cell and the estimated osmolarity in the extracellular and intracellular solutions as a function of time during concentration change from 0.15M to 0.5M NaCl solution at 23°C. Symbols show the measured cell volume, the solid line shows the best-fit curve obtained with an appropriate L_p and V_b values, the broken line shows the osmolarity in the extracellular solution, the dotted chain line shows the estimated osmolarity in the intracellular solution, and the dotted line shows the difference in the osmolarity between extracellular and intracellular solutions. The cell volume is shown by the value normalized by the initial value before contraction. Reproduced from [Zhao *et al.*, 2012] with permission.

barrier for water transport. The estimated largest difference in the intra- and extracellular osmolarities was 0.4–0.5 Osmol/L.

When the temperature was decreased, the change of cell volume became considerably slower (Figure 21). While the time required for a cell to become equilibrium was approximately 40 s at 23°C, it was ~80 s at 11°C and ~200 s at 4°C. Even at the lower temperatures, the best-fit curve calculated for the

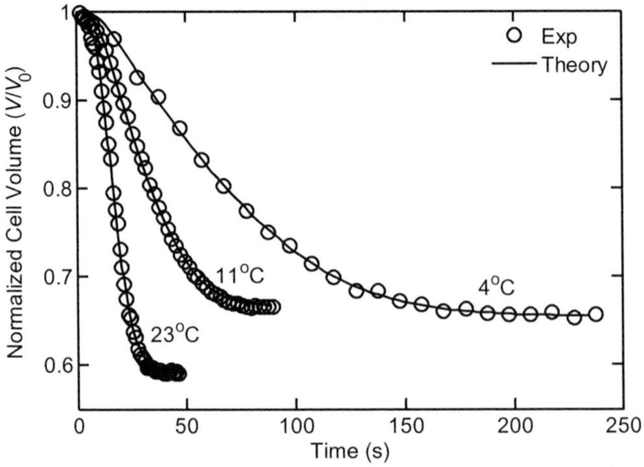

Figure 21. Change of normalized cell volume at different temperatures: 23°C, 11°C, and 4°C; symbols show the experimental data and the lines show the calculation. Reproduced from [Zhao *et al.*, 2012] with permission.

appropriate hydraulic conductivity agreed well with the experimental results. The fact that the final values are different from each other in Figure 21 is attributed not to the difference in temperature but to the difference in the osmotically inactive volume V_b relative to V_0 of selected cells.

The size of cells that were picked up for measurement ranged from ~12 μm to ~23 μm in diameter. However, there was no significant correlation between L_p value for individual cells and the size of cells (Figure 22). The hydraulic conductivity and the other parameters determined from at least 10 individual cells at each temperature are summarized in Table 4. While there was no significant difference in V_0 between the cells measured at different temperatures, significant differences were found in V_b/V_0 between 23°C and 11°C and between 23°C and 4°C.

The mean L_p values at different temperatures are correlated well with the following Arrhenius relationship [Levin *et al.*, 1976] (Figure 23):

$$L_p(T) = L_{pg} \cdot \exp\left[\frac{E_{Lp}}{R}\left(\frac{1}{T_0} - \frac{1}{T}\right)\right] \tag{16}$$

where E_{Lp} is the activation energy (kJ/mol) and T_0 is a reference temperature. The least-square linear fit of Equation 16 to the data points showed that the value of L_{pg} at $T_0 = 273.15$ K was 0.84×10^{-14} m/Pa/s (0.051 μm/atm/min) and the activation energy E_{Lp} was 47.6 kJ/mol.

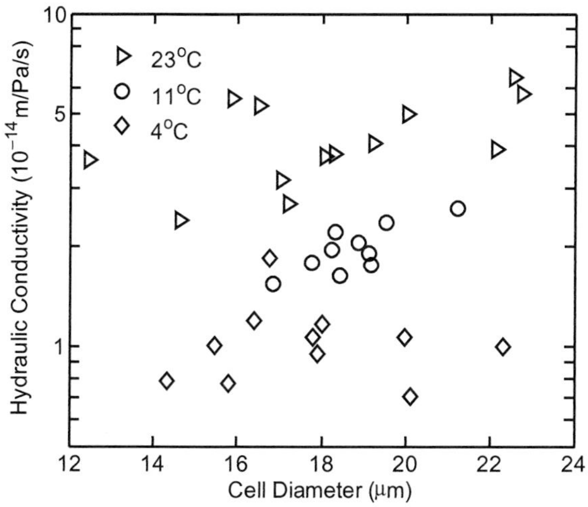

Figure 22. Measured hydraulic conductivity shown as a function of the diameter of cells in the isotonic condition. Reproduced from [Zhao *et al.*, 2012] with permission.

Table 4. Cryobiological parameters for BCAECs [Zhao *et al.*, 2012].

	Temperature		
Parameter	**23°C (n = 13)**	**11°C (n = 10)**	**4°C (n = 11)**
V_0 [μm^3]	3416 ± 1661	3487 ± 669	3054 ± 1253
d_0 [μm]	18.2 ± 3.1	18.7 ± 1.2	17.7 ± 2.3
V_b/V_0 [%]	53.8 ± 9.2*#	43.0 ± 7.3*	43.3 ± 7.2#
L_p [10^{-14} m/Pa/s]	4.27 ± 1.24	1.98 ± 0.33	1.05 ± 0.31
L_p [μm/atm/min]	0.26 ± 0.08	0.12 ± 0.02	0.06 ± 0.02

Note: *,# $p < 0.05$

The L_{pg} and E_{Lp} values determined for BCAECs were compared with available data in Table 4. These include vascular endothelial cells and several types of cell from cornea as well as smooth muscle cells. The comparison was also made in two different figures: one was Figure 24 that shows L_p calculated from Equation 16 with given L_{pg} and E_{Lp} values as a function of temperature, and the other was a map that shows the relationship between L_{pg} and E_{Lp} (Figure 25).

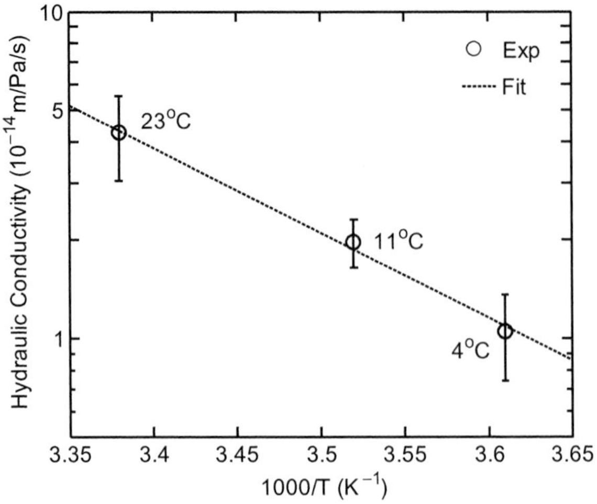

Figure 23. Hydraulic conductivity as a function of the reciprocal of temperature; symbols show the mean value and the bars show the standard deviation. Reproduced from [Zhao *et al.*, 2012] with permission.

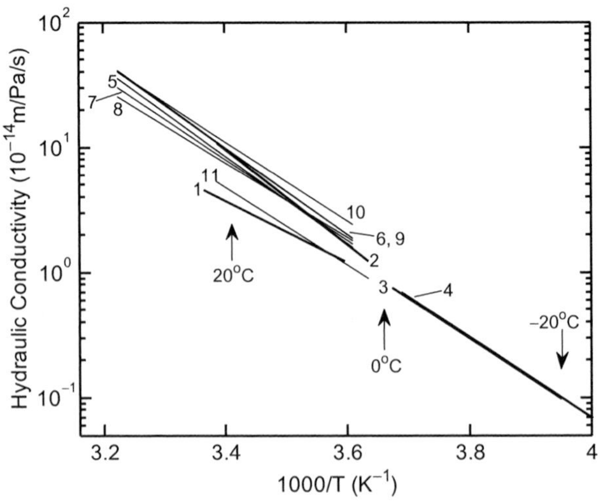

Figure 24. Hydraulic conductivity calculated with the L_{pg} and E_{Lp} in Table 5 as a function of the reciprocal of temperature; the bold lines represent vascular endothelial cells, the thin lines represent the other cells; the numbers on the lines indicate the cell types listed in Table 5. Reproduced from [Zhao *et al.*, 2012] with permission.

Figure 24 allows us to compare L_p data directly. Collected data shown in this figure indicate that the cells from cornea, including endothelial cells, have similar L_p values. They also suggest that the L_p value is not significantly affected by CPAs, while it should be confirmed experimentally with the same

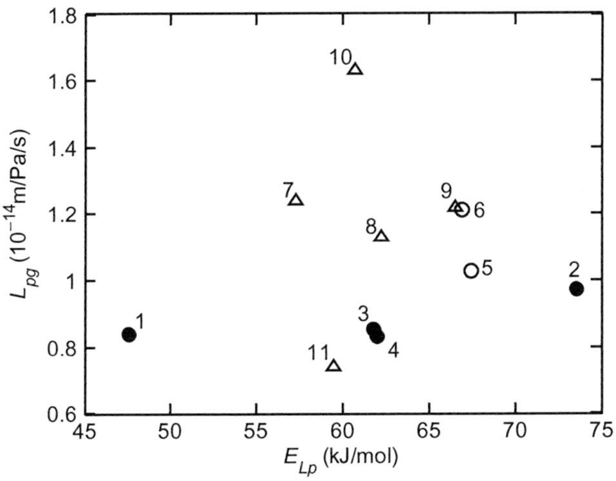

Figure 25. Relationship between L_{pg} and E_{Lp} of the biological cells listed in Table 5; the circles represent endothelial cells and the triangles represent the other cells; the numbers indicate the cell types listed in Table 5. Reproduced from [Zhao *et al.*, 2012] with permission.

type of cells. The porcine aorta endothelial cell has L_p that is similar to that of cornea cells, while the temperature dependence is larger. Our data for BCAECs are smaller than that of these cells, smaller by ~15% at 0°C and ~60% at 20°C than porcine aorta endothelial cells (#2), but are approximately the same as those for porcine smooth muscle cells. Direct comparison with human vascular endothelial cells is not available because these data have been determined for only sub-zero temperatures. However the extrapolation of the lines suggests that all the lines for vascular endothelial cells intersect at ~–6°C ($1000/T = 3.75\ K^{-1}$).

Following the discussion of He and Bischof [2003], the L_{pg} value was shown against E_{Lp} in Figure 24. No significant correlation between L_{pg} and E_{Lp} was observed as in the case of their review [He and Bischof, 2003]. However, it is clear from Figure 24 and Table 5 that the vascular endothelial cells have approximately the same L_{pg} value of ~0.85×10^{-14} m/Pa/s, while E_{Lp} ranges from 47.6 to 73.5 kJ/mol. This resulted in the possible intersection of the lines in Figure 23 at the temperature slightly lower than 0°C. The fact that endothelial cells have similar L_p at the temperature near 0°C is interesting, because it could be a complex result of the temperature effect on the membrane lipids, which depends on the composition and their phase, as well as the water channels expressed in many endothelial cell types.

During freezing of tissues with multiple cell types, slower dehydration of cells with smaller L_p could increase the probability of intracellular ice

Table 5. Hydraulic conductivity of biological cells [Zhao *et al.*, 2012].

Cell type		CPA	L_{pg}	E_{Lp}	T	Reference
			10^{-14} m/Pa/s (μm/atm/min)	kJ/mol (kcal/mol)	°C	
Vascular endothelial cells	Bovine carotid artery	—	0.84 (0.051)	47.6 (11.4)	4~23	Zhao *et al.* [2012]
	Porcine aorta	Me_2SO	0.97 (0.059)	73.5 (17.6)	2~22	[Wusteman and Pegg, [2001]
	Human umbilical vein	—	0.86 (0.052)	61.8 (14.8)	−2~−40	Zhang *et al.* [2006]
	Human microvasculature	—	0.84 (0.05)	62.0 (14.8)	−0.6, −0.7~−20	Berrada and Bischof [2001]
Human cornea	Endothelial cells	Me_2SO	1.03 (0.062)	67.4 (16.1)	4~37	Ebertz and McGann [2004]; Chuenkhum and Cui [2006]
		PG	1.21 (0.074)	66.9 (16.0)		
	Keratocytes	Me_2SO	1.24 (0.075)	57.3 (13.7)		
		PG	1.13 (0.069)	62.2 (14.9)		
	Epithelial cells	Me_2SO	1.22 (0.074)	66.5 (15.9)		
		PG	1.63 (0.099)	60.7 (14.5)		
Porcine smooth muscle cells		DMSO	0.74 (0.045)	59.5 (14.2)	2~22	Yoshimori and Takamatsu [2009]

formation, which is supposed to be lethal for cells. This is one of the difficulties encountered in successful cryopreservation. It is therefore a good news that the L_p obtained in this studies agreed fairly well with smooth muscle cells as shown in Figure 24, while the L_p may depends on the species. The result that showed larger V_b at 23°C than the other temperatures is interesting. However, more critical examination is needed to know if the difference really depended on the temperature because there are other important factors to be taken into account, such as the stage of a cell cycle for instance.

5.3. *Membrane Transport Properties of SF21 Cells*

The Sf21 cell line, derived from the ovary of the fall armyworm (*Spodoptera Frugiperda*) by Vaughn *et al.* [1977], is a host of many viruses; when combined with baculoviruses, it is a powerful platform technology that is widely used for the manufacture of viral particles and heterologous recombinant proteins [Ikonomou *et al.*, 2003; Estrada-Mondaca *et al.*, 2005; Drugmand *et al.*, 2012]. This robust expression system is able to produce large amounts of different proteins, with applications from basic research (protein-protein interaction, multiple-affinity protein purification, ultrasensitive mass spectrometry, and protein structures) [Berger *et al.*, 2004] to clinical medicine (routine diagnostic tests, therapeutic protein drugs for various diseases, etc.) [Gómez-Sebastián *et al.*, 2008; Hancock *et al.*, 2008; Metz and Pijlman, 2011].

By using the improved perfusion microscope developed by Liu *et al.* [2015], incorporated with the self-adaptive time-varying function for extracellular concentration, $C'_{e,p}(t)$, Wang *et al.* [2013] experimentally studied the membrane osmotic properties of the Sf21 cells (IPLB-Sf21AE, a continuous cell line developed from ovaries of the Fall Army worm, *Spodoptera frugiperda*). The membrane permeability coefficient to water (L_p), the membrane permeability coefficient to glycerol of different concentration (P_s), and the activation energies of water and glycerol permeability coefficients at 25°C, 15°C, 5°C and −2°C have been investigated.

Figure 26(a) shows the Sf21 cells adhered to the micro-channel in the initial state at −2°C, and the transient response of a representative cell (indicated by a red box in the left figure) during the osmotic shift from PBS (291 mOsm) to 3× PBS (784 mOsm) at −2°C was shown in Figure 26(b) at a 10 second intervals. Table 6 summarizes the relative volume (V_b/V_0) and the hydraulic conductivity at four different temperatures. The V_b of the Sf21 is similar to mammalian ovarian tissues (0.5 V_0) [Rubin, 1993; Stevens, 1996; Devireddy *et al.*, 2006].

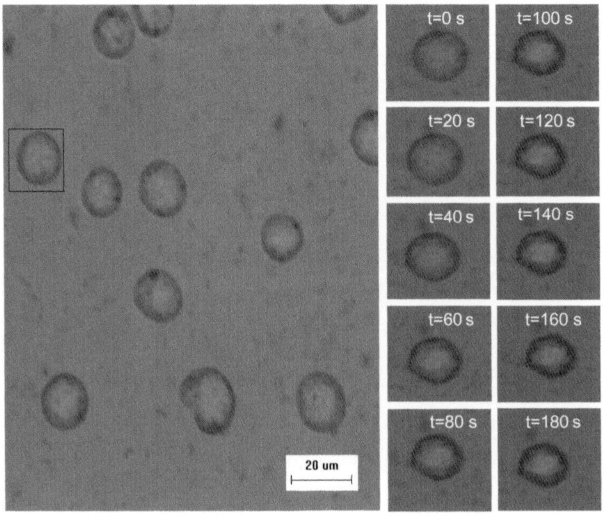

Figure 26. (a) The Sf21 cells were immobilized on the micro-channel. (b) The volume changes of a typical cell during the osmotic shift from PBS to 3× PBS solution at −2°C. Reproduced from [Wang *et al.*, 2013] with permission.

Table 6. Cryobiological parameters for Sf21 [Wang *et al.*, 2013].

		L_p	
Temperature (°C)	V_b/V_0 (%)	(10^{-14}m/Pa/s)	(μm/atm/min)
25(n = 12)	54.6 ± 9.1	1.112 ± 0.334	0.068 ± 0.020
15(n = 8)	50.6 ± 11.8	0.910 ± 0.277	0.055 ± 0.017
5(n = 11)	51.3 ± 4.1	0.449 ± 0.090	0.027 ± 0.006
−2(n = 7)	50.8 ± 6.1	0.368 ± 0.101	0.023 ± 0.006

5.4. *Membrane Transport Properties of Porcine Adipose-derived Stem Cells*

Stem cells are characterized by the ability to self-renew and migrate to locations where tissues have suffered damage from injury or sickness and differentiate into various somatic cells to repair damaged tissue, thus helping to maintain homeostasis [Estrov, 2009; Lazzeri *et al.*, 2012]. Porcine Adipose-derived Stem Cells (ADSCs) can self-renew and differentiate into various cell lineages, and may be used as a potential tool for drug screening to investigate the pharmacodynamic effects on differentiation, development, regeneration, and toxicity [Nirmalanandhan and Sittampalam, 2009; Lea-Currie *et al.*, 2011]. Although ADSCs can be isolated from adipose tissue by liposuction or a needle biopsy [Bastard *et al.*, 1994], fleshly isolated ADSCs cannot meet the needs of tissue engineering and regenerative medicine. Thus, to vastly improve the applications of ADSCs, new techniques need to meet the increasing demands for sufficient numbers of stem cells as well as long-term preservation.

By using the improved perfusion microscope developed by Liu *et al.* [2015], Wang *et al.* [2014] measured the transport properties of the porcine ADSC (pADSC) membrane. The self-adaptive time-varying function for extracellular concentration, $C'_{e,p}(t)$, was also used during curve-fittings to the experimental data.

The transient volume response of the pADSCs after slight adherence to the silicified glass-coated with the poly-L-lysine is presented in Figure 27. To measure the L_p, the normalized volume changes at 28°C, 18°C, 8°C, and −2°C were fitted with the *2-p* model (Figure 28(a)), indicating that the change in the pADSC volume was dependent on temperature. The permeability coefficient fits the *2-p* model as shown in Table 7. A least-squares linear regression ($R^2 = 0.994$) indicated that the temperature dependence of the L_p fit well with the Arrhenius relationship (Figure 28(b)). E_{Lp} and L_{pg} were calculated to be 14.3 kJ/mol and 0.57×10^{-14} m/Pa/s, respectively.

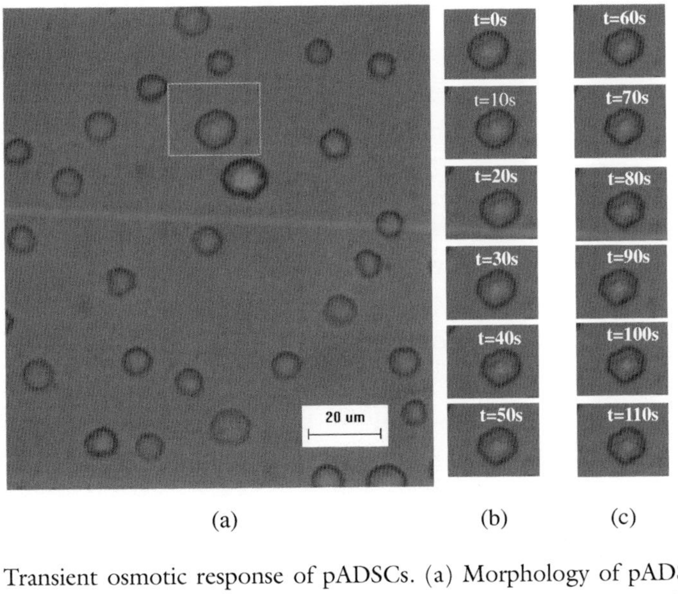

Figure 27. Transient osmotic response of pADSCs. (a) Morphology of pADSCs immobilized on the silicified glass. (b) Volume changes of pADSC during the osmotic shift from PBS to 3× PBS at 8°C. Reproduced from [Wang *et al.*, 2014] with permission.

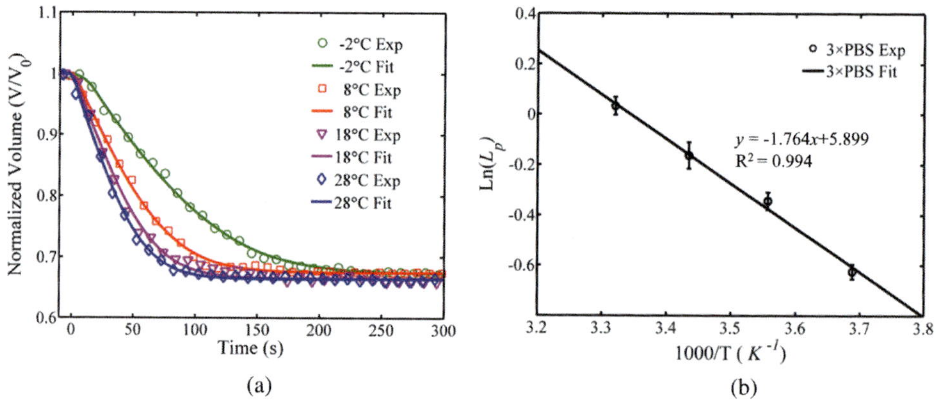

Figure 28. pADSC permeability coefficients in hypertonic saline. (a) Relative cell volume changes as a function of time after the osmotic shift from PBS (291 mOsm) to 3× PBS(781 mOsm) at −2°C, 8°C, 18°C, and 28°C. The theoretical curves (Fit, solid line) were fitted to the experimental data points (Exp, symbols dot). (b) Arrhenius plots for hydraulic conductivity. The y-values were obtained by taking the natural log of 10^{14} times the hydraulic conductivity. Reproduced from [Wang *et al.*, 2014] with permission.

Table 7. Cryobiological parameters for pADSC [Wang *et al.*, 2014].

		L_p	
Temperature (°C)	V_b	**(10^{-14} m/Pa/s)**	**(μm/atm/min)**
28 (n = 19)	0.480 ± 0.064	1.046 ± 0.173	0.064 ± 0.011
18 (n = 15)	0.526 ± 0.065	0.866 ± 0.181	0.053 ± 0.011
5 (n = 11)	0.501 ± 0.044	0.717 ± 0.099	0.044 ± 0.006
−2(n = 7)	0.515 ± 0.052	0.540 ± 0.068	0.033 ± 0.004

The osmotically inactive cell volume was obtained simultaneously in the fitting process. The value of V_b was 0.502 ± 0.057, which was used to determine the membrane permeability coefficients in CPAs.

The transient cell volume changed because of water efflux as Me_2SO was transported into the cell and water followed to equilibrate the chemical potential. Figure 29(b) shows theoretical curves fitted to the experimental data points at −2°C, 8°C, 18°C, and 28°C, indicated that permeation of Me_2SO through the cell membrane occurred rapidly with increasing temperatures. Table 8 shows the membrane permeability coefficients of pADSC in 1.5M Me_2SO after theoretical fitting at 28°C, 18°C, 8°C, and −2°C. The goodness-of-fit parameter indicated that L_p and P_s of Me_2SO obeyed the

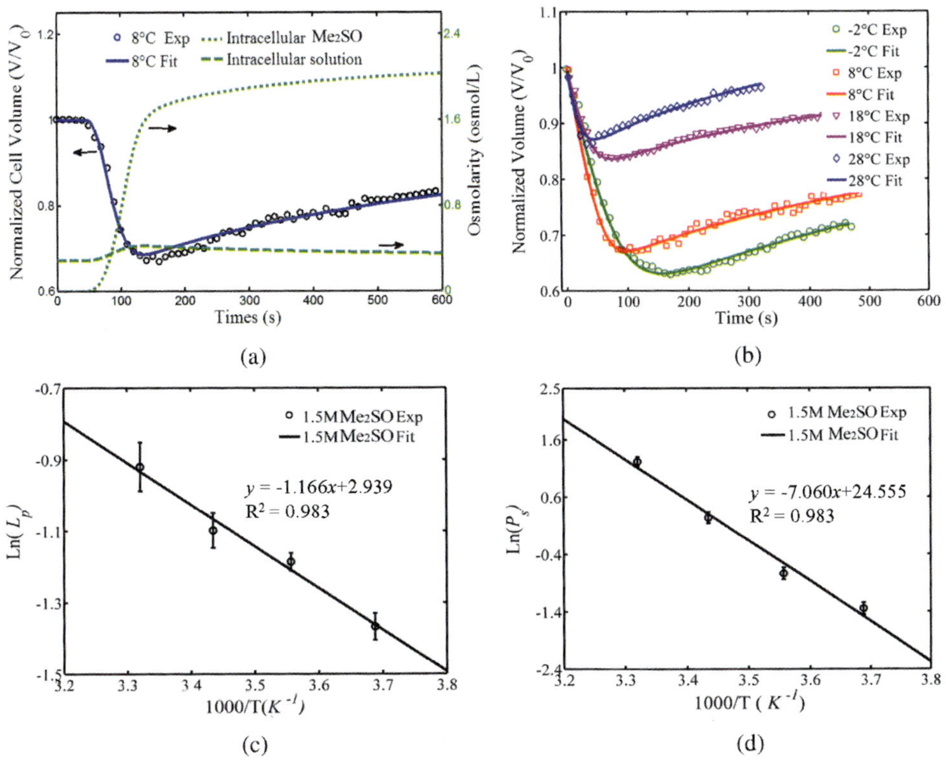

Figure 29. Transport properties of pADSCs in 1.5M Me_2SO. (a) The dynamic equilibrium after 1.5M Me_2SO was added to the isotonic solution at 8°C. The theoretical curves (solid line) were fitted to the experimental data points (o). The osmolarities of intracellular Me_2SO (dotted line) and the intracellular solution are shown. (b) Volume changes of pADSCs with the osmotic shift from PBS to 1.5M Me_2SO (2220 mOsm) at –2°C, 8°C, 18°C, and 28°C. (c) (d) Arrhenius plots for hydraulic conductivity and Me_2SO permeability. Reproduced from [Wang *et al.*, 2014] with permission.

Arrhenius relationship (Figure 29(c) and (d)) and the linear fit yielded E_{Lp} and E_{Ps} to be 9.5 kJ/mol and 57.4 kJ/mol Me_2SO, respectively, as well as L_{pg} and P_{sg} to be 0.26×10^{-14} m/Pa/s and 0.28×10^{-8} m/s, respectively. Furthermore, the L_p and P_s of pADSCs in 1.5M EG were determined by the *2-p* model at 28°C, 18°C, 8°C, and –2°C. Figure 30(a) and (b) shows the procedure for fitting the experimental data and Table 6 presents the fitted values for L_p and P_s in EG. Figure 30(c) and (d) indicate the temperature dependence of L_p and P_s on EG in the Arrhenius plot. The linear fit resulted in an E_{Lp} of 14.2 kJ/mol, E_{Ps} of 68.0 kJ/mol, L_{pg} of 0.30×10^{-14} m/Pa/s, and P_{sg} of 0.24×10^{-8} m/s.

Glycerol as a common CPA is widely used in cell and tissue cryopreservation. The simulation indicated that extracellular glycerol was slowly

Table 8. The cell membrane permeability coefficients of pADSC for 1.5M Me_2SO, EG and glycerol [Wang *et al.*, 2014].

	Temperature	L_p		P_s	
CPA	(°C)	(10^{-14} m/Pa/s)	(μm/atm/min)	(10^{-8} m/s)	(10^{-4} cm/min)
1.5M Me_2SO	28 (n = 12)	0.409 ± 0.092	0.025 ± 0.007	3.502 ± 0.953	2.101 ± 0.572
	18 (n = 12)	0.338 ± 0.056	0.021 ± 0.003	1.344 ± 0.461	0.806 ± 0.277
	8 (n = 13)	0.307 ± 0.028	0.019 ± 0.002	0.510 ± 0.183	0.306 ± 0.110
	−2(n = 12)	0.257 ± 0.033	0.016 ± 0.002	0.276 ± 0.010	0.165 ± 0.074
1.5M EG	28 (n = 12)	0.581 ± 0.193	0.035 ± 0.012	4.844 ± 1.146	2.906 ± 0.688
	18 (n = 12)	0.426 ± 0.093	0.026 ± 0.006	1.435 ± 0.361	0.861 ± 0.217
	8 (n = 13)	0.376 ± 0.091	0.023 ± 0.006	0.555 ± 0.181	0.333 ± 0.109
	−2(n = 12)	0.287 ± 0.067	0.017 ± 0.004	0.226 ± 0.087	0.136 ± 0.052
1.5M Glycerol	32 (n = 11)	0.389 ± 0.069	0.024 ± 0.004	0.139 ± 0.124	0.084 ± 0.075
	28 (n = 12)	0.371 ± 0.107	0.023 ± 0.007	0.032 ± 0.018	0.019 ± 0.011
	18 (n = 12)	0.337 ± 0.120	0.021 ± 0.007	0.008 ± 0.020	0.005 ± 0.012
	8 (n = 13)	0.278 ± 0.033	0.017 ± 0.002	0.002 ± 0.007	0.001 ± 0.005
	−2(n = 12)	0.225 ± 0.031	0.014 ± 0.002	0.002 ± 0.006	0.001 ± 0.004

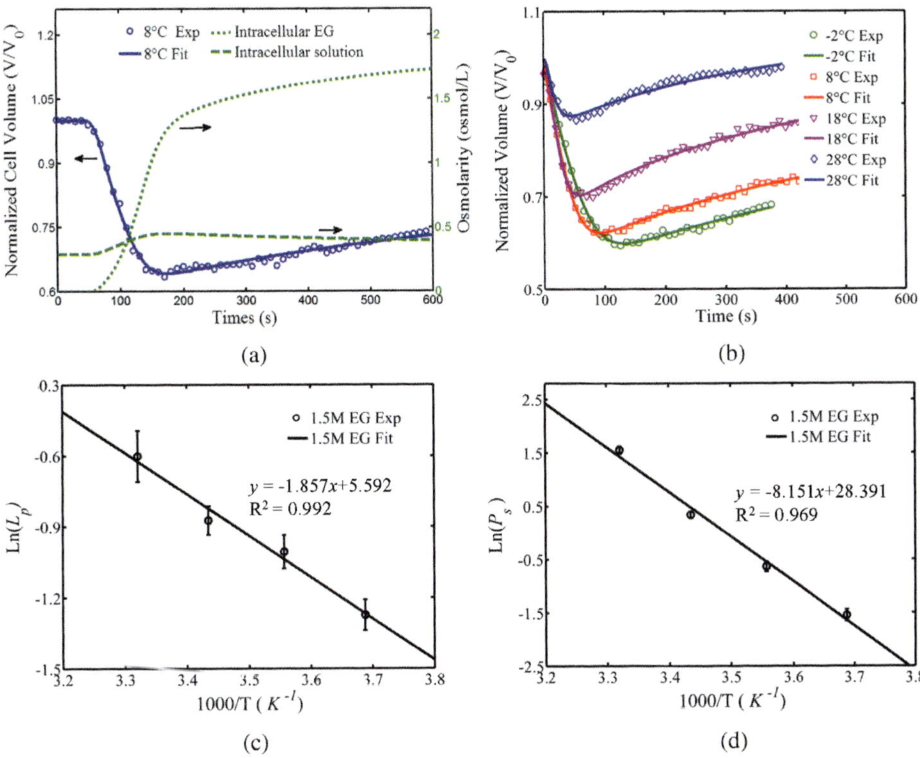

Figure 30. Transport properties of pADSCs in 1.5M EG. (a) The dynamic equilibrium in the isotonic solution containing 1.5M EG at 8°C. (b) Volume changes of pADSCs with the osmotic shift from PBS to 1.5M EG (2080 mOsm) at –2°C, 8°C, 18°C, and 28°C. (c) (d) Arrhenius plots for hydraulic conductivity and EG permeability. Reproduced from [Wang *et al.*, 2014] with permission.

transported into pADSCs even at 28°C (Figure 31(a)). The transient volume change as a function of time is presented in Figure 31(b) at 28°C, 18°C, 8°C, and –2°C. Interestingly, the final volume of the pADSCs did not increase after the pADSCs had shrunk because of exposure to 1.5M glycerol for 10 min at 8°C and –2°C. The increase in the final volume was too small to be ignored, and it indicated that glycerol had hardly entered the pADSCs and the P_s of pADSCs was difficult to accurately measure at low temperatures. Figure 31(c) shows the temperature dependence of L_p on glycerol in the Arrhenius plot. However, the P_s in glycerol was independent of temperatures at 28°C, 18°C, 8°C, and –2°C (Figure 31(d)). An electronic particle was used to rapidly measure the volume distributions in the suspension solution and determine the rapid cell response [Lea-Currie *et al.*, 2011]. However, there are no reports of precise measurement of the slow cell response. We obtained the cell volume change in 1.5M glycerol at 32°C, and the value of

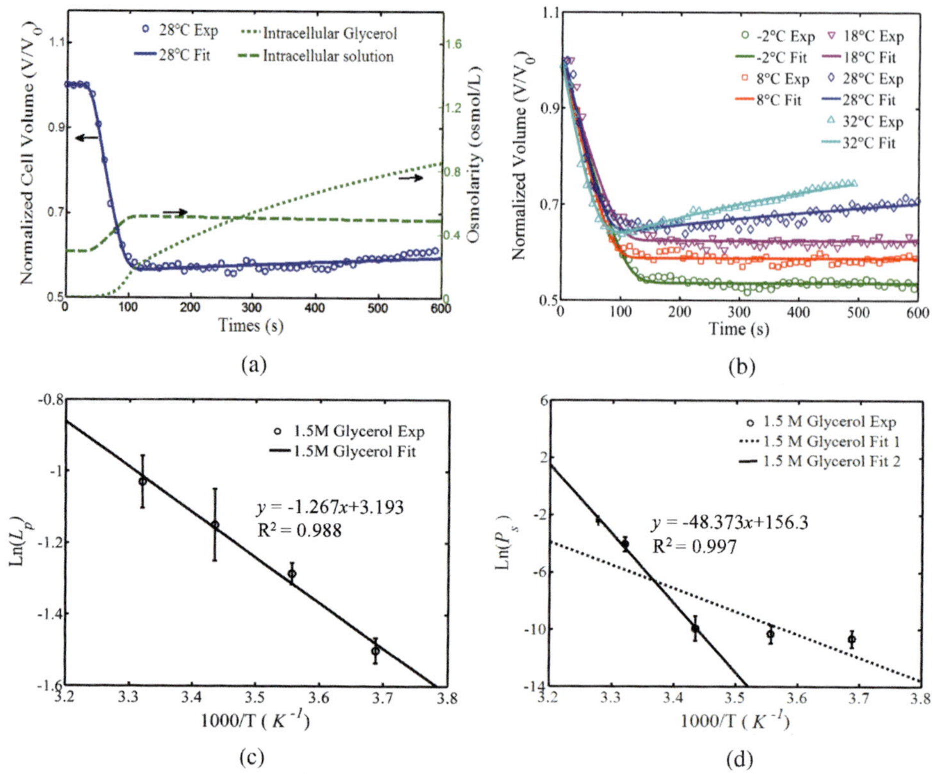

Figure 31. Transport properties of pADSCs in 1.5 M glycerol. (a) The dynamic equilibrium after 1.5M glycerol was added to the isotonic solution. (b) Volume changes of pADSCs with the osmotic shift from PBS to 1.5M glycerol (2204 mOsm) at −2°C, 8°C, 18°C, 28°C, and 32°C. (c) Arrhenius plots for hydraulic conductivity. (d) Arrhenius plots for glycerol permeability at various temperatures. The dotted line (Fit 1) represents the linear curve fitting glycerol permeability at −2°C, 8°C, 18°C, and 28°C. The solid line (Fit 2) represents the linear curve fitting glycerol permeability at 18°C, 28°C, and 32°C. Reproduced from [Wang *et al.*, 2014] with permission.

the least squares linear fit for P_s at 1.5M glycerol was 0.997 at 18°C, 28°C, and 32°C (Figure 31(d)). E_{Ps} was 393.3 kJ/mol and P_{sg} was 9.33×10^{-18} m/s.

The values of L_{pg}, P_{sg}, E_{Lp}, and E_{Ps} in the presence of 1.5M Me_2SO, EG, and glycerol solutions are shown in Table 9. In addition, the cryobiological parameters of some other cell types are shown. The E_{Ps} of pADSCs was similar to that of other cell types, but L_{pg}, P_{sg}, and E_{Lp} in Me_2SO and EG were much smaller than those of other cell types. The permeability coefficient of pADSCs in glycerol was significantly different to the common cell types, and P_{sg} and E_{Ps} were considerably outside of the usual activity energy. These results indicated that glycerol had barely diffused into the pADSCs at low temperatures (below 15°C) and the transported rate of glycerol into the cell at high

Table 9. The permeability coefficients of pADSCs compared with those of other cell types [Wang *et al.*, 2014].

Cell type	CPA	L_{pg} 10^{-14} m/s/Pa	E_{Lp} kJ/mol	P_{sg} 10^{-8}m /s	E_{ps} kJ/mol	Reference
Porcine adipose derived-stem cell	Me_2SO	0.26	9.5	0.27	57.4	Wang *et al.* [2014]
	EG	0.30	14.2	0.24	68.0	
	Glycerol	0.24	10.3	9.3×10^{-10}	393.3	
Porcine smooth muscle cells	Me_2SO	0.74	59.5	0.91	80.3	Wusteman and Pegg [2001]
Porcine aorta endothelial cells	Me_2SO	0.97	73.5	1.36	78.7	
Human cornea endothelial cells	Me_2SO	1.03	67.4	2.25	66.5	Ebertz and McGann [2004]; Chuenkhum and Cui [2006]
	PG	1.21	66.9	2.26	75.7	
Human cornea keratocytes	Me_2SO	1.24	57.3	0.17	62.7	
	PG	1.13	62.2	1.72	64.0	
Human cornea epithelial cells	Me_2SO	1.22	66.5	1.07	72.3	
	PG	1.63	60.7	1.15	75.7	

temperatures was much faster than that at low temperatures. This interesting membrane transportation phenomenon and the significant temperature-dependent membrane permeabilities may provide an ideal model to study transmembrane transportation processes of CPAs.

5.5. *Membrane Transport Properties of Pig Iliac Endothelial Cells*

In order to fully explore the potential applications of nanoparticles in biopreservation, Yue *et al.* [2014] investigated the effects of hydroxyapatite nanoparticles (HA-NPs) on the osmotic properties of PIEC membrane using the microperfusion system, including the cell membrane hydraulic conductivity (L_p) at −2°C, 5°C, 14°C and 23°C, and its activation energy. These findings are of potential significance to optimization of nanoparticle-aided cryopreservation.

The HA-NPs (Nanjing Emperor Nano Material Co., Ltd, China) had a non-spherical shape with a declared 20 nm diameter. The suspensions of HA-NPs were prepared in PBS and 3× PBS with 0.01 wt% sodium citrate and 0.005 wt% sodium hexametaphosphate used as the dispersant. In order to obtain a highly even distribution of the NPs, the solutions were exposed to ultrasonic wave irradiation generated using an ultrasonic cleaner (Shanghai KUDOS Ultrasonic Instrument Co., Ltd, China) for 5 mins. The cell osmotic responses were recorded using the CCD camera after they were transferred from isotonic (PBS, 290 mOsm; PBS with dispersant, 294 mOsm; or PBS with dispersant and 0.01wt% HA-NPs, 293 mOsm) to hypertonic solutions (3× PBS, 875 mOsm; 3× PBS with dispersant, 878 mOsm; or 3× PBS with dispersant and 0.01wt% HA-NPs, 877 mOsm) at −2°C, 5°C, 14°C and −2°C, respectively. For convenience, the osmotic shift experiments were divided into three groups: Group I, from PBS to 3× PBS; Group II, from PBS to 3× PBS with dispersant; Group III, from PBS to 3× PBS with dispersant and 0.01 wt% HA-NPs. The isotonic cell volumes, the osmotically inactive cell volumes, and the hydraulic conductivity values were averaged and compared between the three groups by using student's *t*-test. A probability level of ($p < 0.05$) was set to determine significance among groups.

Figure 32 depicts the typical cell dehydration processes for the three groups of experiments at −2°C, 5°C, 14°C and 23°C. As can be seen, the theoretically predicted transient volumes agree well with those of the experiments, as it is obvious that the cell response becomes slower along with the decrease in temperature.

Figure 33 depicts the student *t*-test results between the L_p values obtained in Group I and II or Group I and III at the four temperatures, respectively.

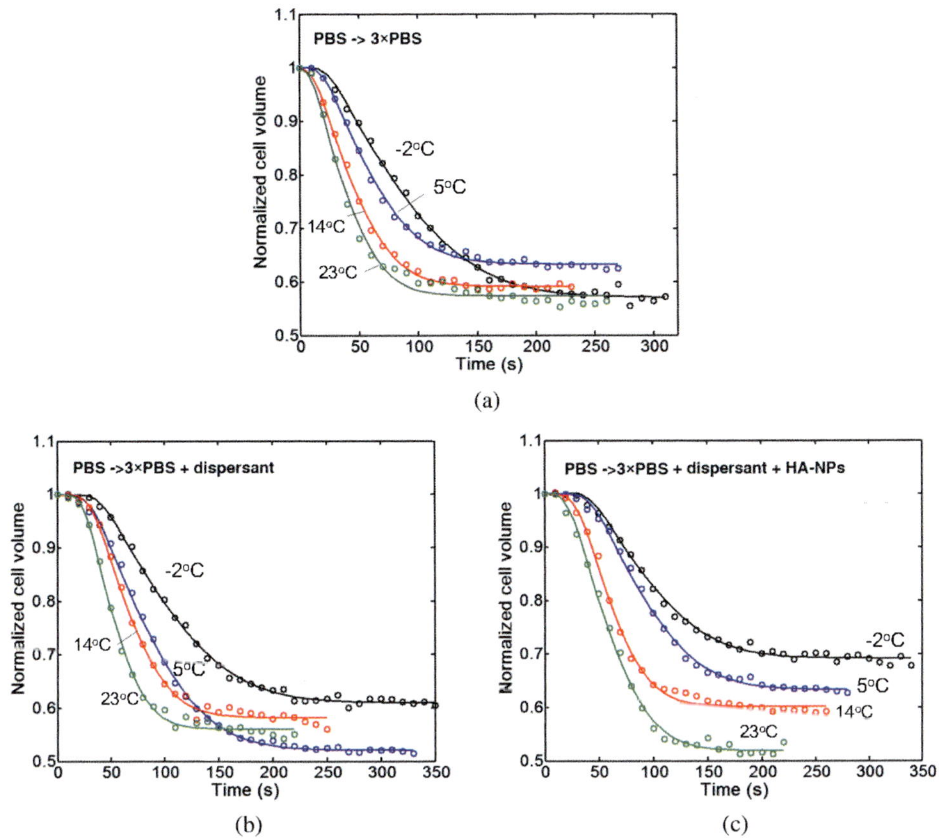

Figure 32. Effect of HA-NPs on cell volume responses during osmotic shifts at –2°C, 5°C, 14°C and 23°C. (a) Osmotic shifts from PBS solution (290 mOsm) to 3× PBS solution (875 mOsm). (b) Osmotic shifts from PBS solution with dispersant (294 mOsm) to 3× PBS solution with dispersant (878 mOsm). (c) Osmotic shifts from PBS solution with dispersant and 0.01 wt% HA-NPs (293 mOsm) to 3× PBS solution with dispersant and 0.01 wt% HA-NPs (877 mOsm). Four typical cells were picked up for each group (Group I, II and III). Reproduced from [Yue *et al.*, 2014] with permission.

The statistical analysis indicates that the introduction of the dispersant does not significantly change the L_p values while the introduction of the NPs (together with the dispersant) does change L_p values significantly. Additionally, the L_p values were also listed in Table 10 for reference.

The logarithms of the L_p values were plotted against the reciprocal of the absolute temperature (Figure 34), and then the least-square linear fits were applied. As can be seen, the experimental data obey the Arrhenius relationship very well, and no discontinuities are visible in the Arrhenius plots of L_p between –2°C and 23°C, while the fit (R-squared) tends to decrease with the introduction of the HA-NPs. The reference values for L_p at the reference

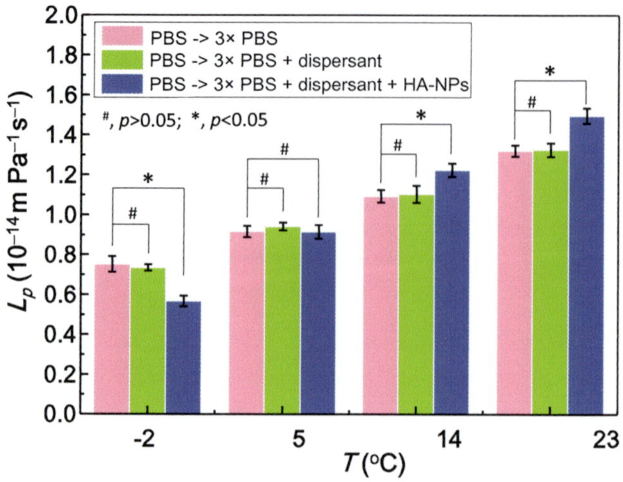

Figure 33. Effect of HA-NPs on the L_p values of PIECs at −2°C, 5°C, 14°C and 23°C. The error bars represent the standard errors (SEM). Reproduced from [Yue *et al.*, 2014] with permission.

Table 10. Hydraulic conductivities of PIECs [Yue *et al.*, 2014].

	L_p (10^{-14}m/Pa/s), [mean ± SD (n)]		
Temperature (°C)	3× PBS	3× PBS + the dispersant	3× PBS + the dispersant + HA-NPS
−2	0.75 ± 0.25 (36)	0.73 ± 0.01 (37)	0.57 ± 0.18 (46)
5	0.92 ± 0.18 (36)	0.94 ± 0.13 (45)	0.91 ± 0.23 (44)
14	1.09 ± 0.20 (40)	1.10 ± 0.28 (43)	1.22 ± 0.20 (37)
23	1.32 ± 0.18 (44)	1.3 ± 0.25 (51)	1.5 ± 0.23 (37)

temperature (T_{ref} = 273.15K), i.e. L_{pg}, are 0.77 (n = 156), 0.78 (n = 176), and 0.65 (n = 164) (10^{-14}m/Pa/s), while the activation energies for L_p, i.e. E_{Lp}, are 15.62 (n = 156), 15.09 (n = 176) and 26.23 (n = 164) (KJ/mol) for Groups I, II and III, respectively. Linear regression analysis on the Arrhenius plots revealed that there was no significant difference between Groups I and II, while there was a significant difference between Groups I and III in the slopes. For the intercepts, no significant differences were found between any two of the groups. The results indicate that the introduction of the dispersant alone has no significant influences on both the values of L_{pg} and E_{Lp}, while the introduction of HA-NPs (together with the dispersant) significantly increases E_{Lp} (meanwhile, L_{pg} is not significantly changed). Although the exact mechanism for the effect of the nanoparticles on water transport

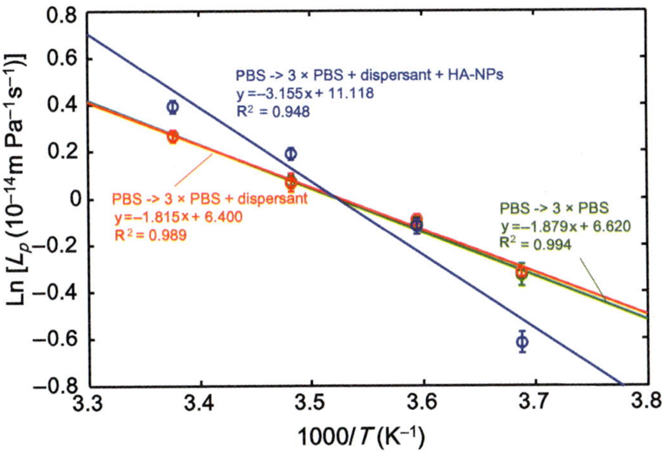

Figure 34. Arrhenius plots of the L_p values of PIECs. L_{pg} and E_{Lp} are 0.77×10^{-14} m/Pa/s and 15.62 kJ/mol ($R^2 = 0.9939$), 0.78×10^{-14} m/Pa/s and 15.09 kJ/mol ($R^2 = 0.9888$), and 0.65×10^{-14} m/Pa/s and 26.23 kJ/mol ($R^2 = 0.9476$) for Group I, II, and III, respectively. Group I: PBS – > 3× PBS; Group II: PBS – > 3× PBS + dispersant; and Group III: PBS – > 3× PBS + dispersant + 0.01 wt% HA-NPs. The error bars represent the standard errors (SEM). Reproduced from [Yue *et al.*, 2014] with permission.

across cell membrane is still unknown, this work provides a pioneering effort on promoting the applications of nanoparticles in cryopreservation.

6. Summary

This chapter summarizes the commonly used microperfusion systems and their applications in measuring kinetic changes of cell volume during addition/removal of CPAs. Both the modern developments on precise description of the concentration and the temperature at the cellular scale in the microperfusion chamber are surveyed. The experimentally or theoretically predetermined profiles for extracellular concentration changes are of primary importance for measurements of cell membrane permeabilities by nonlinear-fitting with high precision.

References

Adams, R. B., W. H. Voelker and E. C. Gregg (1967). Electrical counting and sizing of mammalian cells in suspension — an experimental evaluation. *Physics in Medicine and Biology*, 12(1), pp. 79–92.

Applegate, R. W., J. Squier, T. Vestad, J. Oakey and D. W. M. Marr (2004). Optical trapping, manipulation, and sorting of cells and colloids in microfluidic systems with diode laser bars. *Optics Express*, 12(19), pp. 4390–4398.

Ateya, D. A., F. Sachs, P. A. Gottlieb, S. Besch and S. Z. Hua (2005). Volume cytometry: microfluidic sensor for high-throughput screening in real time. *Analytical Chemistry*, 77(5), pp. 1290–1294.

Bastard, J. P., J. Cuevas, S. Cohen, C. Jardel and B. Hainque (1994). Percutaneous adipose tissue biopsy by mini-liposuction for metabolic studies. *Journal of Parenteral and Enteral Nutrition*, 18(5), pp. 466–468.

Berger, I., D. J. Fitzgerald and T. J. Richmond (2004). Baculovirus expression system for heterologous multiprotein complexes. *Nature Biotechnology*, 22(12), pp. 1583–1587.

Berrada, M. S. and J. C. Bischof (2001). Evaluation of freezing effects on human microvascular-endothelial cells (HMEC). *CryoLetters*, 22(6), pp. 353–366.

Chao, P. H. G., A. C. West and C. T. Hung (2006). Chondrocyte intracellular calcium, cytoskeletal organization, and gene expression responses to dynamic osmotic loading. *American Journal of Physiology-Cell Physiology*, 291(4), pp. C718–C725.

Chen, H. H., J. J. P. Purtteman, S. Heimfeld, A. Folch and D. Gao (2007). Development of a microfluidic device for determination of cell osmotic behavior and membrane transport properties. *Cryobiology*, 55(3), pp. 200–209.

Chen, H. H., H. Shen, S. Heimfeld, K. K. Tran, J. Reems, A. Folch and D. Y. Gao (2008). A microfluidic study of mouse dendritic cell membrane transport properties of water and cryoprotectants. *International Journal of Heat and Mass Transfer*, 51(23–24), pp. 5687–5694.

Chuenkhum, S. and Z. F. Cui (2006). The parameter conversion from the Kedem-Katchalsky model into the two-parameter model. *CryoLetters*, 27(3), pp. 185–199.

Devireddy, R. V., G. Li and S. P. Leibo (2006). Suprazero cooling conditions significantly influence subzero permeability parameters of mammalian ovarian tissue. *Molecular Reproduction and Development*, 73(3), pp. 330–341.

Diller, K. R. and D. A. Bradley (1984). Measurement of the water permeability of single human granulocytes on a microscopic stopped-flow mixing system. *Journal of Biomechanical Engineering*, 106(4), pp. 384–393.

Drugmand, J.-C., Y.-J. Schneider and S. N. Agathos (2012). Insect cells as factories for biomanufacturing. *Biotechnology Advances*, 30(5), pp. 1140–1157.

Duffy, D. C., J. C. McDonald, O. J. A. Schueller and G. M. Whitesides (1998). Rapid prototyping of microfluidic systems in poly(dimethylsiloxane). *Analytical Chemistry*, 70(23), pp. 4974–4984.

Ebertz, S. L. and L. E. McGann (2004). Cryoprotectant permeability parameters for cells used in a bioengineered human corneal equivalent and applications for cryopreservation. *Cryobiology*, 49(2), pp. 169–180.

Elmoazzen, H. Y., J. A. W. Elliott and L. E. McGann (2009). Osmotic transport across cell membranes in nondilute solutions: a new nondilute solute transport equation. *Biophysical Journal*, 96(7), pp. 2559–2571.

Estrada-Mondaca, S., O. T. Ramirez and L. Palomares (2005). Principles and applications of the insect cell-baculovirus expression vector system. In S. S. Ozturk and W.-S. Hu (eds.) *Cell Culture Technology for Pharmaceutical and Cell-Based Therapies* (CRC Press), pp. 627–692.

Estrov, Z. (2009). Stem cells and somatic cells: reprogramming and plasticity. *Clinical Lymphoma and Myeloma*, 9, Supplement 3(0), pp. S319–S328.

Gómez-Sebastián, S., D. M. Pérez-Filgueira, E. Gómez-Casado, M. C. Nuñez, I. Sánchez-Ramos, E. Tabarés and J. M. Escribano (2008). DIVA diagnostic of Aujeszky's disease using an insect-derived virus glycoprotein E. *Journal of Virological Methods*, 153(1), pp. 29–35.

Gao, D. Y., C. T. Benson, C. Liu, J. J. McGrath, E. S. Critser and J. K. Critser (1996). Development of a novel microperfusion chamber for determination of cell membrane transport properties. *Biophysical Journal*, 71(1), pp. 443–450.

Gao, D. Y., J. J. Mcgrath, J. Tao, C. T. Benson, E. S. Critser and J. K. Critser (1994). Membrane-transport properties of mammalian oocytes — a micropipette perfusion technique. *Journal of Reproduction and Fertility*, 102(2), pp. 385–392.

Gilmore, J. A., L. E. Mcgann, J. Liu, D. Y. Gao, A. T. Peter, F. W. Kleinhans and J. K. Critser (1995). Effect of cryoprotectant solutes on water permeability of human spermatozoa. *Biology of Reproduction*, 53(5), pp. 985–995.

Gu, J. H., X. D. Pu, J. L. Yu and W. Qiao, P (2002). Measuring the diffusion coefficient of solution by optical method. *Physics Experiment of University*, 15(4), pp. 34–37.

Hancock, K., S. Narang, S. Pattabhi, M. L. Yushak, A. Khan, S.-c. Lin, R. Plemons, M. J. Betenbaugh and V. C. W. Tsang (2008). False positive reactivity of recombinant, diagnostic, glycoproteins produced in High Five™ insect cells: effect of glycosylation. *Journal of Immunological Methods*, 330(1–2), pp. 130–136.

He, X. and J. C. Bischof (2003). Quantification of temperature and injury response in thermal therapy and cryosurgery. *Critical Reviews in Biomedical Engineering*, 31(5–6), pp. 355–422.

Huh, D., W. Gu, Y. Kamotani, J. B. Grotberg and S. Takayama (2005). Microfluidics for flow cytometric analysis of cells and particles. *Physiological Measurement*, 26(3), pp. R73–R98.

Ikonomou, L., Y. J. Schneider and S. N. Agathos (2003). Insect cell culture for industrial production of recombinant proteins. *Applied Microbiology and Biotechnology*, 62(1), pp. 1–20.

Jacobs, M. H. (1933). The simultaneous measurement of cell permeability to water and to dissolved substances. *Journal of Cellular and Comparative Physiology*, 2(4), pp. 427–444.

Jacobs, M. H. and D. R. Stewart (1932). A simple method for the quantitative measurement of cell permeability. *Journal of Cellular and Comparative Physiology*, 1(1), pp. 71–82.

Kedem, O. and A. Katchalsky (1958). Thermodynamic analysis of the permeability of biological membranes to non-electrolytes. *Biochimica et Biophysica Acta*, 27(2), pp. 229–246.

Kleinhans, F. W. (1998). Membrane permeability modeling: Kedem-Katchalsky vs a two-parameter formalism. *Cryobiology*, 37(4), pp. 271–289.

Kleinhans, F. W., V. S. Travis, J. Y. Du, P. M. Villines, K. E. Colvin and J. K. Critser (1992). Measurement of human sperm intracellular water volume by electron-spin-resonance. *Journal of Andrology*, 13(6), pp. 498–506.

Kubitschek, H. E. (1958). Electronic counting and sizing of bacteria. *Nature*, 182(4630), pp. 234–235.

Lazzeri, E., A. Peired, L. Ballerini and L. Lasagni (2012). Adult stem cells in tissue homeostasis and disease. In P. S. Najman (ed.) *Current Frontiers and Perspectives in Cell Biology*, ISBN: 978-953-51-0544-2.

Lea-Currie, Y. R., D. J. Duffin and B. M. Buehrer (2011). Use of adipose-derived stem cells in high-throughput screening to identify modulators of lipogenesis. *Methods in Molecular Biology*, 702, pp. 359–368.

Leibo, S. P. (1980). Water permeability and its activation-energy of fertilized and unfertilized mouse ova. *Journal of Membrane Biology*, 53(3), pp. 179–188.

Levin, R. L., E. G. Cravalho and C. E. Huggins (1976). Membrane model describing effect of temperature on water conductivity of erythrocyte-membranes at subzero temperatures. *Cryobiology*, 13(4), pp. 415–429.

Li, W. J., X. L. Zhou, H. S. Wang and B. L. Liu (2012). Numerical analysis to determine the performance of different oocyte vitrification devices for cryopreservation. *CryoLetters*, 33(2), pp. 144–150.

Liu, C., C. T. Benson, D. Y. Gao, B. W. Haag, L. E. Mcgann and J. K. Critser (1995). Water permeability and its activation-energy for individual hamster pancreatic-islet cells. *Cryobiology*, 32(5), pp. 493–502.

Liu, W., G. Zhao, Z. Shu, T. Wang, K. Zhu and D. Gao (2015). High-precision approach based on microfluidic perfusion chamber for quantitative analysis of biophysical properties of cell membrane. *International Journal of Heat and Mass Transfer*, 86, pp. 869–879.

Lucke, B. and M. McCutcheon (1932). The living cell as an osmotic system and its permeability to water. *Physiological Reviews*, 12(1), pp. 68–139.

Macey, R. I. and R. E. L. Farmer (1970). Inhibition of water and solute permeability in human red cells. *Biochimica et Biophysica Acta*, 211(1), pp. 104–106.

McDonald, J. C., D. C. Duffy, J. R. Anderson, D. T. Chiu, H. K. Wu, O. J. A. Schueller and G. M. Whitesides (2000). Fabrication of microfluidic systems in poly(dimethylsiloxane). *Electrophoresis*, 21(1), pp. 27–40.

McGann, L. E., A. R. Turner and J. M. Turc (1982). Microcomputer interface for rapid measurements of average volume using an electronic particle counter. *Medical & Biological Engineering & Computing*, 20(1), pp. 117–120.

McGrath, J. J. (1985). A microscope diffusion chamber for the determination of the equilibrium and non-equilibrium osmotic response of individual cells. *Journal of Microscopy-Oxford*, 139(Sep), pp. 249–263.

McGrath, J. J. (1997). Quantitative measurement of cell membrane transport: technology and applications. *Cryobiology*, 34(4), pp. 315–334.

Metz, S. W. and G. P. Pijlman (2011). Arbovirus vaccines; opportunities for the baculovirus-insect cell expression system. *Journal of Invertebrate Pathology*, 107(Supplement), pp. S16–S30.

Nirmalanandhan, V. S. and G. S. Sittampalam (2009). Stem cells in drug discovery, tissue engineering, and regenerative medicine: emerging opportunities and challenges. *Journal of Biomolecular Screening*, 14(7), pp. 755–768.

Paganelli, C. V. and A. K. Solomon (1957). The rate of exchange of tritiated water across the human red cell membrane. *Journal of General Physiology*, 41(2), pp. 259–277.

Philibert, J. (2005). One and a half century of diffusion: Fick, Einstein, before and beyond. *Diffusion Fundamentals*, 2(1), pp. 1–10.

Rubin SC, S. G. (1993). *Ovarian Cancer* (McGraw-Hill Education, London).

Sidel, V. W. and A. K. Solomon (1957). Entrance of water into human red cells under an osmotic pressure gradient. *The Journal of General Physiology*, 41(2), pp. 243–257.

Smith, A. U. and C. Polge (1950). Survival of spermatozoa at low temperatures. *Nature*, 166(4225), pp. 668–669.

Stevens A, L. J. (1996). *Human Histology*, 2nd edn. (Elsevier Health Science, The Netherlands).

Takamatsu, H., Y. Komori, S. Zawlodzka and M. Fujii (2004). Quantitative examination of a perfusion microscope for the study of osmotic response of cells. *Journal of Biomechanical Engineering-Transactions of the ASME*, 126(4), pp. 402–409.

Vaughn, J. L., R. H. Goodwin, G. J. Tompkins and P. McCawley (1977). The establishment of two cell lines from the insect Spodoptera frugiperda (Lepidoptera; Noctuidae). *In Vitro*, 13(4), pp. 213–217.

Walcerz, D. B. and K. R. Diller (1991). Quantitative light-microscopy of combined perfusion and freezing processes. *Journal of Microscopy-Oxford*, 161, pp. 297–311.

Wang, J., G. Zhao, P. Zhang, Z. Wang, Y. Zhang, D. Gao, P. Zhou and Y. Cao (2014). Measurement of the biophysical properties of porcine adipose-derived stem cells by a microperfusion system. *Cryobiology*, 69(3), pp. 442–450.

Wang, J., K. Zhu, G. Zhao, J. Ren, C. Yue and D. Gao (2013). Dual dependence of cryobiogical properties of sf21 cell membrane on the temperature and the concentration of the cryoprotectant. *PLoS ONE*, 8(9), p. e72836.

Woods, E., J. Liu, J. Gilmore, T. Reid, D. Gao and J. Critser (1999). Determination of human platelet membrane permeability coefficients using the Kedem-Katchalsky formalism: Estimates from two- vs three-parameter fits. *Cryobiology*, 38(3), pp. 200–208.

Wu, Z. C., Y. Awakura, S. Ando and H. Majima (1990). Determination of the diffusion-coefficients of Cucl2, Fecl3, Cuso4, and Fe2(So4)3 in aqueous-solutions. *Materials Transactions JIM*, 31(12), pp. 1065–1071.

Wusteman, M. C. and D. E. Pegg (2001). Differences in the requirements for cryopreservation of porcine aortic smooth muscle and endothelial cells. *Tissue Engineering*, 7(5), pp. 507–518.

Xiao, B., C. Wu, Y. Sun and Z. Jin (2009). Conductance method for the determination of ion diffusion coefficient through a nanochannel. *Micro & Nano Letters*, 4(4), pp. 192–197.

Yoshimori, T. and H. Takamatsu (2009). 3-D measurement of osmotic dehydration of isolated and adhered PC-3 cells. *Cryobiology*, 58(1), pp. 52–61.

Yue, C., G. Zhao, J. Yi, C. Gao, L. Shen, Y. Zhang, Z. Wang and W. Liu (2014). Effect of hydroxyapatite nanoparticles on osmotic responses of pig iliac endothelial cells. *Cryobiology*, 69(2), pp. 273–280.

Zhang, A., L. X. Xu, G. A. Sandison and S. Cheng (2006). Morphological study of endothelial cells during freezing. *Physics in Medicine and Biology*, 51(23), pp. 6047–6060.

Zhao, G. (2010). A universal self-adaptive time-varying function for extracellular concentration during osmotic shift for curve-fitting permeability coefficients of cell membrane. *CryoLetters*, 31(3), pp. 279–290.

Zhao, G. (2012). Perfusion Microscope. *SIPO*. S. I. P. O. o. t. P.R.C. (China), Pacific China Intellectual Property Co., Ltd.

Zhao, G., K. Kurata and H. Takamatsu (2012). Measurement of membrane hydraulic conductivity of bovine carotid artery endothelial cells using a perfusion microscope. *CryoLetters*, 33(3), pp. 232–240.

Section 3

Nanoscale

Chapter 8

Polymeric Nanoparticles for Cryobiological Applications

Xiaoming He, Wei Rao and Hai Wang

1. Introduction

Nanoparticles show great promise for biomedical applications, due to their capability of effectively transporting and delivering theranostic agents to the desired location within biological systems for therapy and diagnosis. In particular, polymeric nanoparticles can be readily designed in their structures, compositions, and surface properties to achieve responsiveness to environmental stimuli including but not limited to temperature, pH, and enzymes [Mura *et al.*, 2013; Wang *et al.*, 2015]. Controlled polymerization and chemical modification have enabled the engineering of advanced multifunctional polymeric nanoparticles with precise control over size, surface charge, and functionalization. Due to these advantages of polymeric nanoparticles, they have recently been explored for applications in the field of cryobiology, such as cryotherapy and cryopreservation [Zhang *et al.*, 2009; Rao *et al.*, 2014; Rao *et al.*, 2015], as well as hibernation [Colonna *et al.*, 2011]. In this chapter, the fundamentals and recent advancement of polymeric nanoparticles for cryobiological applications are summarized first, followed by a brief discussion of the potential challenges and approaches for overcoming the challenges to further advance this exciting field.

2. Important Properties of Polymeric Nanoparticles

Polymeric nanoparticles offer unique physical, chemical, and biological properties [Petros and DeSimone, 2010] and the size, shape, and surface characteristics of nanoparticles dictate their bioavailability, intracellular

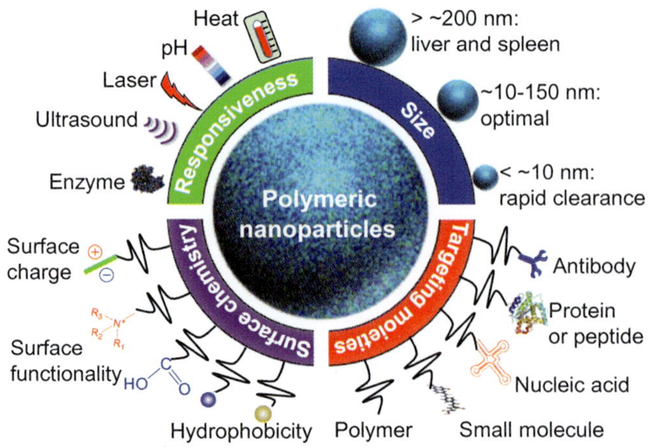

Figure 1. A schematic illustration of important properties of polymeric nanoparticles including size, surface chemistry, targeting moiety, and stimulus-responsiveness.

distribution, and stability (Figure 1). In addition, the presence of nanoparticles can change the thermal and mechanical properties of targeted tissues or cells as a result of their interactions with the environment. In the following section, the characteristics of polymeric nanoparticles will be highlighted with emphasis on the important properties that can effect their applications in cryobiology.

2.1. *Size*

The size of polymeric nanoparticles can greatly influence their uptake by cells, intracellular distribution, circulation time or half-life in blood, biodistribution *in vivo*, and mechanisms of clearance (Figure 1). Large particles (> 0.2 μm) easily accumulate in the liver and spleen, while small nanoparticles (< 10 nm) are cleared rapidly from the blood by extravasation or renal clearance [Elsabahy and Wooley, 2012]. Polymeric nanoparticles of intermediate size (10–150 nm) are advantageous for biomedical applications [Gao *et al.*, 2005; Cabral *et al.*, 2011; Chauhan *et al.*, 2012; Sykes *et al.*, 2014]. Furthermore, polymeric nanoparticle-encapsulated drug within this size range are believed to preferentially accumulate in diseased tissues (e.g., tumor) with leaky vasculature or high vascular permeability and without lymphatic drainage, known as the enhanced permeability and retention (EPR) effect [Farokhzad and Langer, 2009; Cabral *et al.*, 2011; Fang *et al.*, 2011; Chauhan *et al.*, 2012].

2.2. *Surface Charge*

The surface charge can greatly affect the fate of nanoparticles at different stages, such as physical stability, interactions with extracellular matrix, blood stability, cellular uptake, and endocytosis [He *et al.*, 2010; Verma and Stellacci, 2010]. A negatively charged surface is often thought to have good half-life in blood circulation although nanoparticles with a highly negatively charged surface may be difficult for cells to take up. This is because living cells have a net negatively charged plasma membrane. Nanoparticles with a neutral surface may aggregate in aqueous solution if the surface is hydrophobic. Highly positive charge may induce occlusion of blood vessels as a result of their absorption of the many negatively charged macromolecules in blood to form large aggregates [Bertrand *et al.*, 2014]. In addition, highly positively charged nanoparticles interacts with negatively charged components of extracellular matrix, which may hinder their penetration depth in tissue [Stylianopoulos *et al.*, 2010]. At the cellular level, a positive surface charge facilitates the binding between cells and nanoparticles because the negatively charged plasma membrane of living cells has high electrostatic affinity to the positively charged nanoparticles. However, highly positive charge may induce cytotoxicity of nanoparticles and initiate immunological reactions [He *et al.*, 2010; Frohlich, 2012].

2.3. *Targeting Moiety*

Polymeric nanoparticles can be readily modified with moieties to enhance their selective binding to specific receptors overexpressed on the target cells (Figure 1). Targeting moieties that have been used in drug delivery include antibodies, proteins and derivatives, peptides, carbohydrates and derivatives, aptamers, and small molecules [Yu *et al.*, 2012; Steichen *et al.*, 2013]. For example, antibodies can bind with their corresponding antigens with extremely high affinity and specificity [Köhler and Milstein, 1975]. Therefore, nanoparticles surface-decorated with an antibody (or ligand) of a specific antigen (or receptor) can be used to specifically deliver therapeutic agents into the cells via surface receptor-mediated endocytosis. For example, to target cancer cells overexpressed with the variant CD44 receptors, its ligand known as hyaluronan (HA, hydrophilic) has been conjugated with a hydrophobic polymer to form an amphiphilic product for synthesizing nanoparticles [Bulpitt and Aeschlimann, 1999; YoungáChoi *et al.*, 2009]. This modification of polymeric nanoparticles is beneficial in terms of enhancing their uptake by the target cells and reducing toxicity to non-target cells.

2.4. *Stimuli-responsiveness*

Stimuli-responsive polymeric nanoparticles can respond to the stimuli, such as temperature, pH, enzymatic, optical, and electrical signals in their surrounding environment and regulate the release of the encapsulated cargo [Stuart *et al.*, 2010].

Thermally responsive nanoparticles are among the most investigated stimuli-responsive platforms and have been widely explored in cancer therapy. Thermally responsive nanoparticles with a lower critical solution temperature (LCST) are one kind of temperature sensitive nanoparticles that can retain the encapsulated cargo at body temperature (37°C) and rapidly release the cargo at a temperature below the LCST. This characteristic is very useful for encapsulating small therapeutic agents to treat cancer in combination with cryotherapy, to allow controlled drug delivery when cryogenic agent is applied at the tumor site and achieve a synergistic effect between chemo and cryotherapy. This controlled release of drug can minimize the drug systemic toxicity to normal tissue without cryogenic treatment, further enhancing the safety and translational value of the combined therapy. In addition, non-permeable cryoprotectant encapsulated in this type of thermally responsive nanoparticles may facilitate its intracellular release after cellular uptake to achieve more effective cryopreservation. For example, Pluronic F127 based thermally responsive nanocapsules were used for efficient delivery of trehalose into mammalian cells [Zhang *et al.*, 2009; Rao *et al.*, 2015].

The use of pH as the stimulus has also been explored to control the delivery of payloads in specific organs (e.g., the gastrointestinal tract) or subcellular organelles (e.g., endosomes or lysosomes), as well as to trigger the release of the drug in response to the pH change associated with pathological situations (e.g., cancer or inflammation). Two main strategies have been performed: using polymers (polyacids or polybases) with ionizable groups that undergo changes in conformation and/or solubility in response to environmental pH variation, and designing polymeric systems with acid-sensitive bonds whose cleavage enables the release of molecules anchored at the polymer backbone, the modification of the charge of the polymer, or the exposure of targeting ligands [Ganta *et al.*, 2008; Mura *et al.*, 2013].

3. Effect of Polymeric Nanoparticles on Thermophysical Properties Important for Cryobiological Applications

Nanoparticles may also be used to modulate the thermal property of fluids [Keblinski *et al.*, 2005]. For example, studies have shown that adding a small amount of metal nanoparticles or nanotubes (1% w/v) into an aqueous solution can significantly increase the thermal conductivity of the solution [Choi *et al.*, 2001; Eastman *et al.*, 2001]. The capability of augmenting cancer cryoablation has also been reported for inorganic nanoparticles including metal and magnetic nanoparticles, which was attributed to the enhancement of heat transfer for better freezing cells and tissue [Yan and Liu, 2008; Di *et al.*, 2012]. However, polymeric nanoparticles have an intrinsic thermal conductivity lower than that of metals and nanotubes and their capability of enhancing heat transfer is unknown at this time. A few studies showed polymeric nanoparticles with addition of graphite, nanotubes, or carbon nanofibers could improve the thermal conductivity of polymer composites [Elgafy and Lafdi, 2005; Ye *et al.*, 2006; Sui *et al.*, 2008; Zhi *et al.*, 2009]. In addition, nanoparticles with a gold core and polymeric shell have been reported to exhibit a relatively high thermal conductivity [Ge *et al.*, 2005]. These investigations demonstrate that polymeric nanoparticles conjugated with metal or carbon-based nanomaterials may enhance thermal transport.

Besides thermal conductivity, polymeric nanoparticles may be designed to modulate the ice nucleation and/or morphology during freezing to induce better destruction of tumor cells, as with other adjuvants reported in the literature [Koushafar and Rubinsky, 1997; Pham *et al.*, 1999; Han and Bischof, 2004; Wang *et al.*, 2008]. However, further studies are needed to confirm this hypothesis.

4. Polymeric Nanoparticles for Cryotherapy

Cryotherapy (also called cryosurgery or cryoablation) is an energy-based, minimally invasive surgical technique that has been investigated to treat a variety of diseases including cancer, by freezing the diseased tissues to subzero temperatures using one or more cryoprobes [Gage and Baust, 2007; Geha and Abdelhady, 2008; Sabel, 2008; Callstrom *et al.*, 2013]. The advantages of cryotherapy include focal treatment, minimal invasiveness, negligible side effect, and ice ball formation that could be monitored with the widely used imaging techniques in the operation room including ultrasonography (US), computed topography (CT), and magnetic resonance imaging (MRI). During freezing, direct cellular injury occurs as a result of intracellular ice formation (IIF) that could compromise the integrity of the cell membrane and sub-cellular organelles. Significant IIF has been recognized as a lethal event to cells during freezing. In addition, the formation of extracellular ice results in increase in solute concentration (freeze-concentration), leading to cell dehydration, protein denaturation, and lipid phase transition within the cells [Gage and Baust, 1998; Hoffmann and Bischof, 2002; He and Bischof,

2003; Bischof and He, 2005; He, 2011]. Although freeze-concentration and IIF induced cell injury are dependent on all the freezing parameters including cooling rate, end temperature, and holding time at the end temperature experienced by cells, the end temperature is usually considered as the most important parameter in determining cryoinjury during cryotherapy [Gage and Baust, 1998; Hoffmann and Bischof, 2002; He and Bischof, 2003; He, 2011]. An end temperature not higher than –20°C is generally required to kill cancer cells [Gage and Baust, 1998; Hoffmann and Bischof, 2002; He and Bischof, 2003; Bischof and He, 2005; Gage and Baust, 2007; Geha and Abdelhady, 2008; Gilstrap *et al.*, 2011; He, 2011; Callstrom *et al.*, 2013]. However, the temperature at the outer peripheral of a frozen tumor iceball created during cryoablation is higher than –20°C [He and Bischof, 2005; Gilstrap *et al.*, 2011], which may compromise the safety and efficacy of cryotherapy for cancer treatment based on the aforementioned intraoperative imaging techniques. This is because it is the iceball edge that is monitored by the imaging techniques.

To improve the efficacy of cryoablation at high subzero temperatures (> –20°C), various strategies have been explored recently to augment cryoinjury to cancer cells. For example, the modification of ice morphology by adding glycine (a nonessential amino acid) or an excess of electrolytes (e.g., sodium and potassium chloride) has been found to enhance cryoinjury to prostate and breast tumor cells [Han and Bischof, 2004; Wang *et al.*, 2008]. Antifreeze proteins used by arctic fish to resist freezing have been shown to actually enhance cryoinjury to prostate tumor cells once the cells are frozen, via a mechanism also related to the change of ice morphology [Koushafar and Rubinsky, 1997; Pham *et al.*, 1999]. A significant enhancement in cryoablation of prostate tumor has been achieved by inducing pre-treatment inflammation in the tumor using tumor necrosis factor-alpha (TNF-a) either in its free form or tethered to gold nanoparticles [Chao *et al.*, 2004; Shenoi *et al.*, 2013]. Inorganic nanoparticles (without drug encapsulation) of high thermal conductivity have also been shown to improve the outcome of cryoablation, which was attributed to the enhancement of heat transfer in biological samples by the nanoparticles [Yan and Liu, 2008; Di *et al.*, 2012]. In addition, enhanced suppression of tumor growth has been observed by combining cryoablation with the local injection of immunopotentiator or mild systemic chemotherapy [Forest *et al.*, 2006]. Lastly, anticancer drugs for chemotherapy such as peplomycin, 5-fluorouracil, bleomycin, and doxorubicin in their free form have been shown to enhance cryoinjury to prostate, liver, and lung tumor cells under mild freezing conditions, which was attributed to their capability of enhancing apoptosis in the treated cells [Clarke *et al.*, 2001; Mir and Rubinsky, 2002].

Enhancement of cryotherapy using anticancer drugs is of particular interest because both cryotherapy and anticancer drugs have been used for treating cancer in the clinic. In particular, polymeric nanodrug is considered to be promising for further improving the safety and efficacy of cryotherapy for cancer treatment. As illustrated in Figure 2, this combinational therapy could be designed to achieve specific release of therapeutic agents (e.g., small molecular drugs, proteins, peptides, siRNAs, microRNAs, and plasmid DNAs) and

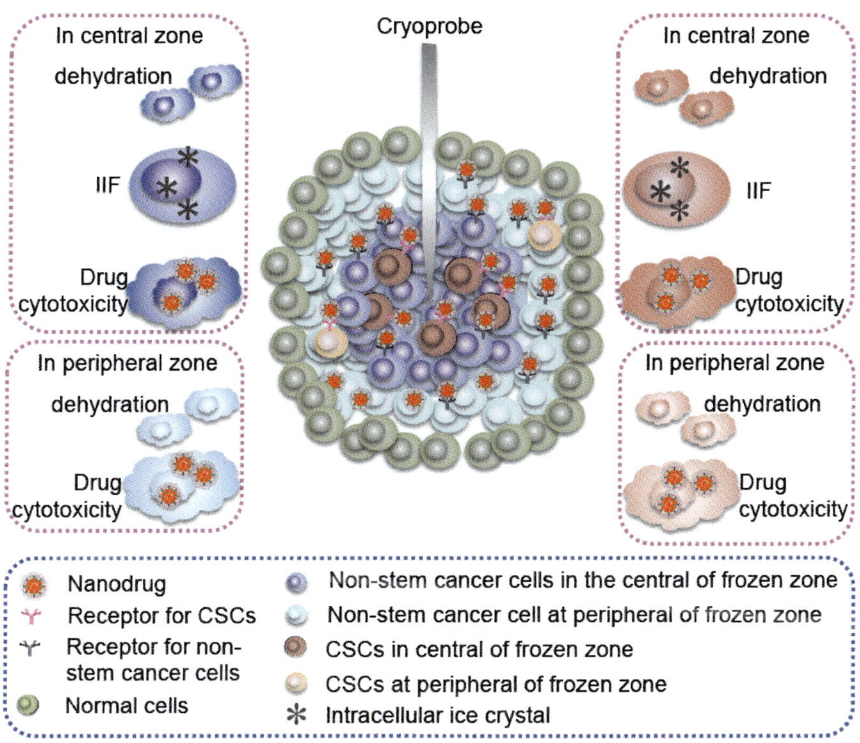

Figure 2. A schematic illustration of polymeric nanodrug-mediated cryotherapy. Nanodrug surface-modified with targeting moieties could selectively bind to non-stem cancer cells and cancer stem cells (CSCs), to potentially achieve synergistic destruction of tumor. Cancer cells including CSCs in the central frozen zone of an frozen tissue iceball could be effectively killed by the combined effect of anticancer drug, intracellular ice formation (IIF), and cell dehydration while cancer cells in the peripheral frozen zone could be effectively killed by the combined effects of anticancer drug and cell dehydration.

deposition of freezing in tumor simultaneously. This is achieved by utilizing the capability of passive targeting of nanomaterials (20–150 nm) as a result of the aforementioned EPR effect of tumor (as compared to normal tissue) vasculature and by modifying the nanomaterials with moieties to actively target tumor. In addition, the synergistic effect between cryo and chemotherapy may allow for a lower freezing threshold and drug dose for complete tumor destruction.

Moreover, by appropriately modifying the nanodrug surface, this combinational therapy could be further designed to specifically target and destroy the cancer stem cells (CSCs), rare cell subpopulations among cancer cells that are capable of self-renewal to maintain their "stemness" and differentiation to create the heterogeneity in tumor [Jordan *et al.*, 2006; Visvader and Lindeman, 2008; Frank *et al.*, 2010]. There is mounting evidence from both basic and clinical studies that the CSCs are responsible for cancer recurrence and metastasis associated with conventional chemotherapy, due at least in part to their high resistance to free anticancer drugs [Al-Hajj *et al.*, 2003; Jordan *et al.*, 2006; Visvader and Lindeman, 2008; Frank *et al.*, 2010]. More recently, it has been shown that the drug resistance of CSCs and multidrug resistant cells can be overcome by delivering anticancer drug using either polymeric or inorganic nanoparticles [Kabanov *et al.*, 2002; Gao *et al.*, 2011; Wang *et al.*, 2011; Rao *et al.*, 2014]. Compared to inorganic nanoparticles, appropriately designed polymeric nanoparticles may be advantageous in terms of bioavailability, controlled release, and minimization of potential toxicity or side effect [Kumari *et al.*, 2010].

A recent study has demonstrated that cryoablation combined with biocompatible Pluronic F127-chitosan nanoparticle-encapsulated doxorubicin (nDOX) can enhance destruction of 3D mammosphere cells enriched with the CSCs [Rao *et al.*, 2014]. This 3D mammosphere model mimics the 3D tumor microenvironment in human patients more closely than the conventional 2D preparation of cancer cells. Typical micrographs showing morphologic differences at day 10 between the samples with various treatments and the corresponding quantitative cell survival data are given in Figures 3a and 3b, respectively. As controls for the nDOX treatment, the corresponding morphologic and quantitative data for samples treated with empty nanoparticles (NP) are also shown in the figure. Clearly, the combined treatments with cryoablation at –2.6 or –8.5°C were not significantly different from the corresponding drug (free doxorubicin, fDOX or nDOX) treatment alone at 37°C. This observation holds true for the combination of NP and cryoablation at –2.6 or –8.5°C and the cell survival was not significantly different from the no drug (ND) control at 37°C, indicating the nontoxic nature of

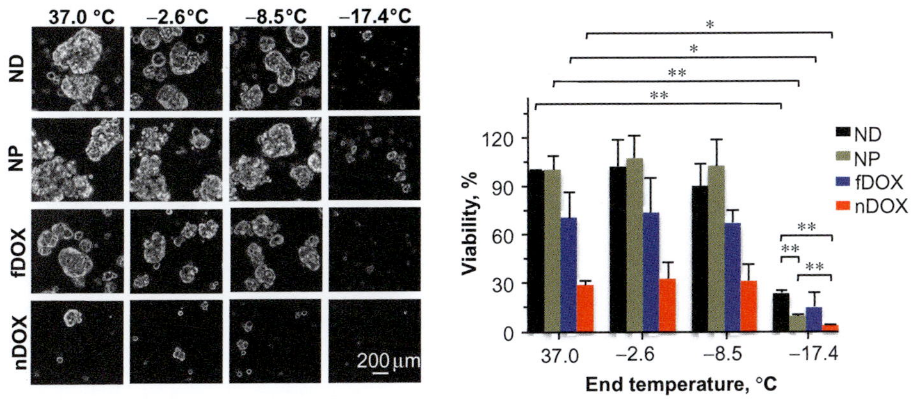

(a) (b)

Figure 3. Morphologic and quantitative survival of mammosphere cells enriched with CSCs after 10 days incubation in culture medium containing no drug (ND), empty nanoparticles (NP), free doxorubicin (fDOX, 1 μg/ml), and nanoparticle-encapsulated doxorubicin (nDOX, 1 μg/ml of DOX) either without (37°C) or with cryoablation (at −2.6, −8.5, and −17.4°C) between 1 and 1.5 h during the incubation. (a) Typical micrographs showing morphologic survival of the mammosphere cells. (b) The corresponding quantitative cell survival. *: $p < 0.05$ and **: $p < 0.01$. Figure reprinted and redrawn from [Rao *et al.*, 2014] with permission from World Scientific Publishing Co. Pte Ltd.

the empty nanoparticles. As shown in Figure 3b, after cryoablation at −17.4°C, all the combined treatments were significantly different from the drug treatment alone at 37°C. However, the combination of fDOX and cryoablation at −17.4°C was not significantly different from the cryoablation treatment alone. In contrast, the combined treatment of nDOX and cryoablation at −17.4°C was significantly different from the cryoablation treatment alone and the combination of nanoparticles and cryoablation.

The CSCs among the mammosphere cells were identified by the overexpression of CD44 and CD133 receptors. Both receptors have been commonly used to judge the presence of CSCs [Al-Hajj *et al.*, 2003; Shipitsin *et al.*, 2007; Wright *et al.*, 2008; Lorico and Rappa, 2011]. As shown in Figure 4, cryoablation alone at all three subzero temperatures did not significantly alter the percentage of $CD44^+CD133^+$ cells although the percentage did appear to increase and decrease slightly at −8.5 and −17.4°C, respectively. The combination of cryoablation at −2.6 and −8.5°C and anticancer drug (either fDOX or nDOX) was not significantly different from the drug treatment alone (at 37°C) in terms of their effect on the percentage of $CD44^+CD133^+$ cells. This observation holds true for the combination of NP and cryoablation at −2.6 or −8.5°C and the percentage was not significantly different from the no drug (ND) control at 37°C. Interestingly, unlike its impact on the viability of all mammosphere cells shown in Figure 3b, the combination of nanoparticles and cryoablation at –17.4°C was not significantly different from cryoablation alone in altering the percentage of $CD44^+CD133^+$ cells although it is significantly different from the nanoparticles treatment alone (NP at 37°C). In other words, the combination of cryoablation and nanoparticles without anticancer drug is ineffective in preferentially destroying the CSCs.

The combined treatment of cryoablation at –17.4°C with nanoparticles, fDOX, or nDOX could significantly lower the percentage of $CD44^+CD133^+$ cells compared to the nanoparticles or drugs alone treatment at 37°C (15.8–17.5%). However, the combination of fDOX and cryoablation at –17.4°C was not significantly different from the cryoablation treatment at –17.4°C alone (12.4 ± 2.4%). In contrast, the combined treatment of nDOX and cryoablation at –17.4°C (0.6 ± 0.2%) was significantly different from the cryoablation treatment alone and more importantly, it was significantly different from the combination of nanoparticles (NP) treatment and cryoablation at –17.4°C (8.8 ± 2.4%). Therefore, the combined treatment of cryoablation at –17.4°C and nDOX was not only effective in destroying all mammosphere cells (Figure 3) but also good at effectively and preferentially killing the $CD44^+CD133^+$ CSCs (Figure 4).

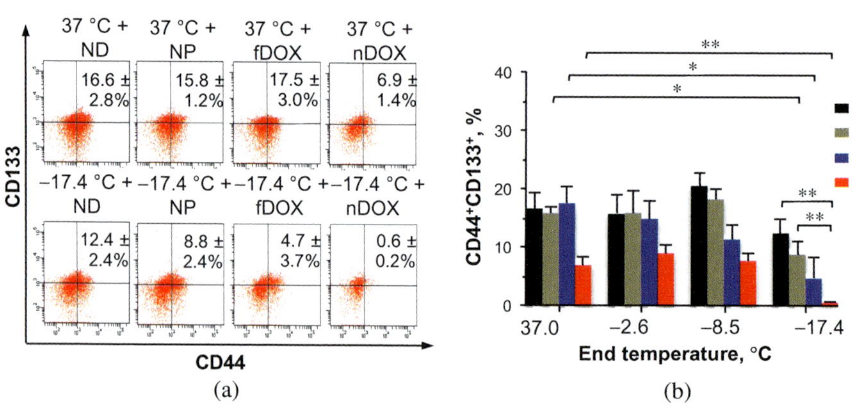

Figure 4. Percentage of CSCs (judged by $CD44^+CD133^+$, two typical CSC markers) among the mammosphere cells after 10 days of incubation in culture medium containing no drug (ND), empty nanoparticles (NP), free doxorubicin (fDOX, 1 μg/ml), and nanoparticle-encapsulated doxorubicin (nDOX, 1 μg/ml of DOX) either without (37°C) or with cryoablation (at –2.6, –8.5, and –17.4°C) between 1 and 1.5 h during the incubation. (a) Typical two-channel flow cytometry data showing the intensity and distribution of mammosphere cells with CD44 and CD133 staining. (b) Quantitative data of the percentage of $CD44^+CD133^+$ cell subpopulations. *: $p < 0.05$ and **: $p < 0.01$. Figure reprinted and redrawn from [Rao *et al.*, 2014] with permission from World Scientific Publishing Co. Pte Ltd.

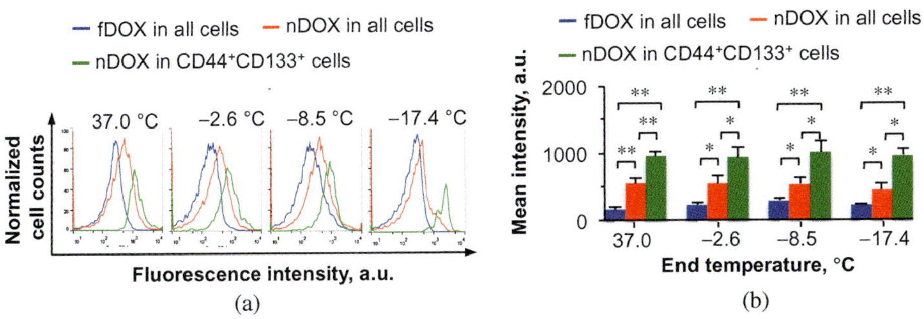

Figure 5. Uptake of free doxorubicin (fDOX, 1 μg/ml) versus nanoparticle-encapsulated doxorubicin (nDOX, 1 μg/ml) in all mammosphere cells and the CSC subpopulations ($CD44^+CD133^+$) either without cryoablation (at 37°C) or with freezing to −2.6, −8.5, or −17.4°C from 1 to 1.5 h during 10-day incubation with the anticancer drugs. (a) Typical flow cytometry peaks. (b) The mean fluorescence intensity of DOX in the mammosphere and $CD44^+CD133^+$ cells from three independent runs. *: $p < 0.05$ and **: $p < 0.01$. Figure reprinted and redrawn from [Rao *et al.*, 2014] with permission from World Scientific Publishing Co. Pte Ltd.

Quantitative analysis of drug uptake by cells in mammospheres was shown in Figure 5. The mean fluorescence intensity of DOX in mammosphere cells treated with nDOX was consistently and significantly higher than that in mammosphere cells treated with fDOX either with or without cryoablation. Moreover, the uptake of nDOX by the $CD44^+CD133^+$ CSCs was consistently and significantly higher than all mammosphere cells, which may render the combined treatment of cryoablation at −17.4°C and nDOX particularly effective in destroying the CSCs.

These results demonstrate the combined treatment of nDOX and cryoablation could minimize cancer recurrence by eliminating the CSCs that are capable of creating the heterogeneity in tumor to reinitiate and further maintain tumor growth post conventional treatments [Jordan *et al.*, 2006; Visvader and Lindeman, 2008; Frank *et al.*, 2010]. It is worth noting that although the 3D mammosphere model used in this study is indeed biomimetic to *in vivo* tumor at the microscale and has been reported to be a more appropriate model than 2D culture for studying the sensitivity of the tumorigenic stem cells in tumor to therapeutic agents and treatment modalities [Grimshaw *et al.*, 2008], it does not completely recapitulate the exact microenvironment *in vivo*. Nonetheless, cancer cells in *in vivo* tumor have been shown to be more susceptible to cryoablation because of several *in vivo* injury mechanisms secondary to direct cell injury including damage due to vascular stasis and potential cryo-immunological responses [Hoffmann and Bischof, 2002; Dong *et al.*, 2009; Sabel, 2009; He, 2011]. Moreover, nanoparticle-encapsulated drug has been extensively reported to enhance the *in vivo* bioavailability of anticancer drug in tumor and minimize the drug systemic toxicity [Gilstrap *et al.*, 2011]. Therefore, even greater efficacy on cancer (stem) cell destruction may occur *in vivo* than *in vitro* for the combined therapy of cryoablation and nDOX, pending further substantiation with *in vivo* studies to monitor tumor relapse. Collectively, these results demonstrate the potential of a novel combined therapy of cryoablation and nanodrug for improving the safety and efficacy of cancer treatment by eliminating the root of cancer resistance associated with conventional chemotherapy — the CSCs.

5. Polymeric Nanoparticles for Cell Cryopreservation

As the living cells hold great promise for wide applications in tissue engineering, regenerative medicine, cell transplantation [Langer and Vacanti, 1993; Malik and Khan, 2013; Emmert *et al.*, 2014], successful long-term storage of living cells for ready availability is crucial for the eventual success of these burgeoning cell-based medicines. The traditional cryopreservation approaches mainly rely on the use of penetrating (i.e., cell membrane permeable) cryoprotectants (CPAs) such as dimethyl sulfoxide (DMSO), glycerol, and ethylene glycol [Fuller, 2004; Hunt, 2011; Ock and Rho, 2011]. However, these CPAs are known to be highly toxic at body temperature. For example, incomplete removal of DMSO could cause intravascular hemolysis, hyperosmolality, and increased serum transaminase levels in clinical applications [Davis *et al.*, 1990; Windrum *et al.*, 2005]. Furthermore, it has been found to induce differentiation of stem cells using more than 25 human stem cell lines [Chetty *et al.*, 2013]. Therefore, it is important to develop better approaches for stem cell cryopreservation without using these toxic CPAs.

Some small carbohydrates (e.g., trehalose) possess excellent capability of stabilizing cellular membranes and proteins during freezing and drying. This is attributed to its exceptional capability of (1) acting as water to form hydrogen bonds with and/or promoting preferential hydration of biomacromolecules so that their functional conformations can be retained during cell dehydration (e.g., by freeze-concentration during cryopreservation) [Clegg *et al.*, 1982; Crowe *et al.*, 1993a, 1993b; Crowe *et al.*, 1998; D'Alfonso *et al.*, 2003; Cottone *et al.*, 2005; Roche *et al.*, 2006; Cottone, 2007; He, 2011], and (2) forming a stable glassy matrix with extremely low molecular mobility to suspend any degradative and metabolic activities in the dehydrated state [Sun *et al.*, 1996; Crowe *et al.*, 1998; Crowe *et al.*, 2001; He *et al.*, 2006; He, 2011]. However, effective delivery of trehalose into a large amount of mammalian cells has been challenging. This is because the small hydrophilic sugar cannot permeate the cell membrane and mammalian cells are not capable of synthesizing trehalose endogenously [Toner and Kocsis, 2002; Wolkers *et al.*, 2002; Acker *et al.*, 2004].

A variety of approaches have been developed to deliver trehalose into living cells, including microinjection [Eroglu *et al.*, 2000; Eroglu *et al.*, 2001; Eroglu *et al.*, 2002; Eroglu *et al.*, 2003; Burgess *et al.*, 2004], engineered pore-forming protein [Acker *et al.*, 2003; Buchanan *et al.*, 2004; Liu *et al.*, 2005; Elliott *et al.*, 2006; Sharp *et al.*, 2013], electroporation [Shirakashi *et al.*, 2002; Zhou *et al.*, 2010], and fluid-phase endocytosis [Wolkers *et al.*, 2003; Oliver *et al.*, 2004; He *et al.*, 2006]. A summary of the major advantages, drawbacks, and the cells studied with these approaches is given in Table 1. However, successful cryo-preservation of mammalian cells with trehalose delivered using these approaches

Table 1. A summary of the major advantages and limitations of commonly used approaches to deliver trehalose for cell preservation

Delivery approaches	Advantages	Limitations	Cells studied
Microinjection	• Allowing delivery of sufficient amount of trehalose	• Small quantity of cells • Cells with large size	Mammalian oocytes
Poration protein/ polymer	• Relative ease of preparation • Easy to operate	• Invasiveness • Toxicity	Fibroblasts, human keratinocytes, human osteosarcoma cells
Electroporation	• High efficacy of delivery	• Electric injury to cells	Hematopoietic cells, human embryonic stem cells, red blood cells
Fluid-phase endocytosis	• Easy to operate	• Inability to deliver enough trehalose	Mesenchymal stem cells
Liposomes	• Non-invasive • Biocompatible • Non-toxic	• Less stable • Leakage and fusion of encapsulated trehalose • Low solubility	Red blood cells
Polymeric nanoparticles	• Non-invasive • Biocompatible, • Non-toxic • Controlled release of trehalose	• Encapsulation efficiency and loading content of trehalose need further improvement	Human adipose derived stem cells

has not been reported in the literature except microinjection which is limited to cryopreserve large cells such as oocytes present in a small quantity.

More recently, nanotechnology has been introduced into the field of cell preservation as an effective strategy to encapsulate and deliver trehalose into living cells. For example, liposomes have been used for delivery of trehalose into red blood cells with promising outcomes [Holovati and Acker, 2007; Holovati *et al.*, 2009]. Compared to liposomes, polymeric nanoparticles have improved biological stability [Date *et al.*, 2007] and have been used to successfully deliver trehalose into fibroblasts [Zhang *et al.*, 2009]. Furthermore, the surface properties and composition of polymeric nanoparticle can be optimized to achieve the desired (e.g., responsive to the low pH in late endosomes and lysosomes) release of the encapsulated trehalose.

However, successful cryopreservation of mammalian cells using trehalose as the sole cryoprotectant has not been reported until a recent study, where polymeric nanoparticles (genipin crosslinked Pluronic-chitosan nanoparticles or GNPs for short) were utilized to efficiently deliver trehalose into primary human adipose-derived stem cells (hADSCs) and the trehalose-laden cells were found to be highly viable after cryopreservation [Rao *et al.*, 2015]. The viability of hADSCs cryopreserved by the nanoparticle-encapsulated trehalose (nTre) was evaluated first by the immediate survival judged by membrane integrity post cryopreservation. As shown in Figures and 6a and 6b, non-encapsulated free trehalose (fTre) or GNPs alone do not help the hADSCs to survive cryopreservation compared to the no pre-treatment (NPT) control. This is probably because fTre cannot enter the cells. By contrast, the nTre could be used to cryopreserve the hADSCs with a high viability similar to that of cells cryopreserved using DMSO and fresh cells without cryopreservation.

Further studies were conducted to assess the viability of the hADSCs one day after cryopreservation by quantifying the attachment efficiency calculated as the number of attached live cells post cryopreservation normalized to that of fresh cells after one day culture. As shown in Figure 6c, only a minimal percentage of cells treated with GNPs or fTre were able to attach-post cryopreservation, which is not significantly different from that of cryopreserved cells with no pre-treatment (NPT). The post-cryopreservation attachment efficiency was much higher for the cells pre-incubated with nTre. Although this attachment efficiency is slightly lower than that of the cells cryopreserved with DMSO, using nTre as the sole cryoprotectant could eliminate the tedious multi-step washing to remove the highly toxic cryoprotectant in DMSO-cryopreserved cells. Besides being time-consuming, the multi-step washing is associated with significant loss of cells (~10% loss in each step). Moreover, the attached cells post-cryopreservation with nTre proliferate similarly to fresh

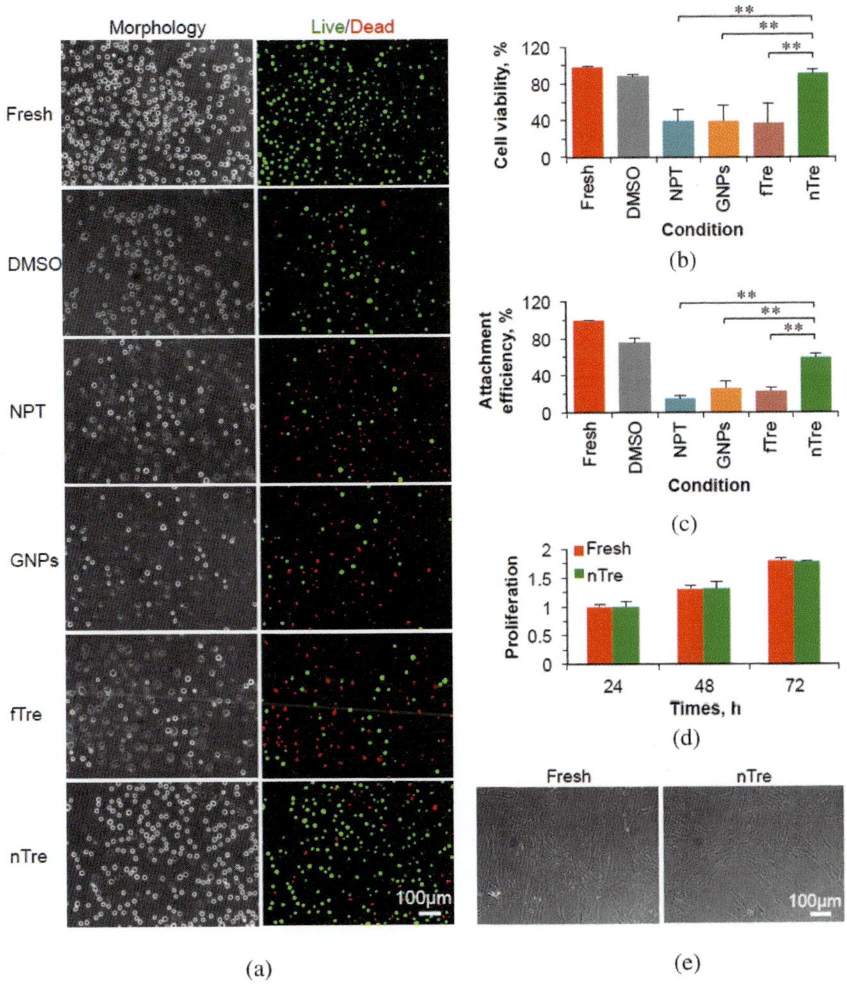

Figure 6. Intracellular delivery of trehalose with polymeric nanoparticles (genipin crosslinked Pluronic-chitosan nanoparticles or GNPs for short) enables successful cryopreservation of hADSCs using trehalose as the sole cryoprotectant. (a) Typical phase and fluorescence micrographs showing high immediate viability of hADSCs after cryopreservation with the nanoparticle-encapsulated trehalose (nTre). (b) Quantitative data showing high immediate cell viability after cryopreservation with nTre similar to that of hADSCs cryopreserved with dimethyl sulfoxide (DMSO). (c) Attachment efficiency after one-day culture post-cryopreservation. (d) Quantitative data showing similar proliferation of cryopreserved (with nTre) to fresh hADSCs. (e) Typical phase micrographs showing similar morphology of fresh to cryopreserved (with nTre) hADSCs. Fresh: without any pre-incubation or cryopreservation; NPT: no pre-incubation with trehalose or GNCs for cryopreservation; GNPs: pre-incubated with empty GNPs for cryopreservation; fTre: pre-incubated with free trehalose for cryopreservation; and nTre: pre-incubated with nTre for cryopreservation. *: $p < 0.05$; **: $p < 0.01$. Figure reprinted and redrawn from [Rao *et al.*, 2015] with permission from American Chemical Society.

hADSCs (Figure 6d). The morphology of hADSCs post-cryopreservation with nTre is also similar to that of fresh hADSCs (Figure 6e).

To ascertain functional survival of the hADSCs post-cryopreservation with nTre as the sole cryoprotectant, further studies were conducted to evaluate the stemness and capability of differentiation into multiple lineages of the cryopreserved hADSCs as compared to fresh hADSCs. As shown in Figure 7a, no significant difference was observed between the cryopreserved and fresh cells

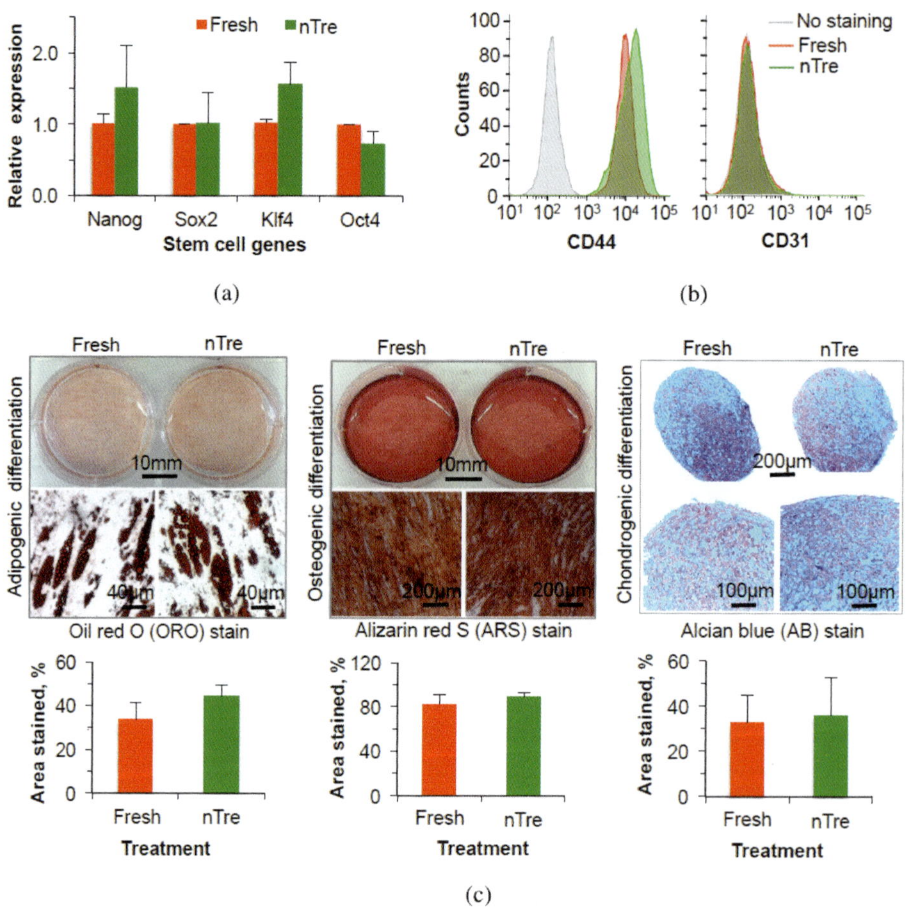

Figure 7. The hADSCs cryopreserved with nTre retain intact stemness and capability of multi-lineage differentiation. (a) Similar level of expression of four typical stem cell genes in fresh and cryopreserved (with nTre) hADSCs. (b) Similar level of expression of CD44(+) and CD31(–) on fresh and cryopreserved (with nTre) hADSCs. (c) Similar capability of adipogenic, osteogenic, and chondrogenic differentiation of fresh and cryopreserved (with nTre) hADSCs indicated by Oil red O (ORO) stain of lipid-filled droplets, Alizarin red S (ARS) stain of calcific depositions, and Alcian blue (AB) stain of sulfated proteoglycan, respectively. Figure reprinted and redrawn from [Rao *et al.*, 2015] with permission from American Chemical Society.

in terms of expression of four common stem cell genes including Nanog, Sox2; Klf4, and Oct4. Further flow cytometry analysis indicates no significant difference in the expression of two typical surface markers ($CD44^+$ and $CD31^-$) on the cryopreserved versus fresh hADSCs (Figure 7b). The data on adipogenic, osteogenic, and chondrogenic differentiation of the cryopreserved versus fresh hADSCs are shown in Figure 7c. For adipogenic differentiation, no significance was observed between the two groups of cells in terms of Oil red O (ORO) staining of adipocyte-like cells with lipid-filled droplets. For osteogenic differentiation, both groups of cells maintained similar osteogenic potential based on Alizarin red S (ARS) staining of calcific deposition, a major function of osteoblasts. For chondrogenic differentiation, Alcian blue (AB) staining of sulfated proteoglycan deposits that are indicative of the presence of functional chondrocytes is not significantly different between the fresh and cryopreserved hADSCs.

In summary, utilizing genipin crosslinking to efficiently encapsulate trehalose in the pH responsive genipin-crosslinked Pluronic-F127 nanoparticles (GNPs) enables successful cryopreservation of hADSCs with high viability and intact function using trehalose as the sole cryoprotectant.

6. Polymeric Nanoparticles for Hibernation

Chitosan-based polymeric nanoparticles were reported to promote sustained intracellular delivery of the synthetic delta opioid, which enables prolonged peptide activity and induces a hibernation-like state in Hela cells [Colonna *et al.*, 2011].

7. Outlook and Conclusion

The investigation of polymeric nanoparticles for cryobiological applications is still at the infant stage. Only a few nanoplatforms have been developed for cryobiological applications. However, the existing data suggest that this strategy holds great promise for both cryoablation and cell preservation. Future work for cryoablation could be (1) to develop cooling/freezing responsive nanoparticle for drug delivery to specifically deliver therapeutic agents in the region covered by iceball that is monitored by intraoperative imaging techniques and (2) to develop nanoparticle system that specifically targets and eliminates the cancer stem cells for combining with cryoablation to minimize cancer recurrence post treatment. For cell preservation, more research is needed to test the capability of trehalose delivered using nanoparticles for cryopreserving many different types of cells. Besides cryopreservation, trehalose ahs been shown to be exceptional to protect cells during drying at ambient temperature. It would be exciting to bank living cells at ambient temperature with aid of nanoparticle-mediated delivery of trehalose. However, cell preservation at ambient temperature is even more challenging than cryopreservation and other protective agents such as the late embryogenesis abundant (LEA) proteins may be needed [Chakrabortee *et al.*, 2007; Li and He, 2009; Hand *et al.*, 2011], in addition to trehalose, to achieve this game-changing goal in the future.

In summary, nanotechnology has demonstrated great potential in advancing the field of cryomedicine. It is expected that it will revolutionize the field in the coming years by (1) improving the safety and efficacy of cryoablation and (2) achieving cell preservation at both cryogenic and ambient temperature, without using non-toxic cryoprotectants.

References

Acker, J. P., T. Chen, A. Fowler and M. Toner (2004). Engineering desiccation tolerance in mammalian cells: tools and techniques. In B. J. Fuller, L. Lane and E. E. Benson (eds.) *Life in the Frozen State* (CRC Press LLC, Boca Raton, FL).

Acker, J. P., X. M. Lu, V. Young, S. Cheley, H. Bayley, A. Fowler and M. Toner (2003). Measurement of trehalose loading of mammalian cells porated with a metal-actuated switchable pore. *Biotechnology and Bioengineering*, 82(5), pp. 525–532.

Al-Hajj, M., M. S. Wicha, A. Benito-Hernandez, S. J. Morrison and M. F. Clarke (2003). Prospective identification of tumorigenic breast cancer cells. *Proceedings of the National Academy of Sciences of the United States of America*, 100(7), pp. 3983–3988.

Bertrand, N., J. Wu, X. Xu, N. Kamaly and O. C. Farokhzad (2014). Cancer nanotechnology: the impact of passive and active targeting in the era of modern cancer biology. *Advanced Drug Delivery Reviews*, 66, pp. 2–25.

Bischof, J. C. and X. He (2005). Thermal stability of proteins. *Annals of the New York Academy of Sciences*, 1066, pp. 12–33.

Buchanan, S. S., S. A. Gross, J. P. Acker, M. Toner, J. F. Carpenter and D. W. Pyatt (2004). Cryopreservation of stem cells using trehalose: Evaluation of the method using a human hematopoietic cell line. *Stem Cells and Development*, 13(3), pp. 295–305.

Bulpitt, P. and D. Aeschlimann (1999). New strategy for chemical modification of hyaluronic acid: preparation of functionalized derivatives and their use in the formation of novel biocompatible hydrogels. *Journal of Biomedical Materials Research*, 47(2), pp. 152–169.

Burgess, C. M., W. B. Wang, D. Powers, M. Alper, M. Toner and T. L. Toth (2004). The survival of human failed to fertilize MII-stage oocytes cryopreserved by microinjection of trehalose. *Fertility and Sterility*, 82, p. S117.

Cabral, H., Y. Matsumoto, K. Mizuno, Q. Chen, M. Murakami, M. Kimura, Y. Terada, M. R. Kano, K. Miyazono, M. Uesaka, N. Nishiyama and K. Kataoka (2011). Accumulation of sub-100 nm polymeric micelles in poorly permeable tumours depends on size. *Nature Nanotechnology*, 6(12), pp. 815–823.

Callstrom, M. R., D. E. Dupuy, S. B. Solomon, R. A. Beres, P. J. Littrup, K. W. Davis, R. Paz-Fumagalli, C. Hoffman, T. D. Atwell, J. W. Charboneau, G. D. Schmit, M. P. Goetz, J. Rubin, K. J. Brown, P. J. Novotny and J. A. Sloan (2013). Percutaneous image-guided cryoablation of painful metastases involving bone: multicenter trial. *Cancer*, 119(5), pp. 1033–1041.

Chakrabortee, S., C. Boschetti, L. J. Walton, S. Sarkar, D. C. Rubinsztein and A. Tunnacliffe (2007). Hydrophilic protein associated with desiccation tolerance exhibits broad protein stabilization function. *Proceedings of the National Academy of Sciences of the USA*, 104(46), pp. 18073–18078.

Chao, B. H., X. He and J. C. Bischof (2004). Pre-treatment inflammation induced by TNF-alpha augments cryosurgical injury on human prostate cancer. *Cryobiology*, 49(1), pp. 10–27.

Chauhan, V. P., T. Stylianopoulos, J. D. Martin, Z. Popovic, O. Chen, W. S. Kamoun, M. G. Bawendi, D. Fukumura and R. K. Jain (2012). Normalization of tumour blood vessels improves the delivery of nanomedicines in a size-dependent manner. *Nature Nanotechnology*, 7(6), pp. 383–388.

Chetty, S., F. W. Pagliuca, C. Honore, A. Kweudjeu, A. Rezania and D. A. Melton (2013). A simple tool to improve pluripotent stem cell differentiation. *Nature Methods*, 10(6), pp. 553–556.

Choi, K. Y., K. H. Min, J. H. Na, K. Choi, K. Kim, J. H. Park, I. C. Kwon and S. Y. Jeong (2009). Self-assembled hyaluronic acid nanoparticles as a potential drug carrier for cancer therapy: synthesis, characterization, and in vivo biodistribution. *Journal of Materials Chemistry*, 19(24), pp. 4102–4107.

Choi, S. U. S., Z. G. Zhang, W. Yu, F. E. Lockwood and E. A. Grulke (2001). Anomalous thermal conductivity enhancement in nanotube suspensions. *Applied Physics Letters*, 79(14), pp. 2252 2254.

Clarke, D. M., J. M. Baust, R. G. Van Buskirk and J. G. Baust (2001). Chemo-cryo combination therapy: an adjunctive model for the treatment of prostate cancer. *Cryobiology*, 42(4), pp. 274–285.

Clegg, J. S., P. Seitz, W. Seitz and C. F. Hazlewood (1982). Cellular responses to extreme water loss: the water-replacement hypothesis. *Cryobiology*, 19(3), pp. 306–316.

Colonna, C., R. Dorati, B. Conti, T. Modena, M. Biggiogera, A. Spedito and I. Genta (2011). Induction of an *in vitro* reversible hypometabolism through chitosan-based nanoparticles. *Journal of Microencapsulation*, 28(4), pp. 229–239.

Cottone, G. (2007). A comparative study of carboxy myoglobin in saccharide-water systems by molecular dynamics simulation. *The Journal of Physical Chemistry B*, 111(13), pp. 3563–3569.

Cottone, G., S. Giuffrida, G. Ciccotti and L. Cordone (2005). Molecular dynamics simulation of sucrose- and trehalose-coated carboxy-myoglobin. *Proteins*, 59(2), pp. 291–302.

Crowe, J. H., J. F. Carpenter and L. M. Crowe (1998). The role of vitrification in anhydrobiosis. *Annual Review of Physiology*, 60, pp. 73–103.

Crowe, J., L. Crowe, and J. Carpenter (1993a). Preserving dry biomaterials: the water replacement hypothesis, part I. *Biopharm*, 6(3), pp. 28–33.

Crowe, J., L. Crowe, and J. Carpenter (1993b). Preserving dry biomaterials: the water replacement hypothesis, part II. *Biopharm*, 6(4), pp. 40–43.

Crowe, J. H., L. M. Crowe, A. E. Oliver, N. Tsvetkova, W. Wolkers and F. Tablin (2001). The trehalose myth revisited: introduction to a symposium on stabilization of cells in the dry state. *Cryobiology*, 43(2), pp. 89–105.

D'Alfonso, L., M. Collini and G. Baldini (2003). Trehalose influence on beta-lactoglobulin stability and hydration by time resolved fluorescence. *European Journal of Biochemistry*, 270(11), pp. 2497–2504.

Date, A. A., M. D. Joshi and V. B. Patravale (2007). Parasitic diseases: Liposomes and polymeric nanoparticles versus lipid nanoparticles. *Advanced Drug Delivery Reviews*, 59(6), pp. 505–521.

Davis, J. M., S. D. Rowley, H. G. Braine, S. Piantadosi and G. W. Santos (1990). Clinical toxicity of cryopreserved bone marrow graft infusion. *Blood*, 75(3), pp. 781–786.

Di, D. R., Z. Z. He, Z. Q. Sun and J. Liu (2012). A new nano-cryosurgical modality for tumor treatment using biodegradable MgO nanoparticles. *Nanomedicine: Nanotechnology, Biology, and Medicine*, 8(8), pp. 1233–1241.

Dong, J., P. Liu and L. X. Xu (2009). Immunologic response induced by synergistic effect of alternating cooling and heating of breast cancer. *International Journal of Hyperthermia*, 25(1), pp. 25–33.

Eastman, J. A., S. U. S. Choi, S. Li, W. Yu and L. J. Thompson (2001). Anomalously increased effective thermal conductivities of ethylene glycol-based nanofluids containing copper nanoparticles. *Applied Physics Letters*, 78(6), pp. 718–720.

Elgafy, A. and K. Lafdi (2005). Effect of carbon nanofiber additives on thermal behavior of phase change materials. *Carbon*, 43(15), pp. 3067–3074.

Elliott, G. D., X. H. Liu, J. L. Cusick, M. Menze, J. Vincent, T. Witt, S. Hand and M. Toner (2006). Trehalose uptake through P2X(7) purinergic channels provides dehydration protection. *Cryobiology*, 52(1), pp. 114–127.

Elsabahy, M. and K. L. Wooley (2012). Design of polymeric nanoparticles for biomedical delivery applications. *Chemistry Society Reviews*, 41(7), pp. 2545–2561.

Emmert, M. Y., R. W. Hitchcock and S. P. Hoerstrup (2014). Cell therapy, 3D culture systems and tissue engineering for cardiac regeneration. *Advanced Drug Delivery Reviews*, 69–70, pp. 254–269.

Eroglu, A., J. A. Lawitts, M. Toner and T. L. Toth (2003). Quantitative microinjection of trehalose into mouse oocytes and zygotes, and its effect on development. *Cryobiology*, 46(2), pp. 121–134.

Eroglu, A., M. J. Russo, R. Bieganski, A. Fowler, S. Cheley, H. Bayley and M. Toner (2000). Intracellular trehalose improves the survival of cryopreserved mammalian cells. *Nature Biotechnology*, 18(2), pp. 163–167.

Eroglu, A., M. Toner and T. L. Toth (2002). Beneficial effect of microinjected trehalose on the cryosurvival of human oocytes. *Fertility and Sterility*, 77(1), pp. 152–158.

Eroglu, A., T. L. Toth and M. Toner (2001). Small amounts of microinjected trehalose protect mouse and human oocytes during freeze-thaw. *Fertility and Sterility*, 76(3), p. S78.

Fang, J., H. Nakamura and H. Maeda (2011). The EPR effect: Unique features of tumor blood vessels for drug delivery, factors involved, and limitations and augmentation of the effect. *Advanced Drug Delivery Reviews*, 63(3), pp. 136–151.

Farokhzad, O. C. and R. Langer (2009). Impact of nanotechnology on drug delivery. *ACS Nano*, 3(1), pp. 16–20.

Forest, V., M. Peoc'h, L. Campos, D. Guyotat and J. M. Vergnon (2006). Benefit of a combined treatment of cryotherapy and chemotherapy on tumour growth and late cryo-induced angiogenesis in a non-small-cell lung cancer model. *Lung Cancer*, 54(1), pp. 79–86.

Frank, N. Y., T. Schatton and M. H. Frank (2010). The therapeutic promise of the cancer stem cell concept. *The Journal of Clinical Investigation*, 120(1), pp. 41–50.

Frohlich, E. (2012). The role of surface charge in cellular uptake and cytotoxicity of medical nanoparticles. *International Journal of Nanomedicine*, 7, pp. 5577–5591.

Fuller, B. J. (2004). Cryoprotectants: the essential antifreezes to protect life in the frozen state. *Cryo Letters*, 25(6), pp. 375–388.

Gage, A. A. and J. Baust (1998). Mechanisms of tissue injury in cryosurgery. *Cryobiology*, 37(3), pp. 171–186.

Gage, A. A. and J. G. Baust (2007). Cryosurgery for tumors. *Journal of the American College of Surgeons*, 205(2), pp. 342–356.

Ganta, S., H. Devalapally, A. Shahiwala and M. Amiji (2008). A review of stimuli-responsive nanocarriers for drug and gene delivery. *Journal of Controlled Release*, 126(3), pp. 187–204.

Gao, H., W. Shi and L. B. Freund (2005). Mechanics of receptor-mediated endocytosis. *Proceedings of the National Academy of Sciences of the United States of America*, 102(27), pp. 9469–9474.

Gao, Y., Y. Chen, X. Ji, X. He, Q. Yin, Z. Zhang, J. Shi and Y. Li (2011). Controlled intracellular release of doxorubicin in multidrug-resistant cancer cells by tuning the shell-pore sizes of mesoporous silica nanoparticles. *ACS Nano*, 5(12), pp. 9788–9798.

Ge, Z., Y. Kang, T. A. Taton, P. V. Braun and D. G. Cahill (2005). Thermal transport in au-core polymer-shell nanoparticles. *Nano Letters*, 5(3), pp. 531–535.

Geha, A. S. and K. Abdelhady (2008). Current status of the surgical treatment of atrial fibrillation. *World Journal of Surgery*, 32(3), pp. 346–349.

Gilstrap, K., X. Hu, X. Lu and X. He (2011). Nanotechnology for energy-based cancer therapies. *American Journal of Cancer Research*, 1(4), pp. 508–520.

Grimshaw, M. J., L. Cooper, K. Papazisis, J. A. Coleman, H. R. Bohnenkamp, L. Chiapero-Stanke, J. Taylor-Papadimitriou and J. M. Burchell (2008). Mammosphere culture of metastatic breast cancer cells enriches for tumorigenic breast cancer cells. *Breast Cancer Research : BCR*, 10(3), p. R52.

Han, B. and J. C. Bischof (2004). Direct cell injury associated with eutectic crystallization during freezing. *Cryobiology*, 48(1), pp. 8–21.

Hand, S. C., M. A. Menze, M. Toner, L. Boswell and D. Moore (2011). LEA proteins during water stress: not just for plants anymore. *Annual Review of Physiology*, 73, pp. 115–134.

He, C., Y. Hu, L. Yin, C. Tang and C. Yin (2010). Effects of particle size and surface charge on cellular uptake and biodistribution of polymeric nanoparticles. *Biomaterials*, 31(13), pp. 3657–3666.

He, X. (2011). Thermostability of biological systems: fundamentals, challenges, and quantification. *The Open Biomedical Engineering Journal*, 5, pp. 47–73.

He, X. M., A. A. Amin, A. Fowler and M. Toner (2006a). Thermally induced introduction of trehalose into primary rat hepatocytes. *Cell Preservation Technology*, 4(3), pp. 178–187.

He, X. and J. C. Bischof (2003). Quantification of temperature and injury response in thermal therapy and cryosurgery. *Critical Reviews in Biomedical Engineering*, 31(5–6), pp. 355–422.

He, X. and J. C. Bischof (2005). Analysis of thermal stress in cryosurgery of kidneys. *Journal of Biomechanical Engineering*, 127(4), pp. 656–661.

He, X. M., A. Fowler and M. Toner (2006b). Water activity and mobility in solutions of glycerol and small molecular weight sugars: Implication for cryo- and lyopreservation. *Journal of Applied Physics*, 100(7), p. 074702.

Hoffmann, N. E. and J. C. Bischof (2002). The cryobiology of cryosurgical injury. *Urology*, 60(2 Suppl 1), pp. 40–49.

Holovati, J. L. and J. P. Acker (2007). Spectrophotometric measurement of intraliposomal trehalose. *Cryobiology*, 55(2), pp. 98–107.

Holovati, J. L., M. I. C. Gyongyossy-Issa and J. P. Acker (2009). Effects of trehalose-loaded liposomes on red blood cell response to freezing and post-thaw membrane quality. *Cryobiology*, 58(1), pp. 75–83.

Hunt, C. J. (2011). Cryopreservation of human stem cells for clinical application: a review. *Transfusion Medicine and Hemotherapy*, 38(2), pp. 107–123.

Jordan, C. T., M. L. Guzman and M. Noble (2006). Cancer stem cells. *The New England Journal of Medicine*, 355(12), pp. 1253–1261.

Kabanov, A. V., E. V. Batrakova and V. Y. Alakhov (2002). Pluronic block copolymers for overcoming drug resistance in cancer. *Advanced Drug Delivery Reviews*, 54(5), pp. 759–779.

Keblinski, P., J. A. Eastman and D. G. Cahill (2005). Nanofluids for thermal transport. *Materials Today*, 8(6), pp. 36–44.

Köhler, G. and C. Milstein (1975). Continuous cultures of fused cells secreting antibody of predefined specificity. *Nature*, 256(5517), pp. 495–497.

Koushafar, H. and B. Rubinsky (1997). Effect of antifreeze proteins on frozen primary prostatic adenocarcinoma cells. *Urology*, 49(3), pp. 421–425.

Kumari, A., S. K. Yadav and S. C. Yadav (2010). Biodegradable polymeric nanoparticles based drug delivery systems. *Colloids and Surfaces B: Biointerfaces*, 75(1), pp. 1–18.

Langer, R. and J. P. Vacanti (1993). Tissue engineering. *Science*, 260(5110), pp. 920–926.

Li, D. and X. He (2009). Desiccation induced structural alterations in a 66-amino acid fragment of an anhydrobiotic nematode late embryogenesis abundant (LEA) protein. *Biomacromolecules*, 10(6), pp. 1469–1477.

Liu, X. H., A. Aksan, M. A. Menze, S. C. Hand and M. Toner (2005). Trehalose loading through the mitochondrial permeability transition pore enhances desiccation tolerance in rat liver mitochondria. *Biochimica et Biophysica Acta-Biomembranes* 1717(1), pp. 21–26.

Lorico, A. and G. Rappa (2011). Phenotypic heterogeneity of breast cancer stem cells. *Journal of Oncology*, 2011, p. 135039.

Malik, A. and W. Khan (2013). Stem cell therapy and tissue engineering applications for bone. *Current Stem Cell Research and Therapy*, 8(3), pp. 183–184.

Mir, L. M. and B. Rubinsky (2002). Treatment of cancer with cryochemotherapy. *British Journal of Cancer*, 86(10), pp. 1658–1660.

Mura, S., J. Nicolas and P. Couvreur (2013). Stimuli-responsive nanocarriers for drug delivery. *Nature Materials*, 12(11), pp. 991–1003.

Ock, S. A. and G. J. Rho (2011). Effect of Dimethyl Sulfoxide (DMSO) on cryopreservation of porcine mesenchymal stem cells (pMSCs). *Cell Transplantation*, 20(8), pp. 1231–1239.

Oliver, A. E., K. Jamil, J. H. Crowe and F. Tablin (2004). Loading human mesenchymal stem cells with trehalose by fluid-phase endocytosis. *Cell Preservation Technology*, 2(1), pp. 35–49.

Petros, R. A. and J. M. DeSimone (2010). Strategies in the design of nanoparticles for therapeutic applications. *Nature Reviews Drug Discovery*, 9(8), pp. 615–627.

Pham, L., R. Dahiya and B. Rubinsky (1999). An *in vivo* study of antifreeze protein adjuvant cryosurgery. *Cryobiology*, 38(2), pp. 169–175.

Rao, W., A. Bellotti, P. J. Littrup, J. Yu, X. Lu and X. He (2014). Nanoparticle-encapsulated doxorubicin enhances cryoablation of cancer stem-like cells. *Technology*, 2, pp. 28–35.

Rao, W., H. Huang, H. Wang, S. Zhao, J. Dumbleton, G. Zhao and X. He (2015). Nanoparticle-mediated intracellular delivery enables cryopreservation of human adipose-derived stem cells using trehalose as the sole cryoprotectant. *ACS Applied Materials and Interfaces*, 7, pp. 5017–5028.

Roche, C. J., F. Guo and J. M. Friedman (2006). Molecular level probing of preferential hydration and its modulation by osmolytes through the use of pyranine complexed to hemoglobin. *Journal of Biological Chemistry*, 281(50), pp. 38757–38768.

Sabel, M. S. (2008). Cryoablation for breast cancer: no need to turn a cold shoulder. *Journal of Surgical Oncology*, 97(6), pp. 485–486.

Sabel, M. S. (2009). Cryo-immunology: a review of the literature and proposed mechanisms for stimulatory versus suppressive immune responses. *Cryobiology*, 58(1), pp. 1–11.

Sharp, D. M. C., A. Picken, T. J. Morris, C. J. Hewitt, K. Coopman and N. K. H. Slater (2013). Amphipathic polymer-mediated uptake of trehalose for dimethyl sulfoxide-free human cell cryopreservation. *Cryobiology*, 67(3), pp. 305–311.

Shenoi, M. M., I. Iltis, J. Choi, N. A. Koonce, G. J. Metzger, R. J. Griffin and J. C. Bischof (2013). Nanoparticle delivered vascular disrupting agents (VDAs): use of TNF-alpha conjugated gold nanoparticles for multimodal cancer therapy. *Molecular Pharmaceutics*, 10(5), pp. 1683–1694.

Shipitsin, M., L. L. Campbell, P. Argani, S. Weremowicz, N. Bloushtain-Qimron, J. Yao, T. Nikolskaya, T. Serebryiskaya, R. Beroukhim, M. Hu, M. K. Halushka, S. Sukumar, L. M. Parker, K. S. Anderson, L. N. Harris, J. E. Garber, A. L. Richardson, S. J. Schnitt, Y. Nikolsky, R. S. Gelman and K. Polyak (2007). Molecular definition of breast tumor heterogeneity. *Cancer Cell*, 11(3), pp. 259–273.

Shirakashi, R., C. M. Kostner, K. J. Muller, M. Kurschner, U. Zimmermann and V. L. Sukhorukov (2002). Intracellular delivery of trehalose into mammalian cells by electropermeabilization. *Journal of Membrane Biology*, 189(1), pp. 45–54.

Steichen, S. D., M. Caldorera-Moore and N. A. Peppas (2013). A review of current nanoparticle and targeting moieties for the delivery of cancer therapeutics. *European Journal of Pharmaceutical Sciences*, 48(3), pp. 416–427.

Stuart, M. A., W. T. Huck, J. Genzer, M. Muller, C. Ober, M. Stamm, G. B. Sukhorukov, I. Szleifer, V. V. Tsukruk, M. Urban, F. Winnik, S. Zauscher, I. Luzinov and S. Minko (2010). Emerging applications of stimuli-responsive polymer materials. *Nature Materials*, 9(2), pp. 101–113.

Stylianopoulos, T., M. Z. Poh, N. Insin, M. G. Bawendi, D. Fukumura, L. L. Munn and R. K. Jain (2010). Diffusion of particles in the extracellular matrix: the effect of repulsive electrostatic interactions. *Biophysical Journal*, 99(5), pp. 1342–1349.

Sui, G., S. Jana, W. H. Zhong, M. A. Fuqua and C. A. Ulven (2008). Dielectric properties and conductivity of carbon nanofiber/semi-crystalline polymer composites. *Acta Materialia*, 56(10), pp. 2381–2388.

Sun, W. Q., A. C. Leopold, L. M. Crowe and J. H. Crowe (1996). Stability of dry liposomes in sugar glasses. *Biophysical Journal*, 70(4), pp. 1769–1776.

Sykes, E. A., J. Chen, G. Zheng and W. C. Chan (2014). Investigating the impact of nanoparticle size on active and passive tumor targeting efficiency. *ACS Nano*, 8(6), pp. 5696–5706.

Toner, M. and J. Kocsis (2002). Storage and translational issues in reparative medicine. *Annals of the New York Academy of Sciences*, 961, pp. 258–262.

Verma, A. and F. Stellacci (2010). Effect of surface properties on nanoparticle-cell interactions. *Small*, 6(1), pp. 12–21.

Visvader, J. E. and G. J. Lindeman (2008). Cancer stem cells in solid tumours: accumulating evidence and unresolved questions. *Nature Reviews Cancer*, 8(10), pp. 755–768.

Wang, C. L., K. Y. Teo and B. Han (2008). An amino acidic adjuvant to augment cryoinjury of MCF-7 breast cancer cells. *Cryobiology*, 57(1), pp. 52–59.

Wang, F., Y. C. Wang, S. Dou, M. H. Xiong, T. M. Sun and J. Wang (2011). Doxorubicin-tethered responsive gold nanoparticles facilitate intracellular drug delivery for overcoming multidrug resistance in cancer cells. *ACS nano*, 5(5), pp. 3679–3692.

Wang, H., S. Zhao, P. Agarwal, J. Dumbleton, J. Yu, X. Lu and X. He (2015). Multi-layered polymeric nanoparticles for pH-responsive and sequenced release of theranostic agents. *Chemical Communications (Cambridge)*, 51(36), pp. 7733–7736.

Windrum, P., T. C. M. Morris, M. B. Drake, D. Niederwieser, T. Ruutu and E. C. L. W. Par (2005). Variation in dimethyl sulfoxide use in stem cell transplantation: a survey of EBMT centres. *Bone Marrow Transplantation*, 36(7), pp. 601–603.

Wolkers, W. F., S. A. Looper, R. A. Fontanilla, N. M. Tsvetkova, F. Tablin and J. H. Crowe (2003). Temperature dependence of fluid phase endocytosis coincides with membrane properties of pig platelets. *Biochimica et Biophysica Acta Biomembranes*, 1612(2), pp. 154–163.

Wolkers, W. F., F. Tablin and J. H. Crowe (2002). From anhydrobiosis to freeze-drying of eukaryotic cells. *Comparative Biochemistry and Physiology Part A: Molecular and Integrative Physiology*, 131(3), pp. 535–543.

Wright, M. H., A. M. Calcagno, C. D. Salcido, M. D. Carlson, S. V. Ambudkar and L. Varticovski (2008). Brca1 breast tumors contain distinct CD44+/CD24– and CD133+ cells with cancer stem cell characteristics. *Breast Cancer Research : BCR*, 10(1), pp. R10.

Yan, J. F. and J. Liu (2008). Nanocryosurgery and its mechanisms for enhancing freezing efficiency of tumor tissues. *Nanomedicine : Nanotechnology, Biology, and Medicine*, 4(1), pp. 79–87.

Ye, C.-M., B.-Q. Shentu and Z.-X. Weng (2006). Thermal conductivity of high density polyethylene filled with graphite. *Journal of Applied Polymer Science*, 101(6), pp. 3806–3810.

Yu, M. K., J. Park and S. Jon (2012). Targeting strategies for multifunctional nanoparticles in cancer imaging and therapy. *Theranostics*, 2(1), pp. 3–44.

Zhang, W., J. Rong, Q. Wang and X. He (2009). The encapsulation and intracellular delivery of trehalose using a thermally responsive nanocapsule. *Nanotechnology*, 20(27), p. 275101.

Zhi, C., Y. Bando, T. Terao, C. Tang, H. Kuwahara and D. Golberg (2009). Towards thermoconductive, electrically insulating polymeric composites with boron nitride nanotubes as fillers. *Advanced Functional Materials*, 19(12), pp. 1857–1862.

Zhou, X. L., J. Yuan, J. F. Liu and B. L. Liu (2010). Loading trehalose into red blood cells by electroporation and its application in freeze-drying. *CryoLetters*, 31(2), pp. 147–156.

Chapter 9

Application of Liposomes in Biopreservation

Luciana da S. Cavalcante, Jelena L. Holovati and Jason P. Acker

1. Introduction

In 1964 when Bangham and Horne first observed and described the structures today known as liposomes, little did they know that their discovery would have such a large impact on a variety of fields. The contribution of liposomes to advances in science, range from pharmaceuticals with targeted drug delivery, vaccines and diagnostic tools to cosmetics, food production, chemistry, toxicology, tissue engineering and cryobiology/biopreservation. Specifically in the cryobiology/biopreservation field, liposomes have contributed to a better understanding of osmotic transport, cryo- and lyo-injury and the mechanisms of action of cryoprotectants. In more practical terms liposomes have also been effectively applied in the biopreservation of a variety of cell types such as oocytes, spermatozoa, stem cells and red blood cells under several different conditions (cryopreservation, lyophilisation, hypothermic storage). However, for the purpose of this review we will focus on the liposomal delivery systems for the cryopreservation and hypothermic storage of red blood cells.

2. Liposome History

Liposomes were first described in 1964 by Bangham and Horne from electron micrographs of negatively stained phospholipids. They observed that lecithin and lecithin/cholesterol dispersions formed rod-shaped lamellar structures in the presence of water [Bangham and Horne, 1964]. In 1965 Alec Bangham and colleagues from the Institute of Animal Physiology at Babraham showed that the lamellar structures, named by him as "smectic mesophases" at the time, were able to maintain an ion concentration gradient similar to the one observed in biological membranes, and that this gradient could be disrupted by the action of surfactants and steroids that caused perturbation of the bilayer structure [Bangham *et al.*, 1965]. Gerald Weissman who was a visitor at Bangham's laboratory coined the term "liposomes" as a reference to his research interest in lysosomes [Bangham, 1995]. The term first appeared in Weisman's 1968 review that addressed the potential for use of liposomes as model membranes [Sessa and Weissmann, 1968]. The use of liposomes as model membranes was particularly important for studying the mechanism of action of anaesthetics [Johnson *et al.*, 1973].

The first clinical application of liposomes came from the collaboration of Bangham and British neonatologist Colin Morley who developed the artificial lung-expanding compound (ALEC) composed of dipalmitoylphosphatidylcholine (DPPC) and phosphatidylglycerol (PG) [Bangham *et al.*, 1984]. Clinical trials showed the success of the synthetic surfactant in treating neonatal respiratory distress syndrome thereby reducing mortality and respiratory support of premature babies [Ten Centre Study Group, 1987]. In 1994, ALEC^{TM} (Pumactant) (Britannia Pharmaceuticals Ltd) was licensed in the UK and in 2000 it was voluntarily withdrawn from the market after a clinical trial showing higher mortality compared to a porcine-derived surfactant [Ainsworth *et al.*, 2000].

The consideration of liposomes as drug delivery vehicles started in the early 1970's and was due to their characteristic low toxicity, biodegradability, small size, and large variety of structural components that made them excellent candidates for drug carriers, as well as a view that a vast number of pathologies would benefit from targeted drug delivery therapies [Gregoriadis, 2008]. The early studies on liposomal delivery systems involved the entrapment and delivery of enzymes and was demonstrated using a lysosomal storage disease model by Gregory Gregoriadis and Brenda Ryman; two of the pioneers in liposomal delivery systems [Gregoriadis and Ryman 1972a, 1972b; Gregoriadis *et al.*, 1971]. Their studies were further extended to antimicrobial and antitumor therapies and current progress points towards the use of liposomes in gene therapy, vaccines and diagnostic tools [Lasic, 1998]. Many studies had to be performed to understand liposome interactions and stability *in vivo*, and it took over 20 years and a lot of persistence between the first study by Gregoriadis and the approval of the first liposomal drugs in 1995 (Doxorubicin) and 1997 (Amphotericin B). Currently there are at least four licensed liposomal drugs in the US market and several others undergoing clinical trials [Allen and Cullis, 2013; Chonn and Cullis, 1995; Lasic, 1998].

Liposomes were introduced to the cosmetics industry in 1986 by L'Oréal (Niosomes) and Christian Dior (Capture) in anti-ageing creams [Meybeck, 1992]. Currently they can be found in several cosmetic products like moisturizers, make up and sunscreen as delivery vehicles of active ingredients, based on studies showing increased skin absorption and hydration compared to regular formulations [Lampen *et al.*, 2003; Strauss, 1989].

In the biopreservation field, liposomes have been used as model membranes to study freezing [McGrath, 1984], as delivery vehicles of cryoprotectants [Holovati *et al.*, 2008] and as membrane stabilizers for freeze, freeze-drying and hypothermic storage of a variety of cell types ranging from spermatozoa and oocytes [Graham and Foote, 1987; Zeron *et al.*, 2002] to stem cells and red blood cells [Holovati and Acker, 2011; Kheirolomoom *et al.*, 2005; Motta *et al.*, 2014].

3. Liposomes

Liposomes are defined as microscopic lipid vesicles composed of a central aqueous core surrounded by one or more phospholipid bilayers (Figure 1). They can be classified in terms of size, lamellarity and preparation method [Immordino *et al.*, 2006].

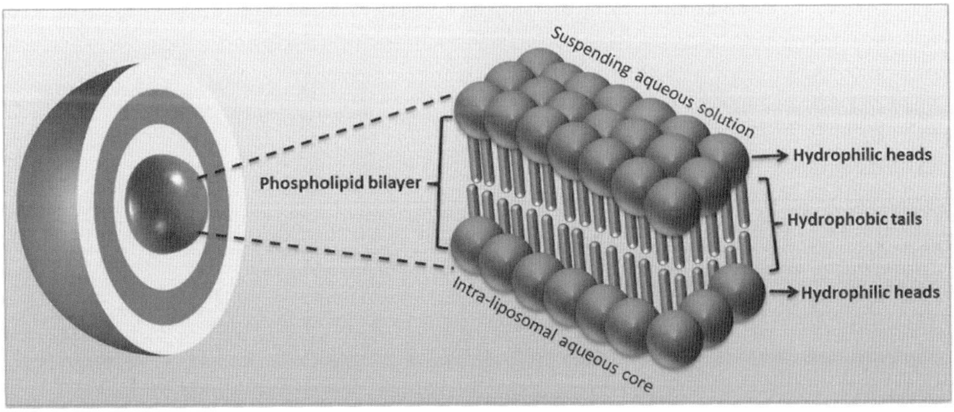

Figure 1. Liposome and phospholipid bilayer structure.

3.1. *Structural Components and Characteristics*

3.1.1. *Composition*

Liposomes are mainly composed of phospholipids and cholesterol. Phospholipid molecules are composed of a hydrophilic head group (e.g. phosphatidylcholine, phosphatidylethanolamine), a glycerol backbone and two hydrophobic fatty acids (e.g. palmitic, oleic, stearic) (Figure 2) [Ulrich, 2002]. They can be extracted from natural sources (i.e. egg, soy, tissues) or artificially synthesized [Lian and Ho, 2001]. Phosphatidylcholine and phosphatidylethanolamine phospholipids are the major constituents of cell membranes and therefore the major constituents of liposomes designed for biological applications [Chandran *et al.*, 1997; Vemuri and Rhodes, 1995]. The fatty acid portion can vary in number of carbon atoms and degree of unsaturation, and are important determinants of the liposome-cell interaction. Liposomes containing phospholipids with short fatty acid chains (12–14 carbons) have been shown to induce cell lysis, probably due to extraction of cholesterol from the cell membrane or rapid uptake and replacement of the native phospholipid by the short chain phospholipid. Either mechanism results in destabilization of the cell membrane with consequent lysis [Mineo and Hara, 2005; 2007; Tamura *et al.*, 1986]. In terms of unsaturation, liposomes composed of unsaturated fatty acids have been

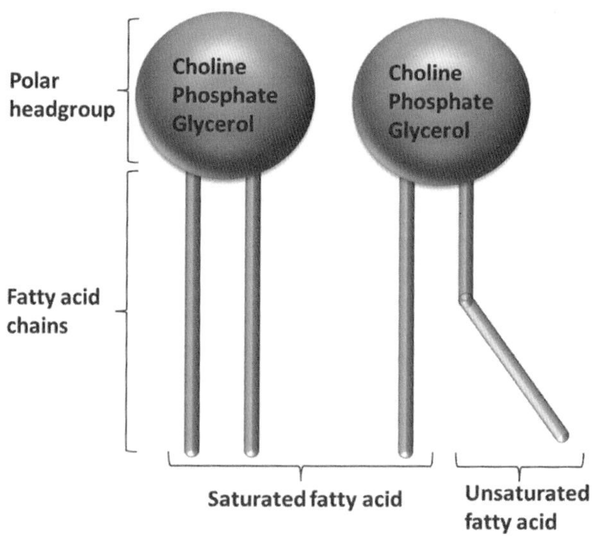

Figure 2. Schematic structure of phospholipid composed of fatty acid chains, glycerol backbone and the headgroup (choline).

shown to have better preservation effects for freeze-dried and freezing of human red blood cells, while liposomes composed of saturated fatty acids confer protection for hypothermic storage (1–6°C) of human red blood cells [Kheirolomoom *et al.*, 2005; Stoll *et al.*, 2011a]. Cholesterol (another major component of cell membranes) is also normally included in the liposome preparation as it improves the fluidity and stability of the bilayer [Woodle and Papahadjopoulos, 1989].

3.1.2. *Size and lamellarity*

Size and lamellarity are the most common properties used to classify liposomes, and less frequently, the preparation method. In terms of size they can be divided into small, large and giant vesicles. In terms of lamellarity they can be divided into unilamellar (ULV), multilamellar (MLV) and multivesicullar (Figure 3) [Akbarzadeh *et al.*, 2013; Laouini *et al.*, 2012]. Size should be considered for delivery purposes, if a substance has protective effects in a dose-dependent manner, large liposomes are able to entrap a larger volume of the substance to be delivered. Liposome size has also been implicated in interactions with the immune system. Liposomes larger than 100 nm can induce secretion of various types of cytokines [Fifis *et al.*, 2004; Mottram *et al.*, 2007; Schöler *et al.*, 2002]. Lamellarity, on the other hand, is related to the release capacity, where more layers means slower release. The characteristics of the molecules to be encapsulated also play a role in deciding between ULVs and MLVs. ULVs are more efficient for encapsulation of hydrophilic molecules due to increased internal aqueous volume, while MLVs having more lipid content are capable of encapsulating larger amounts of hydrophobic molecules [Monteiro *et al.*, 2014]. Approximate size and nomenclature are shown in Table 1.

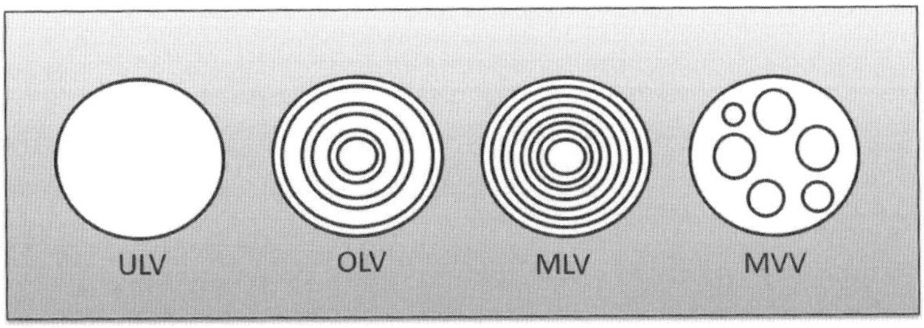

Figure 3. Diagram of liposome type based on lamellae number.

Table 1. Classification of liposomes based on size and lamellarity.

Parameter	Classification	Nomenclature
Size	Small (20–100 nm)	Small Unilamellar Vesicles (SUV)
	Large (100–400 nm)	Large Unilamellar Vesicles (LUV)
	Giant (≥ 1 μm)	Giant Unilamellar Vesicles (GUV)
Lamellarity	Unilamellar (single lamellae)	Unilamellar Vesicles (ULV)
	Oligolamellar (few lamellae)	Oligolamellar Vesicles (OLV)
	Multilamellar (multiple lamellae)	Multilamellar Vesicles (MLV)
	Multivesicullar (multiple lamellae)	Multivesicullar Vesicles (MVV)

3.1.3. *Surface charge*

The head group composition of the phospholipid and pH will determine the surface charge of liposomes. Liposomes composed of phosphatidylcholine (PC), sphingomyelin (SM) and phosphatidylethanolamine (PE) (pH 7–7.4) will have a neutral charge. Liposomes containing phosphatidic acid (PA), phosphatidylserine (PS) and phosphatidylglycerol (PG) (pH 7–7.4) will have a negative charge. Liposomes containing trimethylammonium (i.e. DOTAP) or sterylamine (SA) (pH 7–7.4) will have a positive charge [Szoka Jr and Papahadjopoulos, 1980; Ulrich, 2002].

Surface charge is particularly important for fusion of liposomes and cells, which will increase the delivery of intra-liposomal content. Holovati *et al.* [2008] reported a 100 times increase in the delivery of trehalose from negatively charged liposomes (containing PS) compared to neutral liposomes, under the same incubation conditions [Holovati *et al.*, 2008].

Charge can also determine the toxicity as well as the interaction of liposomes with the immune system *in vivo*, which will determine how long they stay in the circulation [Landesman-Milo and Peer, 2012]. Cationic liposomes interact with plasma proteins causing turbidity and have been shown to cause dose-dependent cytotoxicity [Senior *et al.*, 1991]. They are also rapidly cleared from the circulation by macrophages due to activation of the complement system [Christensen *et al.*, 2009]. Anionic liposomes are taken up by phagocytic cells at faster rates than neutral liposomes in an unclear mechanism [Foged *et al.*, 2004]. Neutral liposomes are reported as not inducing any detectable toxicity. The issue of reduced half-life has been addressed by coating the liposomes with polyethylene glycol (PEGylated or stealth liposomes) to suppress the uptake and increase circulation time [Allen *et al.*, 1991].

3.2. *Synthesis and Characterization*

The preparation of liposomes comprises four general steps which are common to all methods. These are: (a) drying of lipids from organic solvent; (b) dispersion of lipids in aqueous media; (c) purification of formed vesicles (i.e. removal of detergent or organic solvents) and (d) characterization of the final product [Akbarzadeh *et al.*, 2013].

3.2.1. *Synthesis*

Several methods are available for laboratory-scale preparation of liposomes, but many of these methods are unsuitable for scale-up production, therefore industrial manufacturing of liposomes can be very challenging and continued developments in this are needed [Wagner and Vorauer-Uhl, 2011].

3.2.1.1. Small scale vesicle preparation

Lipid film hydration

This method can be used for liposomes composed of one type of lipid or for liposomes containing lipid mixtures. The lipids are dissolved and mixed in an organic solvent to ensure a homogenous mixture. The lipid solutions are mixed in a round bottom flask, the organic solvent is evaporated by rotary evaporation, or by streaming nitrogen or argon gas leaving a thin film on the walls of the flask. An aqueous media (usually buffer) is then added to the dried lipid film at a temperature above the melting temperature (Tm) of the lipid (or above the Tm of the highest melting component in case of a mixture). The lipid mixture is dispersed into the aqueous phase by shaking or vortex forming MLVs. MLVs can be fragmented into smaller and/or unilamellar particles using the methods described below (*see Section 3.2.1.3*) [Bangham *et al.*, 1965; Bangham *et al.*, 1967].

Reverse phase evaporation

The phospholipids are dissolved first in organic solvents or mixtures of organic solvents. The aqueous media is added directly to the phospholipid-solvent mixture. The preparation is then briefly sonicated, forming a homogeneous water in oil emulsion. The organic solvents are removed under reduced pressure, resulting in the formation of a viscous gel. Further removal of the organic solvent under reduced pressure spontaneously forms a liposome dispersion composed of LUVs [Lasic, 1988; Szoka and Papahadjopoulos, 1978].

Detergent removal

Dried lipid mixtures are solubilized with detergents at their critical micelle concentration. Once the detergent is removed, the micelles become richer in phospholipids and combine to form unilamellar vesicles. By controlling the rate of detergent removal either LUVs or small unilamellar vesicles (SUVs) can be produced, although the chemical nature of the detergent and the type of lipids used can also have an influence on the vesicle size. Several techniques can be used for detergent removal, the most common ones are: dialysis, dilution, gel chromatography and adsorption [Alpes *et al.*, 1986; Fleischer and Klouwen, 1961; Kagawa and Racker, 1971].

3.2.1.2. Large scale vesicle preparation

Ether or ethanol injection

Briefly, a solution of lipids dissolved in an organic solvent (ether, ether/methanol mixtures) is injected into an aqueous solution under reduced pressure, the evaporation of the solvent leads to the formation of LUVs, which can be further downsized. In case of ethanol (miscible in water), the dilution of the ethanol in the aqueous phase causes the lipid molecules to precipitate and form SUVs. This method has many advantages as the technique is convenient for controlled hydration of lipids even on large scale, and ethanol is a relatively harmless solvent, accepted as an injectable by the regulatory agencies. Other solvents might also be used, but one has to keep in mind the regulations for residual solvents that are accepted for clinical applications. This is the most common manufacturing technique for industrial liposome production [Batzri and Korn, 1973; Stano *et al.*, 2004; Szebeni *et al.*, 1984; Toh and Chiu, 2013].

3.2.1.3. Small scale vesicle downsizing

Sonication

The use of sound (sonication) to disrupt MLVs or other aqueous dispersions of phospholipids typically produces SUVs. The purpose of sonication is to produce a homogenous dispersion of small vesicles that can be achieved using a bath type or a probe sonicator [Abramson *et al.*, 1964; Akbarzadeh *et al.*, 2013; Saunders *et al.*, 1962].

Low/medium pressure membrane extrusion

Repeated extrusion of MLVs at low pressures (80 psi or less) through polycarbonate filters of decreasing pore size produces LUVs. Using moderate pressure (500–800 psi) and stacked polycarbonate filters the homogeneity of the LUV population can be further improved [Hope *et al.*, 1985; Mayer *et al.*, 1986; Olson *et al.*, 1979].

French press (high pressure)

The French press method involves the extrusion of MLVs through a small orifice under high pressure (20,000 psi, 4°C). Multiple extrusions of the preparation results in decreased size heterogeneity and can be used to produce smaller MLVs or SUV. Due to volume restrictions (maximum 40 ml) it cannot be used for large scale production [Barenholzt and Amselem, 1979; Hamilton *et al.*, 1980].

3.2.1.4. Large scale vesicle downsizing

Microfluidizer

Microfluidization is a scale-up strategy based on the French press technique. This continuous and scalable variation of the French press technique forces downsizing of liposomes by collision of larger vesicles at high pressure in the interaction chamber of the microfluidizer (≈ 10,000 psi) for disruption of multilamellar systems. The starting volume is 50 ml and can reach up to 4000 ml [Brandl *et al.*, 1990; Mayhew *et al.*, 1984; Vemuri *et al.*, 1990].

3.2.2. *Characterization*

After synthesis by any of the preceding methods, liposomes must be characterized for adequate quality control of the product and to assist in the interpretation of results. Many parameters can be used for liposome characterization depending on the formulation type and intended application, including determination of residual solvent, turbidity, surface charge, stability and presence of degradation products [Chatterjee and Banerjee, 2002]. However, for this review we will focus on the ones most frequently used for general characterization.

3.2.2.1. Size

Particle size and distribution are important parameters on the characterization of liposomes especially for intended clinical use. There are many methods that can be used for particle size analysis but most of them can only provide the average size and little or no information about the size distribution. Dynamic light scattering is a simple and reliable technique for size distribution, but it has limitations when applied to a heterogeneous vesicle population. The same can happen with other biophysical techniques such as nuclear magnetic resonance (NMR) and analytical ultracentrifugation, which makes them more suitable for analysing SUVs. For larger and multilamellar vesicles that tend to form more heterogeneous populations, microscopy and gel exclusion chromatography are more reliable. Many techniques using atomic force and electron microscopy (negative stain and freeze-fracture) have been described. Cryoelectron microscopy has also been used with the advantage of not needing any stain, shadow, or replica preparation, although some argue that is a costly method and only a small fraction of the sample is analysed. Other possible methods include fluorescent microscopy and flow cytometry [Chen *et al.*, 2013; Egelhaaf *et al.*, 1996; Kanno *et al.*, 2002; Nozaki *et al.*, 1982; Olson *et al.*, 1979].

3.2.2.2. Lamellarity

The lamellarity of liposomes can be measured by electron microscopy using negative staining, by light microscopy using phase-contrast or by spectroscopic techniques, especially NMR (^{32}P-NMR or ^{19}F-NMR) [Fröhlich *et al.*, 2001; Hargreaves and Deamer, 1978; Jousma *et al.*, 1987].

3.2.2.3. Entrapped volume

The percentage of the compound initially added to the preparation that becomes entrapped in the aqueous space is also referred to as encapsulation efficiency. The encapsulated fraction of the compound is treated with a detergent to lyse the liposomes and release the compound into surrounding aqueous media. The compound can then be assayed by a suitable technique. The trapped volume is usually expressed as the trapped volume per lipid concentration or weight. Fluorescent dyes such as carboxyfluorescein can be used to determine the volume of entrapped water per mole of phospholipid using a UV/VIS spectrophotometer. Column chromatography techniques can also be used [Hope *et al.*, 1985; Kawakami *et al.*, 1999; Oku *et al.*, 1982; Weinstein *et al.*, 1977].

3.2.2.4. Phospholipid concentration

Lipid mixtures can be separated by chromatographic techniques or assayed as whole samples. Phospholipids in lipid extracts or in chromatographic fractions are estimated by phosphorus determination through an acidic digestion. The released inorganic phosphate reacts with ammonium molybdate and ascorbic acid, generating a blue color that can be measured spectrophotometrically at 825 nm [McClare, 1971; Poorthuis *et al.*, 1976].

4. Liposomes as Models Systems to Study Cryobiology

4.1. *Models for Osmotic Transport*

Understanding the movement of water and cryoprotectants across the cell membrane is critical to the development and optimization of cryopreservation protocols. Liposomes have been used as model systems to examine the movement of water and solutes across the cell membrane and to study osmotic injury. The use of liposomes to study water and solute permeability was first described by Bangham in a series of seminal papers examining self-forming "phospholipid liquid crystals" whereby it was shown that liposomes function as perfect osmometers and would be suitable models to study osmotic transport [Bangham *et al.*, 1965; Bangham *et al.*, 1967]. Following the work by Bangham, Reeves and Dowben were one of the first groups to describe the use of multilamellar liposomes to measure the transport of glycine and water [Reeves and Dowben, 1970]. By monitoring changes in the turbidity of the liposome-containing solution, they were able to not only demonstrate that liposomes could be used to roughly calculate the temperature dependence of water and solute permeability, but were able to examine the effect that liposome size and composition have on osmotic transport. Siminovich and Chapman, appreciating the advantages that liposomes provided as a model system where control of the lamellarity, lipid and cholesterol composition and internal composition of the liposome's aqueous core was possible, demonstrated their use as a model system to study osmotic injury during freezing and thawing [Siminovich and Chapman, 1971; 1974]. By studying the osmotic response of multi- and unilamellar liposome systems, McGrath *et al.* were able to develop and validate a thermodynamic model for water transport that has helped create a better understanding of the osmotic response of cells to anisotonic conditions [Callow and McGrath, 1985; McGrath, 1984]. By modifying the composition of the liposome and the structural interaction of phospholipids and cholesterol, the impact of these factors on the selective permeability of the lipid bilayers has been examined [Biondi *et al.*, 1991; de Gier J, 1993]. While liposomes lack the complexity of a cell membrane, the ability to manipulate their structure and composition has been used effectively over the years to better understand the role that water and solute play in the low temperature response of cells.

4.2. *Liposomes as a Model System to Study Cryoinjury*

As the membrane has been known to be a primary site of injury during freezing and thawing [Mazur, 1965; Meryman, 1971], liposomes have been used to examine the biophysical and biomechanical changes that occur to membranes during cryopreservation. By taking advantage of the ability to entrap inside liposomes various markers and then monitoring their leakage as a function of various osmotic and temperature-induced changes, Siminovich and Chapman [1971] demonstrated the osmotic effects of slow cooling on cell dehydration and the formation of intracellular ice during rapid cooling — in line with the "two-factor hypothesis of cryoinjury" [Mazur *et al.*, 1972]. Using a cryomicroscope to examine cell volume changes and the formation of intracellular ice, Morris and McGrath were able to show the relationship of lipid and cholesterol composition to cooling rate and cryoprotectant sensitivity [Morris and McGrath, 1981; Morris, 1981; 1982]. As the response of liposomes to freezing and thawing was shown to mimic that of many cell types [Callow and McGrath, 1985; McGrath, 1984; Morris, 1981], the ability to test a thermodynamic model for freezing injury was actively pursued [Callow and McGrath, 1985] and observations on the interaction of ice and cell-size liposomes were reported [Tandorf *et al.*, 1987]. The influence of the composition on the freezing-induced dehydration of liposome systems has been used to examine the phase transitions that may occur in cells during freezing and thawing [Bronshteyn and Steponkus, 1993; Siow *et al.*, 2007]. With the increased interest in the use of liposomes as drug delivery vehicles, much of the more recent work looking at the freezing and thawing of liposomes has moved away from using them as cell model systems to study cryoinjury and focused more on the developing liposome formulations and preservation methods for the pharmaceutical industry. While there has been direct application of liposome stabilization studies to our understanding of membrane biophysics and cryo- and lyoprotectant mechanisms of action (see Section 4.3), efforts to leverage these studies to better understand cryoinjury have been limited.

4.3. *Liposomes as Models for Understanding Cryo- and Lyoprotectant Mechanisms of Action*

A large focus for cryobiology research has been investigating compounds with the ability to protect membranes from cryo- and lyo-injury. One of the main advantages of liposomes as membrane stabilizers is their non-toxic composition, which, unlike many other protective compounds, does not necessitate removal before clinical application. Crowe *et al.* pioneered the use of liposomes as model membranes to study mechanisms of cryo- and lyo-protection of a novel protective agent — the sugar trehalose [Carpenter *et al.*, 1992; Crowe *et al.*, 1987; *Crowe et al.*, 1989; Oliver *et al.*, 2001]. As reviewed in Crowe *et al.*, liposomes lyophilized in trehalose were reported to remain structurally intact upon rehydration [Crowe *et al.*, 1987]. If the liposomal lipid bilayer was prepared from low melting point lipids, such as egg phosphotidylcholine, the presence of trehalose appeared to both inhibit the fusion between the dry liposomes and depress the phase transition temperature (Tm) in the dry state, maintaining it in liquid crystalline phase [Carpenter *et al.*, 1992; Crowe *et al.*, 1989; Crowe *et al.*, 1992]. This is claimed to be an essential factor for the membrane to remain stable upon rehydration, since the phospholipid phase transition upon rehydration translates into the well described "leaky membrane" phenomenon [Crowe *et al.*, 1989; Koynova and Caffrey, 1998; Rietveld and Simons, 1998]. Furthermore, Crowe *et al.* [1989] reported that the presence of trehalose as a biostabilizing agent was critical — as other potential lyoprotective additives, such as oligosaccharides and polymers, did not exude similar protective action. However, for maximum protection efficiency, the presence of trehalose has proven to be required on both sides of the cell membrane. As mammalian cells are virtually impermeable to trehalose, liposomes have proven to be an invaluable model to explore mechanisms of cryoprotective action of trehalose. Versatility of liposome synthesis allowed for trehalose to be selectively incorporated into the liposome's aqueous core and/or extra-liposomal buffer, thus enabling research into this biopreservative agent.

In addition to being a membrane model to investigate cryoprotective actions of different biostabilizing agents, research has established that liposomes have important properties that can reduce the chilling injury of spermatozoa and oocytes [Healey, 1969; Phillips and Lardy, 1940; Quinn *et al.*, 1980; Zeron *et al.*, 2002]. During cold shock, sperm and oocyte membranes become transiently leaky and release phospholipid into the surrounding medium, leading to coalescence of lipid rafts in membranes, which tend to be irreversible upon warming. After Phillips and Lardy [1940] first reported that egg yolk was beneficial for the preservation of sperm [Phillips and Lardy, 1940], investigation of the lipid and lipoprotein portions of the egg yolk suggested that it is specifically the phospholipid portion which affords protection to chilling injury [Phillips and Lardy, 1940; Quinn *et al.*, 1980]. Furthermore, as it was shown that sperm from species in which membranes contain higher cholesterol to phospholipid ratios are less susceptible to cold shock than sperm from species with lower corresponding ratios [Phillips and Lardy, 1940; Quinn *et al.*, 1980], and the addition of cholesterol to phospholipid membranes lowers the phase transition temperature of cell membranes [Phillips and Lardy, 1940; Quinn *et al.*,1980], liposomes were proposed as vehicles to alter sperm and oocyte compositional membranes in a way that would improve cellular ability to withstand chilling injury.

A key study by Graham and Foote [1987] tested the hypothesis that specific lipids, with potential for lowering membrane phase transition temperatures, when added to sperm could minimize cold shock membrane damage as well as cryoinjury [Graham and Foote, 1987]. This classic study described the effect of sonicated liposomes of purified phosphatidylserine, phosphatidylcholine with varying fatty acyl chain lengths and double bonds and cholesterol alone or in combination, on cold shock and cryoinjury of bull sperm [Graham and Foote, 1987]. One interesting finding is that liposomes containing lipids with fatty acyl chains of less than 12 carbons were toxic to sperm cells. Phosphatidylserine alone or in combination with phosphatidylcholine or cholesterol, but not phosphatidylcholine or cholesterol alone, protected sperm from cold shock as well as did egg yolk lecithin liposomes or egg yolk. Lipid concentrations of 0.5mM were found to be more effective than liposomes at 5mM in protecting sperm during freezing and thawing [Graham and Foote, 1987]. During freezing, phosphatidylserine in combination with phosphatidylcholine partially protected sperm, but only phosphatidylserine/cholesterol liposomes were as effective as egg yolk in protecting sperm from cryoinjury. However, as recently pointed out by Mutalik *et al.*, [2014] using egg yolk in semen cryopreservation medium has multiple disadvantages, including variability of egg yolk composition and a potential to harbor various infectious microorganisms and endotoxins [Mutalik *et al.*, 2014]. Liposome encapsulated soy lecithin and cholesterol has been shown to effectively replace the chicken egg yolk from semen cryopreservation medium without compromising post-thaw quality and function [Mutalik *et al.*, 2014].

Although liposomes have been repeatedly shown to be useful as biostabilizing agents, the underlying mechanism of how liposomes stabilize cells during freezing or drying is still poorly understood. The protective properties of liposomes are often attributed to lipid and cholesterol transfer between liposomes and cells. It has been demonstrated that cholesterol can exchange rapidly between different membrane bilayers [Bruckdorfer *et al.*, 1969; Steck *et al.*, 1988], whereas lipid transfer is a relatively slow process [Phillips *et al.*, 1987]. Stoll *et al.* [2011b] have recently applied FTIR spectroscopy, MALDI-TOF MS, and colorimetric assays to investigate the effects of liposomes composed of a homologous series of linear saturated phosphatidylcholine phospholipids on red blood cell (RBC) membranes [Stoll *et al.*, 2011b]. This study provided direct evidence that liposomes alter RBC membrane properties by cholesterol depletion and lipid addition, as well as novel information about kinetics of the interaction with liposomes of different compositions [Stoll *et al.*, 2011b]. This new knowledge will further enhance the use of liposomes in biopreservation of red blood cells, as well as other mammalian cells.

5. Liposomes in Cell Preservation

5.1. *Liposomal Delivery System for Cryopreservation of Red Blood Cells*

Cryopreservation of RBCs for transfusion is one of the most important applications of cryobiology research. There have been several approaches to use liposomes to improve the biopreservation of red blood cells. A key study by Kheirolomoom *et al.* [2005] reported that the membrane lesion caused by RBC lyophilization can be minimized by addition of liposomes to the lyophilisation buffer [Kheirolomoom *et al.*, 2005]. During lyophilization and rehydration, RBC membranes lose phospholipids through microvesiculation, resulting in the deterioration of RBC membrane composition, morphology, and rheology [Kheirolomoom *et al.*, 2005; Sowemimo-Coker *et al.*, 1993; Torok *et al.*, 2005]. In their study, liposome-treated, lyophilized RBCs were shown to synthesize 2,3-DPG, have low levels of methemoglobin, and have preserved morphology [Kheirolomoom *et al.*, 2005]. This phenomenon was shown to be due to direct interactions between liposomes and RBC membranes, requiring vesicles of defined size and composition: unsaturated, unilamellar liposomes were found to be most effective as a lyoprotective agent [Kheirolomoom *et al.*, 2005]. Similar RBC membrane damage has been reported by Yamaguchi *et al.* [1993] when RBCs were subjected to high-pressure and hypotonic conditions [Yamaguchi *et al.*, 1993]. This study also showed that the addition of aminophospholipid vesicles, in combination with amphipathic drugs and hypertonic conditions, markedly reduced pressure-induced RBC membrane damage [Yamaguchi *et al.*, 1993].

As similar stress conditions are associated with ice formation that accompanies low temperatures, our group has proposed that delivering of liposome phospholipids through liposome incorporation into RBC membranes may provide an effective approach for reducing cell cryoinjury. Our studies show that liposome treatment significantly improves recovery and membrane integrity of human RBCs following low temperature exposure [Holovati *et al.*, 2009; Stoll *et al.*, 2011b]. Liposomes modulate the RBC freezing response, resulting in cell cryoprotection and, thus, behaving as a novel cryoprotectant. The extent of liposome-RBC interaction was shown to play an essential role for cryoprotection as charged trehalose-containing liposomes were more effective at protecting RBCs against cryoinjury than uncharged liposomes [Holovati *et al.*, 2009; Stoll *et al.*, 2011b]. While both trehalose- and saline-containing liposomes improved RBC post-thaw recovery, there was no cryoprotection when negatively charged lipids in the free and multilamellar form were incubated with RBCs [Holovati *et al.*, 2009]. These results suggest that liposomes in the form of defined unilamellar vesicles are necessary to confer cryoprotection to RBCs. This study also demonstrated that different extracellular solutions and cooling rates play an important role in liposome cryoprotective effect [Holovati *et al.*, 2009]. Liposome-treated RBCs showed an improved post-thaw survival if higher cooling rates were applied and extracellular freezing solutions contained trehalose or liposomes. Furthermore, we provided important evidence that the liposome cryoprotective effect is not due to the delivery of intracellular trehalose, but is likely related to modification and/or preservation of the RBC membrane [Holovati *et al.*, 2009].

In our follow-up study, we have evaluated the effects of DMPC, DOPC, and DPPC liposomes on *in vitro* quality of RBCs cryopreserved with known extracellular cryoprotectants, hydroxyethyl starch (HES) and trehalose [Stoll *et al.*, 2012]. The idea of combining intra- and extracellular cryoprotectant for maximum biostabilization effect is not new. In 1965, Strumia and Strumia published a well-known study investigating the effects of lactose, dextran and albumin on recovery and survival of frozen red cells [Strumia and Strumia, 1965]. This study demonstrated a synergistic protective effect of lactose and dextran or albumin on cryopreserved RBCs, resulting in post-thaw recovery rate of 98–99% and 69% RBC *in vivo* survival 24-hours post-transfusion [Strumia and Strumia, 1965]. However, nephro- and hepatoxic effect of lactose has prevented further clinical translation of this experimental protocol. Our approach also involved combining cryoprotectants with different mechanisms of action to maximize potential cryoprotection: large unilamellar liposomes composed of DPPC, DMPC or DOPC lipids, trehalose and HES [Stoll *et al.*, 2012]. This study demonstrated that DMPC liposomes caused hemolysis even before the cryopreservation step, negatively affecting hemorheologic properties of RBCs. RBCs treated with DMPC and DOPC liposomes significantly increased cell membrane permeability, resulting in loading of intracellular trehalose [Stoll *et al.*, 2012]. In addition to the membrane stabilizing benefits of the liposome treatment, HES conferred additional cryoprotection during cryopreservation [Stoll *et al.*, 2012]. One advantage of this approach is that the cryoprotective compounds used in this study are already approved for human pharmaceutical use, with well-established safety records. Unlike the current gold standard method for RBC clinical cryopreservation, which uses 40% (wt/vol) glycerol as cryoprotectants, using liposomes, trehalose or HES would not require time-consuming, extensive washing procedures before transfusion. The synergistic protective effects of trehalose, DOPC liposomes and HES potentially hold promise for the development of new cryopreservation methods for RBCs.

5.2. *Liposomal Delivery System for Hypothermic Storage of Red Blood Cells*

Although the use of additive solutions extends storage length of red blood cell (RBC) units, the quality of stored RBCs progressively decreases during hypothermic storage (1–6°C). RBCs undergo a series of biochemical and biomechanical changes, collectively known as "hypothermic storage lesion" (HSL). Characteristics of the HSL include RBC membrane remodeling, decreased metabolites, loss of intracellular potassium, oxidative injury of proteins and lipids, membrane loss, microparticle release and ultimately hemolysis [Blasi *et al.,* 2012; Gevi *et al.,* 2012; Rinalducci *et al.,* 2011; Van De Watering, 2011]. Alternate additive solutions have been investigated in order to decrease cell injury during storage, such as rejuvenating solutions, anaerobic storage and the use of plasticizers [Draper *et al.,* 2002; Koshkaryev *et al.,* 2009; Yoshida and Shevkoplyas, 2010]. Since the cell membrane plays an important role in post-transfusion survival and oxygen delivery capacity of RBCs, our research group has focused on the use of liposomes to mitigate RBC membrane injury during hypothermic storage [Holovati *et al.,* 2008; Kiraly *et al.,* 2009; Stoll *et al.,* 2011b].

Liposomes constitute an important tool for preservation of RBCs not only because of their delivery capacity, but also because the lipids present in the liposomes themselves can act as membrane stabilizers. Mature RBCs lose the capacity to synthesize phospholipids, so they rely on lipid exchange with plasma lipoproteins to renew their membrane phospholipids and acylation of lysophospholipids [Mulder *et al.,* 1965; Mulder and Van Deenen, 1965a; 1965b]. In the blood bank processing of whole blood into a packed RBC unit, the lipoproteins remain in the plasma fraction, which is separated from RBCs. So during the next 42 days of storage following manufacturing, RBCs no longer have a source of lipid exchange relying exclusively on acylation reactions that are normally reduced during storage [Rusnak *et al.,* 2000].

Numerous studies using RBCs as membrane models have utilized liposomes as means for modifying the composition of membranes and have documented the effect of these manipulations [Eytan *et al.,* 1982; Lange and Slayton, 1982; Rindlisbacher and Zahler, 1983]. Liposomes containing only phospholipids or cholesterol-poor formulations (less than 1.0 C/P ratio) are known to deplete cholesterol from RBC membranes, destabilizing the membrane, increasing the permeability to ions and resulting in hemolysis. The addition of cholesterol to the formulation can prevent these effects. However, cholesterol-rich formulations are also not ideal (more than 1.0 C/P ratio) since cholesterol will then be transferred from liposomes to RBC membrane and can result in decreased red cell deformability [Cooper *et al.*, 1975]. Short-chain phospholipids and lysophospholipids are also known to cause hemolysis probably due to a rapid insertion and accumulation into the RBC membrane that leads to destabilization [Kobayashi *et al.*, 1983; Mineo and Hara, 2005; 2007]. Kuypers and colleagues [1984] observed that shape changes were induced by replacement of native phosphadylcholine (PC) of human RBCs by liposomes composed of other PC species containing different fatty acid compositions. They concluded that the RBC membrane structure and the overall discoid cell shape are optimally stabilized by PC species that contain one saturated and one mono- or diunsaturated fatty acid, and that the cell tolerates only limited variations in the species composition of its PC [Kuypers *et al.*, 1984]. Our group examined the effect of liposome treatment on RBC membrane quality during hypothermic storage and the results also indicated that RBCs react differently to liposomes with varied fatty acid compositions. RBCs incubated with liposomes containing phospholipids with higher degree of unsaturation and shorter acyl chain length had higher percent hemolysis after 42 days of storage. In terms of deformability, RBCs treated with liposomes containing phospholipids with saturated acyl chains maintained better membrane deformability during hypothermic storage [Acker *et al.*, 2010]. A curious fact is that RBCs from different species will have different preferences in terms of fatty acid saturation. Our recent experience with liposome treatment of rat RBCs for an animal study on RBC preservation yielded the exact opposite results. Rat RBCs had lower hemolysis levels and better deformability during hypothermic storage when treated with liposomes containing unsaturated phospholipids [Cavalcante *et al.*, 2013].

In the previous works mentioned above, the liposomes were not loaded with any active compounds, so we attribute the improvements observed in hemolysis and deformability solely to lipid exchange or adsorption of liposomes to the RBC membrane that could result in either membrane lipid renewal or a "coating" effect [Holovati *et al.*, 2009]. However one aspect of the liposome approach remains to be explored as part of the efforts to improve RBC hypothermic storage, which is the loading of liposomes with antioxidants, cryoprotectants and rejuvenating solutions for delivery during storage. Because the hypothermic lesion results from a combination of different mechanisms (oxidative and metabolic), the association of liposomes with metabolic restoration or antioxidant protection might have a synergistic effect in protecting RBCs during hypothermic storage. A study by Marathe and Mishra [2002] showed that the presence of antioxidants; alpha-tocopherol (0.15 mole %) in the bilayer membrane or ascorbic acid (0.1 mM) in the aqueous region significantly protected DPPC vesicles from radiation damage [Marathe and Mishra, 2002]. In a similar way, antioxidants added to a unit of RBCs have been shown to protect them against oxidative damage of gamma irradiation [Sharifi *et al.*, 2000]. Since evidence suggests that liposome-encapsulated antioxidants can be superior to the corresponding free antioxidants against oxidative damage [Yang *et al.*, 2008], we believe that the combined action of liposomes and antioxidants could provide a new alternative for mitigating RBC storage lesion.

Future investigations into the optimal liposome composition and optimal active compounds to be delivered could result in effective liposome-based strategies to mitigate the RBC hypothermic storage lesion.

6. Conclusions and Future Perspectives

Liposomes have been applied in a broad number of ways to the biopreservation / cryobiology field. Liposomes have been used extensively by cryobiologists as model systems to study the osmotic, thermal and physical stresses that cells encounter during freezing, thawing or drying and as a tool to modify the membrane of cells or delivery protectants into the cell. As an artificial biological system, the ability to carefully control the composition and structure of liposomes at the nano-scale to have the specific properties of interest, has been used to great effect. Looking forward, new techniques will emerge to integrate protein structures with the liposome to add structural elements more alike a cell and to enhance the targeting ability of the liposomes to specific cells. This will lead to new opportunities to use liposomes in both discovery and applied aspects of the preservation sciences. While first described more than 50 years ago, there is an exciting future for liposomes in biopreservation.

References

Abramson, M.B., R. Katzman, and H. P. Gregor (1964). Aqueous dispersions of phosphatidylserine ionic properties. *Journal of Biological Chemistry,* 239, pp. 70–76.

Acker, J., C. Stoll, H. Stadnick, J. Holovati, and W. Wolkers (2010). Application of liposomes to improve the hypothermic storage of red blood cells, *Vox Sanguinis,* 99, pp. 67–68.

Ainsworth, S., M. Beresford, D. Milligan, N. Shaw, J. Matthews, A. Fenton and M. W. Platt (2000). Pumactant and poractant alfa for treatment of respiratory distress syndrome in neonates born at 25–29 weeks gestation: a randomised trial. *Lancet,* 355 (9213), pp. 1387–1392.

Akbarzadeh, A., R. Rezaei-Sadabady, S. Davaran, S. W. Joo, N. Zarghami, Y. Hanifehpour, M. Samiei, M. Kouhi and K. Nejati-Koshki (2013). Liposome: classification, preparation, and applications. *Nanoscale Research Letters,* 8(1), p. 102.

Allen, T., C. Hansen, F. Martin, C. Redemann and A. Yau-Young (1991). Liposomes containing synthetic lipid derivatives of poly (ethylene glycol) show prolonged circulation half-lives *in vivo. Biochimica et Biophysica Acta,* 1066(1), pp. 29–36.

Allen, T.M. and P. R. Cullis (2013). Liposomal drug delivery systems: from concept to clinical applications. *Advance Drug Delivery Reviews,* 65(1), pp. 36–48.

Alpes, H., K. Allmann, H. Plattner, J, Reichert, R. Rick and S. Schulz (1986). Formation of large unilamellar vesicles using alkyl maltoside detergents, *Biochimica et Biophysica Acta—Biomembranes,* 862(2), pp. 294–302.

Bangham, A.D. and R.W. Horne (1964). Negative staining of phospholipids and their structural modification by surface-active agents as observed in the electron microscope. *Journal of Molecular Biology,* 8(5), pp. 660–668.

Bangham, A.D., M. M. Standish and J. C. Watkins (1965). Diffusion of univalent ions across the lamellae of swollen phospholipids, *Journal of Molecular Biology,* 13(1), pp. 238–252.

Bangham, A.D., J. De Gier and G. D. Greville (1967). Osmotic properties and water permeability of phospholipid liquid crystals. *Chemistry and Physics of Lipids,* 1(3), pp. 225–246.

Bangham, A.D., N. G. A. Miller, R. J. Davies, A. Greenough and C. J. Morley (1984). Introductory remarks about artificial lung expanding compounds (ALEC). *Colloids and Surfaces,* 10, pp. 337–341.

Bangham, A.D. (1995). Surrogate cells or trojan horses. The discovery of liposomes. *Bioessays,* 17(12), pp. 1081–1088.

Barenholzt, Y. and S. Amselem (1979). A new method for preparation of phospholipid vesicles (liposomes)—French press. *FEBS Letters,* 99(1), pp. 210–214.

Batzri, S. and E. D. Korn (1973). Single bilayer liposomes prepared without sonication. *Biochimica et Biophysica Acta—Biomembranes,* 298(4), pp. 1015–1019.

Biondi, A.C., M. R. Feliz and E. A. Disalvo, E.A. (1991). Surface changes induced by osmotic stress and its influence on the glycerol permeability in lipid bilayers. *Biochim et Biophysica Acta,* 1069(1), pp. 5–3.

Blasi, B., A. D'Alessandro, N. Ramundo and L. Zolla (2012). Red blood cell storage and cell morphology. *Transfusion Medicine,* 22(2), pp. 90–96.

Brandl, M., D. Bachmann, M. Drechsler and K. Bauer (1990). Liposome preparation by a new high pressure homogenizer Gaulin Micron Lab 40. *Drug Development and Industrial Pharmacy,* 16(14), pp. 2167–2191.

Bronshteyn, V.L. and P. L. Steponkus (1993). Calorimetric studies of freeze-induced dehydration of phospholipids. *Biophysical Journal,* 65(5), pp. 1853–1865.

Bruckdorfer, K.R., R. A. Demel, J. De Gier and L. L. van Deenen (1969).The effect of partial replacements of membrane cholesterol by other steroids on the osmotic fragility and glycerol permeability of other erythrocytes. *Biochimica et Biophysica Acta,* 183(2), pp. 334–345.

Callow, R. A. and J. J. McGrath (1985). Thermodynamic modeling and cryomicroscopy of cell-size unilamellar and paucilamellar liposomes. *Cryobiology,* 22, pp. 251–267.

Carpenter, J. F, T. Arakawa and J. H. Crowe (1992).Interactions of stabilizing additives with proteins during freeze-thawing and freeze-drying. *Developments in Biological Standardization,* 74, pp. 225–238.

Cavalcante, L. S., J. P. Acker and J. L. Holovati (2013). Membrane changes in liposome-treated rat red blood cells upon hypothermic storage. *Cryobiology,* 67(3), pp. 429–430.

Chandran, S., A. Roy and B. Mishra (1997). Recent trends in drug delivery systems: liposomal drug delivery system—preparation and characterisation. *Indian Journal of Experimental Biology,* 35(8), pp. 801–809.

Chatterjee, S. and D. Banerjee (2002). Preparation, isolation, and characterization of liposomes containing natural and synthetic lipids. In S. Basu and M. Basu (eds.) *Methods in Molecular Biology* 199: *Liposome Methods and Protocols* (Humana Press Inc., Totowa, NJ), Chap. 1, pp. 3–16.

Chen, C., S. Zhu, T. Huang, S. Wang and X. Yan (2013). Analytical techniques for single-liposome characterization. *Analytical Methods,* 5(9), pp. 2150–2157.

Chonn, A. and P. R. Cullis (1995). Recent advances in liposomal drug-delivery systems. *Current Opinion in Biotechnology,* 6(6), pp. 698–708.

Christensen, D., E. M. Agger, L. V. Andreasen, D. Kirby, P. Andersen and Y. Perrie (2009) Liposome-based cationic adjuvant formulations (CAF): past, present, and future. *Journal of Liposome Research,* 19(1), pp. 2–11.

Cooper, R. A., E. C. Arner, J. S. Wiley and S. J. Shattil (1975). Modification of red cell membrane structure by cholesterol-rich lipid dispersions. A model for the primary spur cell defect. *Journal of Clinical Investigation,* 55(1), pp. 115–126.

Crowe, J. H., L. M. Crowe, J. F. Carpenter and W. C. Aurell (1987). Stabilization of dry phospholipid bilayers and proteins by sugars. *Biochemical Journal,* 242(1), pp. 1–10.

Crowe, J. H., L. M. Crowe and F. A. Hoekstra (1989). Phase transitions and permeability changes in dry membranes during rehydration. *Journal of Bioenergetics and Biomembranes,* 21(1), pp. 77–91.

Crowe, J. H., F. A. Hoekstra and L. M. Crowe (1992). Anhydrobiosis. *Annual Review of Physiology,* 54, pp. 579–599.

De Gier, J. (1993). Osmotic behaviour and permeability properties of liposomes. *Chemistry and Physics of Lipids,* 64(1–3), pp. 187–196.

Draper, C., T. Greenwalt and U. Dumaswala (2002). Biochemical and structural changes in RBCs stored with different plasticizers: the role of hexanol. *Transfusion,* 42(7), pp. 830–835.

Egelhaaf, S., E. Wehrli, M. Adrian and P. Schurtenberger (1996). Determination of the size distribution of lecithin liposomes: a comparative study using freeze fracture, cryoelectron microscopy and dynamic light scattering. *Journal of Microscopy,* 184(3), pp. 214–228.

Eytan, G. D., R. Broza, B. Notsani, D. Dachir and A. E. Gad (1982). Interaction of acidic liposomes with red blood cells: induction of endocytosis and shedding of particles. *Biochimica et Biophysica Acta,* 689(3), pp. 464–474.

Fifis, T., A. Gamvrellis, B. Crimeen-Irwin, G. A. Pietersz, J. Li, P. L. Mottram, I. F. McKenzie and M. Plebanski (2004). Size-dependent immunogenicity: therapeutic and protective properties of nano-vaccines against tumors. *Journal of Immunology,* 173(5), pp. 3148–3154.

Fleischer, S. and H. Klouwen (1961). The role of soluble lipid in mitochondrial enzyme systems. *Biochemical and Biophysical Research Communications,* 5(5), pp. 378–383.

Foged, C., C. Arigita, A. Sundblad, W. Jiskoot, G. Storm and S. Frokjaer (2004). Interaction of dendritic cells with antigen-containing liposomes: effect of bilayer composition. *Vaccine,* 22(15), pp. 1903–1913.

Fröhlich, M., V. Brecht and R. Peschka-Süss (2001). Parameters influencing the determination of liposome lamellarity by 31P-NMR. *Chemistry and Physics of Lipids,* 109(1), pp. 103–112.

Gevi, F., A. D'Alessandro, S. Rinalducci and L. Zolla (2012). Alterations of red blood cell metabolome during cold liquid storage of erythrocyte concentrates in CPD–SAGM. *Journal of Proteomics,* 76, pp. 168–180.

Graham, J. K. and R. H. Foote (1987). Effect of several lipids, fatty acyl chain length, and degree of unsaturation on the motility of bull spermatozoa after cold shock and freezing. *Cryobiology,* 24(1), pp. 42–52.

Gregoriadis, G., P. D. Leathwood and B. E. Ryman (1971). Enzyme entrapment in liposomes. *FEBS Letters,* 14(2), pp. 95–99.

Gregoriadis, G. and B. E. Ryman (1972a). Fate of protein-containing liposomes injected into rats. An approach to the treatment of storage diseases. *European Journal of Biochemistry,* 24(3), pp. 485–491.

Gregoriadis, G. and B. E. Ryman (1972b). Lysosomal localization of beta-fructofuranosidase-containing liposomes injected into rats. Some implications in the treatment of genetic disorders. *Biochemical Journal,* 129, pp. 123–133.

Gregoriadis, G. (2008). Liposome research in drug delivery: the early days. *Journal of Drug Targeting,* 16(7), pp. 520–524.

Hamilton, R. L., J. Jr, Goerke, L. S. Guo, M. C. Williams and R. J. Havel (1980). Unilamellar liposomes made with the French pressure cell: a simple preparative and semiquantitative technique. *Journal of Lipid Research,* 21(8), pp. 981–992.

Hargreaves, W. R. and D. W. Deamer (1978). Liposomes from ionic, single-chain amphiphiles. *Biochemistry,* 17(18), pp. 3759–3768.

Healey, P. (1969). Effects of freezing on ultrastructure of spermatozoa of some domestic animals. *Journal of Reproduction and Fertility,* 18, pp. 21–27.

Holovati, J. L., M. I. C. Gyongyossy-Issa and J. P. Acker (2008). Effect of liposome charge and composition on the delivery of trehalose into red blood cells. *Cell Preservation Technology,* 6(3), pp. 207–218.

Holovati, J. L., M. I. Gyongyossy-Issa and J. P. Acker (2009). Effects of trehalose-loaded liposomes on red blood cell response to freezing and post-thaw membrane quality. *Cryobiology,* 58(1), pp. 75–83.

Holovati, J. L. and J. P. Acker (2011). Emerging role for use of liposomes in the biopreservation of red blood cells. *Transfusion Medicine and Hemotherapy,* 38(2), pp. 99–106.

Hope, M., M. Bally, G. Webb and P. Cullis (1985). Production of large unilamellar vesicles by a rapid extrusion procedure. Characterization of size distribution, trapped volume and ability to maintain a membrane potential. *Biochimica et Biophysica Acta,* 812(1), pp. 55–65.

Immordino, M. L., F. Dosio and L. Cattel (2006). Stealth liposomes: review of the basic science, rationale, and clinical applications, existing and potential. *International Journal of Nanomedicine,* 1(3), pp. 297–315.

Johnson, S. M., K. W. Miller and A. D. Bangham (1973). The opposing effects of pressure and general anaesthetics on the cation permeability of liposomes of varying lipid composition. *Biochimica et Biophysica Acta,* 307(1), pp. 42–57.

Jousma, H., H. Talsma, F. Spies, J. Joosten, H. Junginger and D. Crommelin (1987). Characterization of liposomes. The influence of extrusion of multilamellar vesicles through polycarbonate membranes on particle size, particle size distribution and number of bilayers. *International Journal of Pharmaceutics,* 35(3), pp. 263–274.

Kagawa, Y. and E. Racker (1971). Partial resolution of the enzymes catalyzing oxidative phosphorylation. XXV. Reconstitution of vesicles catalyzing 32Pi-adenosine triphosphate exchange. *Journal of Biological Chemistry,* 246(17), pp. 5477–5487.

Kanno, T., T. Yamada, H. Iwabuki, H. Tanaka, S. Kuroda, K. Tanizawa and T. Kawai (2002). Size distribution measurement of vesicles by atomic force microscopy. *Analytical Biochemistry,* 309(2), pp. 196–199.

Kawakami, K., Y. Nishihara and K. Hirano (1999). Determination of the entrapped volume of liposomes: dilution method. *Analytical Biochemistry,* 269(1), pp. 139–142.

Kheirolomoom, A., G. R. Satpathy, G.R., Török, Z., Banerjee, M., Bali, R., Novaes, R.C., Little, E., Manning, D.M., Dwyre, D.M., Tablin, F., Crowe, J.H. and Tsvetkova, N.M. (2005). Phospholipid vesicles increase the survival of freeze-dried human red blood cells. *Cryobiology,* 51(3), pp. 290–305.

Kiraly, L. N., S. Underwood, J. A. Differding and M. A. Schreiber (2009). Transfusion of aged packed red blood cells results in decreased tissue oxygenation in critically injured trauma patients. *Journal of Trauma,* 67(1), pp. 29–32.

Kobayashi, T., K. Takahashi, A. Yamada, S. Nojima and K. Inoue (1983). Lysis of erythrocytes by phosphatidylcholine containing polyunsaturated fatty acid. *Journal of Biochemistry,* 93(3), pp. 675–680.

Koshkaryev, A., O. Zelig, N. Manny, S. Yedgar and G. Barshtein (2009). Rejuvenation treatment of stored red blood cells reverses storage-induced adhesion to vascular endothelial cells. *Transfusion,* 49(10), pp. 2136–2143.

Koynova, R. and M. Caffrey (1998). Phases and phase transitions of phosphatidylcholines. *Biochimica et Biophysica Acta,* 137, pp. 91–145.

Kuypers, F. A., B. Roelofsen, W. Berendsen, J. A. Op den Kamp and L. L. van Deenen (1984). Shape changes in human erythrocytes induced by replacement of the native phosphatidylcholine with species containing various fatty acids. *Journal of Cell Biology,* 99(6), pp. 2260–2267.

Lampen, P., W. Pittermann, H. M. Heise, M. Schmitt, H. Jungmann and M. Kietzmann (2003). Penetration studies of vitamin E acetate applied from cosmetic formulations to the stratum corneum of an *in vitro* model using quantification by tape stripping, UV spectroscopy, and HPLC. *Journal of Cosmetic Science,* 54(2), pp. 119–131.

Landesman-Milo, D. and D. Peer (2012). Altering the immune response with lipid-based nanoparticles. *Journal of Controlled Release,* 161(2), pp. 600–608.

Lange, Y. and J. M. Slayton (1982). Interaction of cholesterol and lysophosphatidylcholine in determining red cell shape. *Journal of Lipid Research,* 23(8), pp. 1121–1127.

Laouini, A., C. Jaafar-Maalej, I. Limayem-Blouza, S. Sfar, C. Charcosset and H. Fessi (2012). Preparation, characterization and applications of liposomes: state of the art, *Journal of Colloid Science and Biotechnology,* 1(2), pp. 147–168.

Lasic, D. D. (1988). The mechanism of vesicle formation. *Biochemical Journal,* 256(1), pp. 1–11.

Lasic, D. D. (1998). Novel applications of liposomes. *Trends in Biotechnology,* 16(7), pp. 307–321.

Lian, T. and R. J. Ho (2001). Trends and developments in liposome drug delivery systems. *Journal of Pharmaceutical Science,* 90(6), pp. 667–680.

Marathe, D. and K. Mishra (2002). Radiation-induced changes in permeability in unilamellar phospholipid liposomes. *Radiation Research,* 157(6), pp. 685–692.

Mayer, L., M. Hope and P. Cullis (1986). Vesicles of variable sizes produced by a rapid extrusion procedure. *Biochimica et Biophysica Acta,* 858(1), pp. 161–168.

Mayhew, E., R. Lazo, W. Vail, J. King and A. Green (1984). Characterization of liposomes prepared using a microemulsifier. *Biochimica et Biophysica Acta,* 775(2), pp. 169–174.

Mazur, P. (1965). The role of cell membranes in the freezing of yeast and other single cells. *Annals of the New York Academy of Science,* 125, pp. 658–676.

Mazur, P., S. P. Leibo and E. H. Chu (1972). A two-factor hypothesis of freezing injury: evidence from Chinese hamster tissue-culture cells. *Experimental Cell Research,* 71(2), pp. 345–355.

McClare, C. W. F. (1971). An accurate and convenient organic phosphorus assay. *Analytical Biochemistry,* 39(2), pp. 527–530.

McGrath, J. J. (1984). Cryomicroscopy of liposome sytems as simple models to study cellular freezing response. *Cryobiology,* 21(1), pp. 81–92.

Meryman, H. (1971). Osmotic stress as a mechanism of freezing injury. *Cryobiology,* 8(5), pp. 489–500.

Meybeck, A. (1992). Past, present and future of liposome cosmetics. In O. Braun–Falco, H. Korting and H. Maibach (eds.) *Liposome Dermatics—Griesbach Conference* (Springer, Berlin Heidelberg), pp. 341–345.

Mineo, H. and H. Hara (2005). Structure-dependent and receptor-independent increase in osmotic fragility of rat erythrocytes by short-chain fatty acids. *Biochimica et Biophysica Acta,* 1713(2), pp. 113–117.

Mineo, H. and H. Hara (2007). Chemical specificity in short-chain fatty acids and their analogues in increasing osmotic fragility in rat erythrocytes in vitro. *Biochimica et Biophysica Acta,* 1768(6), pp. 1448–1453.

Monteiro, N., A. Martins, R. L. Reis and N. M. Neves (2014). Liposomes in tissue engineering and regenerative medicine. *Journal of the Royal Society Interface,* 11: 20140459, pp. 1–24.

Morris, G. J. and J. J. McGrath (1981). The response of multilamellar liposomes to freezing and thawing. *Cryobiology,* 18(4), pp. 390–398.

Morris, G. J. (1981). Liposomes as a model system for investigating freezing injury. In G. J. Morris and A. Clarke (eds.) *Effects of Low Temperatures on Biological Membrane* (Academic Press, New York), pp. 241–262.

Morris, G. J. (1982). The response of liposomes to various rates of cooling to −196°C: effect of phospholipid cholesterol ratio. *Cryobiology,* 19, pp. 215–218.

Motta, J. P. R., F. H. Paraguassú-Braga, L. F. Bouzas and L. C. Porto (2014). Evaluation of intracellular and extracellular trehalose as a cryoprotectant of stem cells obtained from umbilical cord blood. *Cryobiology,* 68(3), pp. 343–348.

Mottram, P. L., D. Leong, B. Crimeen-Irwin, S. Gloster, S. D. Xiang, J. Meanger, R. Ghildyal, N. Vardaxis and M. Plebanski (2007). Type 1 and 2 immunity following vaccination is influenced by nanoparticle size: formulation of a model vaccine for respiratory syncytial virus. *Molecular Pharmaceutics,* 4(1), pp. 73–84.

Mulder, E. and L. Van Deenen (1965a). Metabolism of red-cell lipids: I. Incorporation *in vitro* of fatty acids into phospholipids from mature erythrocytes. *Biochimica et Biophysica Acta,* 106(1), pp. 106–117.

Mulder, E., J. Van Den Berg and L. Van Deenen (1965). Metabolism of red-cell lipids: II. Conversions of lysophosphoglycerides. *Biochimica et Biophysica Acta,* 106(1), pp. 118–127.

Mulder, E. and L. Van Deenen (1965b). Metabolism of red-cell lipids: III. Pathways for phospholipid renewal. *Biochimica et Biophysica Acta,* 106(2), pp. 348–356.

Mutalik, S., S. R. Salian, K. Avadhani, J. Menon, H. Joshi, A. R. Hegde, P. Kumar, G. Kalthur and S. K. Adiga (2014). Liposome encapsulated soy lecithin and cholesterol can efficiently replace chicken egg yolk in human semen cryopreservation medium, *Systems Biology in Reproductive Medicine,* 60(3), pp. 183–188.

Nozaki, Y., D. D. Lasic and J. A. Tanford (1982). Size analysis of phospholipid vesicle preparations. *Science,* 217(4557), pp. 366–367.

Oku, N., D. A. Kendall and R. C. MacDonald (1982). A simple procedure for the determination of the trapped volume of liposomes. *Biochimica et Biophysica Acta — Biomembranes,* 691(2), pp. 332–340.

Oliver, A. E., O. Leprince, W. F. Wolkers, D. K. Hincha, A. G. Heyer and J. H. Crowe (2001). Non-disaccharide-based mechanisms of protection during drying. *Cryobiology,* 43(2), pp. 151–167.

Olson, F., C. Hunt, F. Szoka, W. Vail and D. Papahadjopoulos (1979). Preparation of liposomes of defined size distribution by extrusion through polycarbonate membranes. *Biochimica et Biophysica Acta — Biomembranes,* 557(1), pp. 9–23.

Phillips, M. C., W. J. Johnson and G. H. Rothblat (1987). Mechanisms and consequences of cellular cholesterol exchange and transfer. *Biochimica et Biophysica Acta,* 906(2), pp. 223–276.

Phillips, P. H. and Lardy, H. A. (1940). A yolk-buffer pablum for the preservation of bull semen. *Journal of Dairy Science,* 23, pp. 399–404.

Poorthuis, B. J., P. J. Yazaki and K. Y. Hostetler (1976). An improved two dimensional thin-layer chromatography system for the separation of phosphatidylglycerol and its derivatives. *Journal of Lipid Research,* 17(4), pp. 433–437.

Quinn, P.J., P. Y. W. Chow and I. G. White (1980). Evidence that phospholipid protects spermatozoa from cold shock at a plasma membrane site. *Journal of Reproduction and Fertility,* 60, pp. 403–407.

Reeves, J. P. and R. M. Dowben (1970). Water permability of phospholipid vesicles. *Journal of Membrane Biology,* 3, pp. 123–141.

Rietveld, A. and K. Simons (1998). The differential miscibility of lipids as the basis for the formation of functional membrane rafts. *Biochimica et Biophysica Acta,* 1376, pp. 467–479.

Rinalducci, S., G. M. D'Amici, B. Blasi, S. Vaglio, G. Grazzini and L. Zolla (2011). Peroxiredoxin-2 as a candidate biomarker to test oxidative stress levels of stored red blood cells under blood bank conditions. *Transfusion,* 51(7), pp. 1439–1449.

Rindlisbacher, B. and P. Zahler (1983). Interaction of phosphatidylcholine liposomes and plasma lipoproteins with sheep erythrocyte membranes: preferential transfer of phosphatidylcholine containing unsaturated fatty acids. *Biochimica et Biophysica Acta,* 732(3), pp. 485–491.

Rusnak, A., G. Coghlan, T. Zelinski and G. M. Hatch (2000). Incorporation of fatty acids into phosphatidylcholine is reduced during storage of human erythrocytes: evidence for distinct lysophosphatidylcholine acyltransferases. *Molecular and Cellular Biochemistry,* 213(1–2), pp. 137–143.

Saunders, L., J. Perrin and D. Gammack (1962). Ultrasonic irradiation of some phospholipid sols. *Journal of Pharmacy and Pharmacology,* 14(1), pp. 567–572.

Schöler, N., H. Hahn, R. Müller and O. Liesenfeld (2002). Effect of lipid matrix and size of solid lipid nanoparticles (SLN) on the viability and cytokine production of macrophages. *International Journal of Pharmaceutics,* 231(2), pp. 167–176.

Senior, J. H., K. R. Trimble and R. Maskiewicz (1991). Interaction of positively-charged liposomes with blood: implications for their application in vivo. *Biochimica et Biophysica Acta,* 1070(1), pp. 173–179.

Sessa, G. and G. Weissmann (1968). Phospholipid spherules (liposomes) as a model for biological membranes. *Journal of Lipid Research,* 9(3), pp. 310–318.

Sharifi, S., W. Dzik and S. Sadrzadeh (2000). Human plasma and tirilazad mesylate protect stored human erythrocytes against the oxidative damage of gamma-irradiation. *Transfusion Medicine,* 10(2), pp. 125–130.

Siminovich, D. and D. Chapman (1971). Liposome bilayer model systems of freezing living cells. *FEBS Letters,* 16, pp. 207–212.

Siminovich, D. and D. Chapman (1974). The simulation of osmotic stresses of plant cells by lipid liposome membrane systems. *Cryobiology,* 11, pp. 552–553.

Siow, L. F., T. Rades and M. H. Lim (2007). Characterizing the freezing behavior of liposomes as a tool to understand the cryopreservation procedures. *Cryobiology,* 55(3), pp. 210–221.

Sowemimo-Coker, S. O., R. P. Goodrich, C. R. Zerez and K. R. Tanaka (1993). Refrigerated storage of lyophilized and rehydrated, lyophilized human red cells. *Transfusion,* 33, pp. 322–329.

Stano, P., S. Bufali, C. Pisano, F. Bucci, M. Barbarino, M. Santaniello, P. Carminati and P. L. Luisi (2004). Novel camptothecin analogue (gimatecan)-containing liposomes prepared by the ethanol injection method. *Journal of Liposome Research,* 14(1–2), pp. 87–109.

Steck, T.L., F. J. Kezdy and Y. Lange (1988). An activation–collision mechanism for cholesterol transfer between membranes. *Journal of Biological Chemistry,* 263(26), pp. 13023–13031.

Stoll, C., J. L. Holovati, J. P. Acker and W. F. Wolkers (2011a). Liposomes composed of unsaturated lipids for membrane modification of human erythrocytes. *Molecular Membrane Biology,* 28(7–8), pp. 454–461.

Stoll, C., H. Stadnick, O. Kollas, J. L. Holovati, B. Glasmacher, J. P. Acker and W. F. Wolkers (2011b). Liposomes alter thermal phase behavior and composition of red blood cell membranes. *Biochimica et Biophysica Acta,* 1808(1), pp. 474–481.

Stoll, C., J. L. Holovati, J. P. Acker and W. F. Wolkers (2012). Synergistic effects of liposomes, trehalose, and hydroxyethyl starch for cryopreservation of human erythrocytes. *Biotechnology Progress,* 28(2), pp. 364–371.

Strauss, G. (1989). Liposomes: from theoretical model to cosmetic tool. *Journal of the Society of Cosmetic Chemists,* 40(1), pp. 51–60.

Strumia, M. M., and P. V. Strumia (1965). Effect of lactose, dextran and albumin on recovery and survival of frozen red cells. *Transfusion,* 5(5), pp. 399–411.

Szebeni, J., J. H. Breuer, J. G. Szelenyi, G. Bathori, G. Lelkes and S. R. Hollan (1984). Oxidation and denaturation of hemoglobin encapsulated in liposomes. *Biochimica et Biophysica Acta,* 798(1), pp. 60–67.

Szoka, F. Jr. and D. Papahadjopoulos (1978). Procedure for preparation of liposomes with large internal aqueous space and high capture by reverse-phase evaporation. *Proceedings of the National Academy of Sciences USA,* 75(9), pp. 4194–4198.

Szoka, F. Jr. and D. Papahadjopoulos (1980). Comparative properties and methods of reparation of lipid vesicles (liposomes). *Annual Review of Biophysics and Bioengineering,* 9(1), pp. 467–508.

Tamura, A., K. Yoshikawa, T. Fujii, K. Ohki, Y. Nozawa and Y. Sumida (1986). Effect of fatty acyl chain length of phosphatidylcholine on their transfer from liposomes to erythrocytes and transverse diffusion in the membranes inferred by TEMPO-phosphatidylcholine spin probes. *Biochimica et Biophysica Acta,* 855(2), pp. 250–256.

Tandorf, S., J. J. McGrath and C. R. Olien (1987). On the adhesive interaction between ice and cell-size liposomes. *CryoLetters,* 8, pp. 322–333.

Ten Centre Study Group (1987). Ten centre trial of artificial surfactant (artificial lung expanding compound) in very premature babies. Ten Centre Study Group, *British Medical Journal. (Clinical Research Edition),* 294(6578), pp. 991–996.

Toh, M. and G. N. Chiu (2013). Liposomes as sterile preparations and limitations of sterilisation techniques in liposomal manufacturing. *Asian Journal of Pharmaceutical Sciences,* 8(2), pp. 88–95.

Torok, Z., G. R. Satpathy, M. Banerjee, R. Bali, E. Little, R. C. Novaes, H. V. Ly, D. M. Dwyre, A. Kheirolomoom, F. Tablin, J. H. Crowe and N. M. Tsvetkova (2005). Preservation of trehalose-loaded red blood cells by lyophilization. *Cell Preservation Technology,* 3(2), pp. 96–111.

Ulrich, A. S. (2002). Biophysical aspects of using liposomes as delivery vehicles. *Bioscience Reports,* 22(2), pp. 129–150.

Van De Watering, L. (2011). Red cell storage and prognosis. *Vox Sanguinis,* 100(1), pp. 36–45.

Vemuri, S., C. Yu, V. Wangsatorntanakun and N. Roosdorp (1990). Large-scale production of liposomes by a microfluidizer. *Drug Development and Industrial Pharmacy,* 16(15), pp. 2243–2256.

Vemuri, S. and C. T. Rhodes (1995). Preparation and characterization of liposomes as therapeutic delivery systems: a review. *Pharmaceutical Acta Helvetiae,* 70(2), pp. 95–111.

Wagner, A. and K. Vorauer-Uhl (2011). Liposome technology for industrial purposes. *Journal of Drug Delivery,* 591325, pp.1–9.

Weinstein, J.N., Yoshikami, S., Henkart, P., Blumenthal, R. and Hagins, W.A. (1977). Liposome-cell interaction: transfer and intracellular release of a trapped fluorescent marker, *Science,* 195 (4277), pp. 489–492.

Woodle, M. C. and D. Papahadjopoulos (1989). Liposome preparation and size characterization. *Methods in Enzymology,* 171, pp. 193–217.

Yamaguchi, T., K. Kuranoshita, T. Harano and E. Kimoto, E. (1993). Effects of drugs, salts, and phospholipid vesicles on hemoglobin release from hydrostatic pressure-treated human erythrocytes. *Journal of Biochemistry,* 113, pp. 513–518.

Yang, H., V. Paromov, M. Smith and W. L. Stone (2008). Preparation, characterization, and use of antioxidant-liposomes. In D. Armstrong (ed.) *Advanced Protocols in Oxidative Stress I- Methods in Molecular Biology* 477 (Humana Press Inc., Totowa, NJ), pp. 277–292.

Yoshida, T. and S. S. Shevkoplyas (2010). Anaerobic storage of red blood cells. *Blood Transfusion,* 8(4), pp. 220–236.

Zeron, Y., M. Tomczak, J. Crowe and A. Arav (2002). The effect of liposomes on thermotropic membrane phase transitions of bovine spermatozoa and oocytes: implications for reducing chilling sensitivity. *Cryobiology,* 45(2), pp. 143–152.

Chapter 10

Thermal Physical Properties of CPAs with HA Nanoparticles and their Application in Cryopreservation

Baolin Liu, Fukou Lv and Weijie Li

1. Introduction

Water, a main constituent in many biological systems, may crystallize upon cooling to certain sub-zero or ultra-low temperatures. This process, termed as crystallization, consists mainly of two stages: nucleation and crystal growth. The interaction between these two steps determines the crystal characteristics, i.e. size, distribution and morphology of the crystals. The nucleation step is defined as the formation of a new crystal and happens either in a crystal-free solution, which is called primary nucleation, or in the presence of formerly created crystals, which is defined as secondary nucleation. The behavior of crystallization of aqueous solutions has been a subject of extensive study in cryopreservation. For the cryopreservation of cells, tissues, organs and other biological materials, the uncontrolled formation of ice crystals is often the major cause of freeze damage [Smith *et al.*, 1988; Gao and Critser, 2000; Pegg, 2010].

There is significant evidence for a correlation between ice formation and injury of biomaterials including cell suspension, tissues, and even organs [Mazur, 1977; 1984; Muldrew and McGann, 1990]. As is well known, the thawing process is also the key step in determining the final condition of the specimen. That is, devitrification and recrystallization during warming can also play a major role in determining the quality of the frozen product in the same way as crystallization does during freezing. Recent work suggests that the warming process is more important than cooling in cryopreserving biological function [Seki and Mazur, 2009]. Therefore, the control and prediction of the crystallization or vitrification, devitrification, recrystallizing process and related phenomena in regard to the crystal characteristics are very essential for the improvement of freezing and thawing processes.

Vitrification is considered as the most feasible method for successful cryopreservation of cells, tissues and organs because the ice-induced damages during freezing are totally avoided. This technique employs high concentration of low toxic cryoprotectants (CPAs) that suppress the freezing point of water in the sample and freezes them at ultra-fast cooling rates to its glass transition temperature or below. As a result the water in the system is converted from liquid to glass state without crystallization. In regular conditions, two factors (the concentration of CPAs and the cooling rate) affect the probability of vitrification for a solution. Achieving cooling rates of high orders is almost impractical in large sized samples due to the large thermal resistance. Amongst the several ways to increase the freezing and warming rate during cryopreservation, use of nanoparticles holds most promise.

Nanoparticles in solution offer unique electrical, mechanical and thermal properties due to their physical presence and interaction with the state of dispersion. The early inspiration of enhancing the thermal conductivity of base fluid by adding conductive particles goes back to more than a century ago when it was first theoretically proposed by Maxwall amd Maxwell [1954]. However, the inherent problems resulted from the relatively large size of colloidal particles, such as their sedimentation, clogging, and abrasion, have been a major obstacle to realize the concept in practice. In order to alleviate the pertinent problems, Choi [1995] pioneered the use of nanofluids based on advanced manufacturing techniques for nanosized metallic particles with high thermal conductivities. Since Lee *et al.* [1999] made the first measurement of thermal conductivity of nanofluids, a series of nanofluids have been studied with regard to the aspect of thermal conductivity [Eastman *et al.*, 2001; Wang *et al.*, 1999; Xie *et al.*, 2002]. Most of the research on nanofluids' thermal conductivity was performed above room temperature. There were only a few studies on the thermal parameters of nanofluids below their phase change temperature and on the application of nanoparticles in cryogenics.

Han *et al.* [2008] investigated the nucleation and devitrification temperatures of polyol cryoprotectant solutions with diamond, gold, copper–nickel, and silicon nanoparticles at cryogenic temperature by differential scanning calorimetry (DSC). To explore the cryobiological applications of hydroxyapatite (HA) nanoparticles, the phase transition temperatures [Hao *et al.*, 2009; Hao *et al.* 2011; Xu *et al.*, 2011], specific heat [Xu *et al.*, 2010; Gao *et al.*, 2011; Liu *et al.*, 2010], density [Hao *et al.*, 2009; Hao *et al.* 2011], crystallization [Xu *et al.*, 2011; Tang *et al.*, 2010], devitrification and recrystallization

[Lv *et al.*, 2014a; Lv *et al.*, 2014b; Lv *et al.*, 2012] of aqueous cryoprotectant solutions with HA nanoparticles during the cooling and warming were investigated by using DSC and cryomicroscopy. The results show that the complex interactions among solution type, nanoparticle size and nanoparticle concentration apparently influence the thermal physical properties of CPAs at cryogenic temperature.

2. Preparation of CPAs with HA Nanoparticles

The ideal CPAs with HA nanoparticles should be stable, with no chemical change and no agglomeration of particles in a certain time period. General preparation protocol for CPAs with HA nanoparticles is: (a) addition of acid or soda to adjust the pH value of the suspension, (b) addition of some surfactants to disperse particles into fluid, and (c) treatment with ultrasonic vibration. Addition of some other materials may have negative effects because of the structural change of solid/liquid interface. So an ultrasonic vibration method was used in our study to prevent structural change of CPAs with HA nanoparticles.

Glycerol, Polyvinyl Pyrrolidone (PVP) and polyethylene glycol (PEG-600) were purchased from Sinopharm Chemical Reagent Co., Ltd and HA nanoparticles from Nanjing Emperor Nano Material Co., Ltd. The average molecular weight of PEG-600 was 570–630 g/mol. HA nanoparticles of different concentrations and sizes were suspended in aqueous cryoprotectants solutions using a two-step method. The HA nanoparticles were weighed out by using a precise electronic balance and added into the solutions. Then these solutions with HA nanoparticles were placed in a water bath at 20±2°C, and constantly agitated for certain time by a JY92-2D ultrasonic vibrator. This procedure resulted in a uniform distribution of nanoparticles in the solutions.

Figure 1 shows the photographs of PVP solutions (60%, w/w) with 20 nm HA nanoparticles treated by ultrasonic vibration for three hours. It can be seen that nanoparticles have a uniform distribution within two hours. As the time to obtain a glassy state of this solution is far less than two hours, the nanoparticles should have a uniform distribution during the experiment of cryopreservation. The sonication time and vibration power for different nano-cryoprotectants may be different. The specific time and power of sonication for the certain cryoprotectant solutions should be determined by experiments.

3. Thermal Physical Properties of CPAs with HA Nanoparticles

3.1. *Phase Transition Temperatures*

A phase transition is the transformation of a thermodynamic system from one phase or state of matter to another one by heat transfer. During a phase

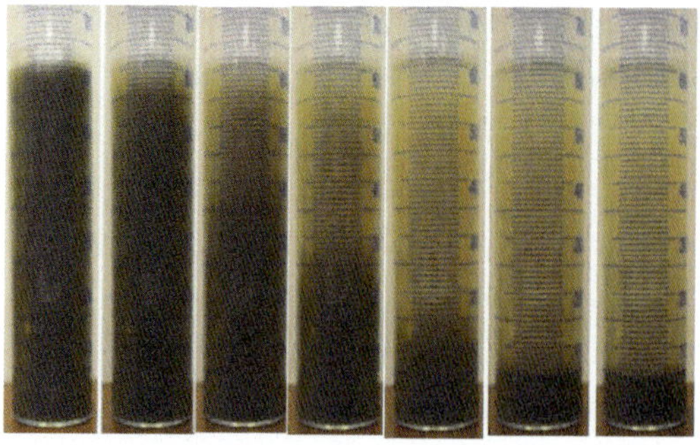

Figure 1. The nanofluids' dispersion stability with time for PVP solutions (60%, w/w) with 20 nm HA nanoparticles. Solutions with HA nanoparticles were placed in a water bath at 20±2°C and constantly agitated by ultrasonic power of 200 W for 3 hours.

Note: "(a) 0h" represents the time after the three hours' of agitation was just completed.

transition of a given medium, certain properties of the medium change, often discontinuously, as a result of the change of some external condition, such as temperature, pressure, or others. The measurement of the external conditions at which the transformation occurs is termed the phase transition. DSC is usually a common technique to determine the phase transition temperatures for cryoprotectants.

The Pyris Diamond DSC (Perkin-Elmer, Waltham, MA, USA) was calibrated using the crystal-crystal transition of cyclopentane at −135.06°C and the melting point of distilled water at 0°C before the experiments. Calorimetric analysis was performed with 10 to 15 μl of HA nanoparticle-containing CPAs. Sample weight was measured with the precision of 1.0 μg using a microbalance (Beijing Sartorius Instrument Co., Beijing, China). The aliquot was hermetically sealed in an aluminum pan and loaded into the DSC under nitrogen purge at 20 ml/min. Both cooling and warming thermograms were analyzed with the instrument's Pyris Software. Each combination of samples was tested thrice and results were averaged.

Glycerol, PVP and PEG-600 based CPA solutions of different concentrations with and without HA nanoparticles were tested. All the sample CPA solutions were cooled from 20°C to −160°C at 150°C /min and warmed up to 20°C at 10°C /min. No exotherms were visible in the cooling DSC curves, indicating that the CPA solutions were vitrified as was explained by Seki and Mazur [2009].

The warming DSC thermograms of nanoparticle-containing glycerol (60%, w/w) and PEG-600 (50%, w/w) solutions are shown in Figures 2 and 3 respectively. Each DSC curve shows one or two exothermic peaks that

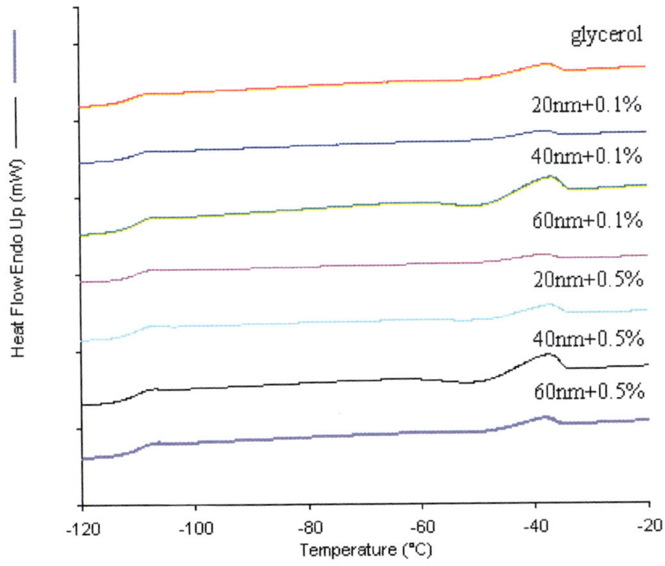

Figure 2. DSC warming curves of quenched glycerol solutions (60% w/w) with and without HA nanoparticles. Nanoparticle size and content are shown in line labels. Glycerol solutions were quenched at 150°C/min to –160°C, and then warmed at 10°C/min. "20 nm + 0.1%" means the solution contained 0.1% 20 nm HA nanoparticles. The same meaning applies for other data. The values of phase transition temperatures (T_g, T_d and T_m) for solutions are tabulated in Table 1.

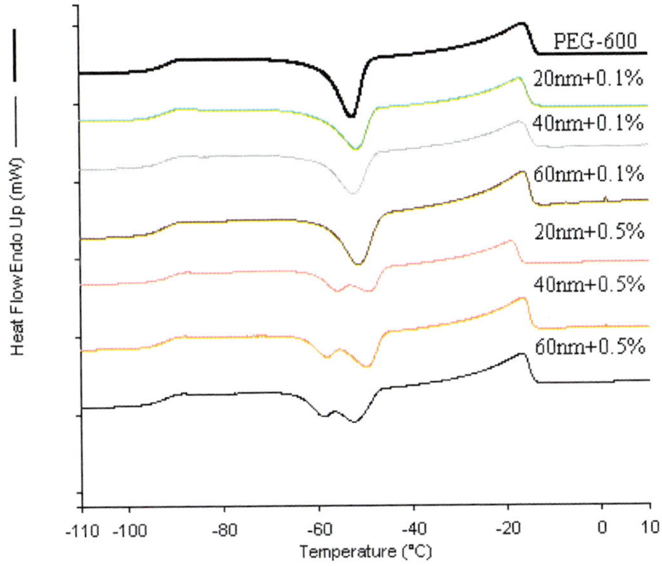

Figure 3. DSC warming curves of quenched PEG-600 solutions (50% w/w) with and without HA nanoparticles. Nanoparticle size and content are shown in line labels. PEG-600 solutions were quenched at 150°C/min to –160°C, and then warmed at 10°C/min. "20 nm + 0.1%" means the solution contained 0.1% 20 nm HA nanoparticles. The same meaning applies for other data. The values of phase transition temperatures (T_g, T_d and T_m) for solutions are tabulated in Table 1.

correspond to devitrification and an endothermic peak that corresponds to the melting of ice crystals. The glass transition temperature (T_g), devitrification temperature (T_d) and melting temperature (T_m) were determined according to the specific DSC curves for each CPA solution, as summarized in Table 1.

Most studies of cryopreservation have focused on the cooling process and its consequences. But there have been few experimental attempts to study the effect of warming process on biological functions. Recent work in vitrification of mouse oocytes indicates that the warming process is possibly more important than the cooling [Seki and Mazur, 2009; Mazur and Seki, 2011; Seki and Mazur, 2012].

Lv *et al.* [2014a; 2014b; 2012] investigated the effect of HA nanoparticles on devitrification, recrystallization and ice crystal characteristics of aqueous glycerol and PEG-600 solutions during warming. Experimental results show that the incorporation of HA nanoparticles can significantly affect the behavior of devitrification and recrystallization of the supercooled dispersions, but the complex interactions among solution type, nanoparticle size and nanoparticle content remain to be understood.

The presence of HA nanoparticles has little effect on the glass transition temperatures and melting temperatures of glycerol and PEG-600 solutions. The addition of HA nanoparticles increases the devitrification temperatures of glycerol solutions. The effect of nanoparticles on devitrification is significant in both solutions at all three sizes, particularly at the higher nanoparticle content (0.5% w/w). The increase of devitrification temperatures demonstrates that the ice nuclei generated during cooling needs more energy for devitrification and recrystallization, indicating that the stability of glass medium is improved with the introduction of HA nanoparticles.

For PEG-600 solution, the devitrification temperatures are depressed with the addition of nanoparticles. Devitrification exhibits as double peaks with the content increased from 0.1% to 0.5% (w/w) (Figure 3), suggesting that the presence of HA nanoparticles might enhance the disturbance of medium and result in domain or phase separation.

Devitrification is not found in 60% PVP solutions, whereas devitrification peak exists in 50% and 55% PVP solutions. As shown in Table 1, the glass transition temperatures and devitrification temperatures of PVP solutions with HA nanoparticles are lower than that without HA nanoparticles. These results also show a nonlinear relationship between T_g, T_d, T_m and the concentrations of HA nanoparticles. A qualitative explanation for the effect of the nanoparticle concentration on phase transition temperatures is that a higher concentration of nanoparticles results in a larger total contact surface between nanoparticles and liquid molecules, which consequently increases the possibility of nucleation [Han *et al.*, 2008].

Table 1. Phase transition temperatures for CPA solutions with and without HA nanoparticles. T_g, glass transition temperature; T_d, devitrification temperature; T_m, melting temperature.

Solutions	HA nanoparticle sizes and concentrations	T_g (°C)	T_d (°C)	T_m (°C)
Glycerol (60% w/w)	No nanoparticle	−111.5±0.4	−64.6±0.6	−37.5±0.5
	0.1% + 20 nm	−111.2±0.3	−58.2±0.4	−38.0±0.7
	0.1% + 40 nm	−111.1±0.4	−59.6±0.5	−37.5±0.6
	0.1% + 60 nm	−111.2±0.3	−64.8±0.7	−38.2±0.8
	0.5% + 20 nm	−111.4±0.3	−61.5±0.6	−37.5±0.4
	0.5% + 40 nm	−111.3±0.3	−61.7±0.6	−37.6±0.6
	0.5% + 60 nm	−111.1±0.4	−59.6±0.7	−38.3±0.5
Glycerol (50% w/w)	without HA	−114.61±0.6	−98.29±0.5	—
	0.1% + 20 nm	−116.50±0.8	−96.46±0.4	—
	0.5% + 20 nm	−116.56±0.3	−96.28±0.6	—
	0.1% + 40 nm	−115.11±0.4	−93.31±0.6	—
	0.5% + 40 nm	−115.18±0.6	−92.02±0.3	—
	0.1% + 60 nm	−114.80±0.5	−94.34±0.2	—
	0.5% + 60 nm	−114.73±0.9	−93.34±0.8	—
PEG-600 (50% w/w)	No nanoparticle	−92.4±0.3	−57.8±0.5	−16.3±0.5
	0.1% + 20 nm	−92.6±0.4	−59.0±0.6	−16.6±0.6
	0.1% + 40 nm	−92.5±0.4	−59.0±0.7	−16.7±0.4
	0.1% + 60 nm	−92.6±0.3	−57.9±0.4	−16.4±0.6
	0.5% + 20 nm	−92.7±0.5	−61.2±0.8	−16.9±0.8
	0.5% + 40 nm	−92.4±0.3	−63.0±0.7	−16.4±0.8
	0.5% + 60 nm	−92.6±0.5	−63.9±0.9	−16.8±0.5
PVP (60% w/w)	No nanoparticle	−43.7±0.7	—	−13.8±0.4
	0.1% + 20 nm	−47.4±0.5	—	−13.1±0.6
	0.2% + 20 nm	−50.3±0.4	—	−12.3±0.5
	0.5% + 20 nm	−52.8±0.5	—	−11.6±0.8
PVP (55% w/w)	No nanoparticle	−60.1±0.6	−19.6±0.7	−8.8±0.1
	0.1% + 20 nm	−63.7±0.2	−22.6±0.9	−8.4±0.5
	0.2% + 20 nm	−66.5±0.7	−24.3±0.4	−8.2±0.6
	0.5% + 20 nm	−68.8±0.8	−27.8±0.5	−7.1±0.7
PVP (50% w/w)	No nanoparticle	−69.2±0.7	−30.6±0.3	−5.7±0.9
	0.1% + 20 nm	−72.6±0.4	−33.5±0.7	−5.5±0.6
	0.2% + 20 nm	−75.1±0.2	−35.9±0.8	−5.4±0.3
	0.5% + 20 nm	−77.3±0.7	−38.4±0.2	−5.3±0.5

Note: "0.1% + 20 nm" represents "0.1% concentration HA of 20 nm diameter". The same meaning applies for other table data.

Sources: Hao and Liu [2011]; Xu *et al.* [2010]; Lv *et al.* [2014].

In general, the smaller the T_m-T_d, the lower the critical warming rate, and the higher the glass medium stability [Boutron, 1986]. The T_m-T_d of PVP (55%, w/w) solutions are smaller than that of PVP (50%, w/w) solutions, indicating that the 55% PVP solutions are more stable than 50% PVP solutions. It can be found that the stability of PVP (55%, w/w) solutions increases with the nanoparticles concentration for all three sizes (20 nm, 40 nm, 60 nm), which is true for 60% glycerol solutions.

3.2. *Specific Heat*

The quantity of heat required to raise the temperature of per gram of a substance by one degree is called the specific heat of the substance. It can be seen that the fast cooling rate could be obtained for solutions with smaller specific heat.

The DSC at 20°C /min cooling rate and 10°C /min warming rate was used to measure the specific heat of PVP solutions with HA nanoparticles of diameters 20 nm, 40 nm, and 60 nm. The density of PVP solutions with and without nanoparticles was determined by using a straw (Shanghai Xiangming Medical Instrument Co., Ltd.). The PVP solution was injected into the straw and inserted into liquid nitrogen for five minutes. The volume change was recorded by a digital video system. The density was calculated by the volume change of solutions [Gao *et al.*, 2011].

Figure 4 demonstrates the endothermal curves of specific heat of PVP (60%, w/w) solutions during warming with HA nanoparticles of different concentrations at the size of 20 nm. Map-like change in the level is the process of glass transition. There is a general consensus that the specific heat changes sharply at the glass transition temperature and varies inversely with temperature. It can be seen that the nanoparticle concentration has great effects on glass transition temperature. There is an endothermic peak in high temperature at the side of the glass transition region, which is caused by structural relaxation of the enthalpy recovery process. The lower the nanoparticle concentration, the greater the endothermal peak.

At unfrozen state, the specific heat of PVP (50%, w/w) solutions reduces with the increment of temperature, as shown in Figure 5. The specific heat of PVP (50%, w/w) solutions also decreases significantly with the increment of nanoparticle concentration. According to the adsorption theory [Gao and Li, 2007], the molecules of solute adhere easily on the surface of solid particles, so the concentration of the solution reduces. In addition, the majority of solute adsorption decreases with increasing temperature. Therefore, HA nanoparticles in PVP solutions absorb water molecules to form a layer of

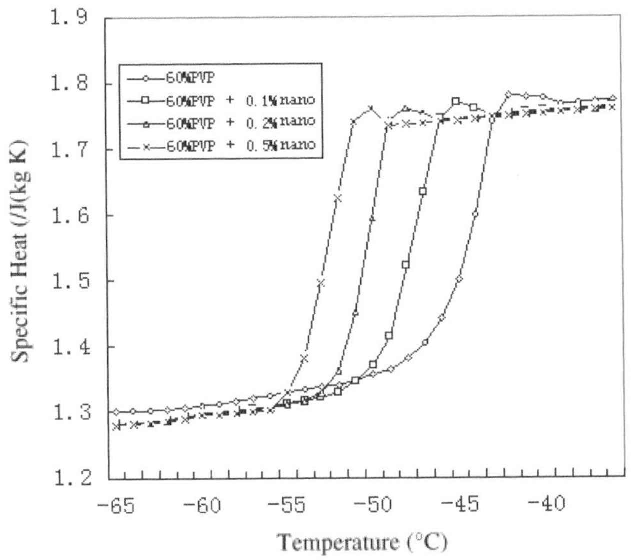

Figure 4. The effect of HA nanoparticles on specific heat of PVP (60%, w/w) solutions.

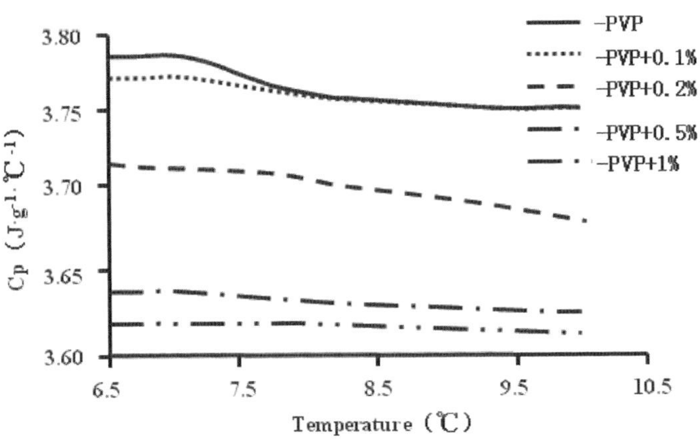

Figure 5. The specific heat of PVP (50%, w/w) solutions with HA nanoparticles (size = 20 nm) at unfrozen state.

water molecules around the particles, so the concentration of solution increases and the specific heat reduces.

At frozen state, the specific heat of PVP (50%, w/w) solutions increases with the increment of temperature, as illustrated in Figure 6. The specific heat of PVP (50%, w/w) solutions decreases with the increment of the concentration of nanoparticles. This phenomenon may be caused by the interaction between nanoparticles and solutions due to the state change of frozen phase.

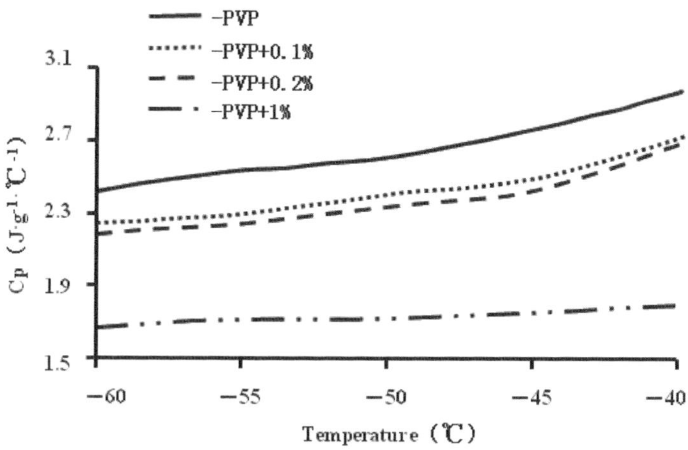

Figure 6. The specific heat of PVP (50%, w/w) solutions with HA nanoparticles (size = 20 nm) at frozen state.

Specific heat of the system reduces with increase in the nanoparticles concentration at both frozen and unfrozen states. In accordance with the configuration entropy theory [Edmund and Marzio, 1997], the specific heat of glassy state is constituted by the specific heat of molecular vibration and specific heat of configuration, and the transition of specific heat caused by the changes of configuration during the glass transition process. The specific heat of configuration of supercooled PVP (60%, w/w) solutions reduces with the increment of nanoparticles concentration. The reduction of specific heat of PVP (60%, w/w) solutions with nanoparticles can be qualitatively analyzed by Equation (1). Therefore it is difficult to quantitatively calculate the reduction of specific heat of PVP with nanoparticles because of the interaction between nanoparticles and PVP molecules.

$$C = \frac{(1-f)(\rho C_p)_{BF} + f(\rho C_p)_p}{(1-f)\rho_{BF} + f\rho_p} \tag{1}$$

where C is specific heat, p is density, the subscript "BF" represents the base fluid matrix and the subscript "p" represents the particles in base fluid.

3.3. *Ice Morphology*

Ice morphology is crucial for cryopreservation, cryosurgery and food manufacturing. Based on the ice formation theory, cells or interstitial structures are more vulnerable to mechanical damage caused by these "sharper" needle-shaped or dendritic crystals. Muldrew *et al.* [2001] reported that the needle-like ice crystal structure significantly increases cellular destruction in the high subzero temperature range. Ice crystal morphology also plays an important role in the textural and physical properties of frozen and frozen-thawed foods [Petzold and Aguilera, 2009].

The cryomicroscopy system consists of a BCS196 Biological Cryo-stage coupled with a Linksys 32 system (Linkam Scientific Instruments Ltd, UK). A thin film of sample solution (~1 μl) was placed between two transparent glass cover slides of 8-mm diameter on a sample holder that was then inserted in the cold-stage. A high resolution camera (Olympus Microscope BX50, Olympus Optical Co., Japan) was used to record the sample images. This system can freeze and warm samples at a ramp rate between 0.01°C/min and 150°C/min from −196°C to 125°C. CCD transmits the image to computer, and the microphotographs can be taken together with a temperature logger. To prevent condensation of the observation window, the sample chamber was purged with purified nitrogen gas during the experiment. The temperature program of the cold-stage was the same as the calorimetrical study. When an individual ice crystal grew to over 4 μm (maximum resolution of the microscope 100×), the temperature of samples was defined as the onset devitrification temperature.

When the ice crystal ceased growing or had grown to the whole field of observation, the temperature of samples was defined as the end devitrification temperature. The average growth rate was determined by the difference in average diameter of ice crystals between two successive images. Since the direction of crystal growth was restricted by coverslips, ice crystals mainly grew two-dimensionally rather than three-dimensionally. Assuming that the ice crystals were uniform in thickness, the fraction of ice crystals on a volume-per-volume (v/v) basis was determined as Ai/A on an area-per-area basis, where Ai and A (2.9 mm^2) represent the area of ice crystal and that of entire field of observation respectively. The size of individual crystals was calculated as the radius of a circle having the equivalent projected area of the crystal as is employed by Hagiwara *et al.* [2006]. The software Digimizer and Photoshop were used for the analysis. In some cases that these two programs did not extract the exact projection area of ice crystals, the outline of ice crystal was traced on the LCD monitor and the projected area information was extracted manually.

The study of ice crystal growth with the cryomicroscope show that the presence of HA nanoparticles does not alter the morphology of ice crystals formed during warming of the quenched solutions. Ice crystals grow in the dendritic form in glycerol solutions (Figure 7), and in the spheric form in PEG-600 solutions (Figure 8).

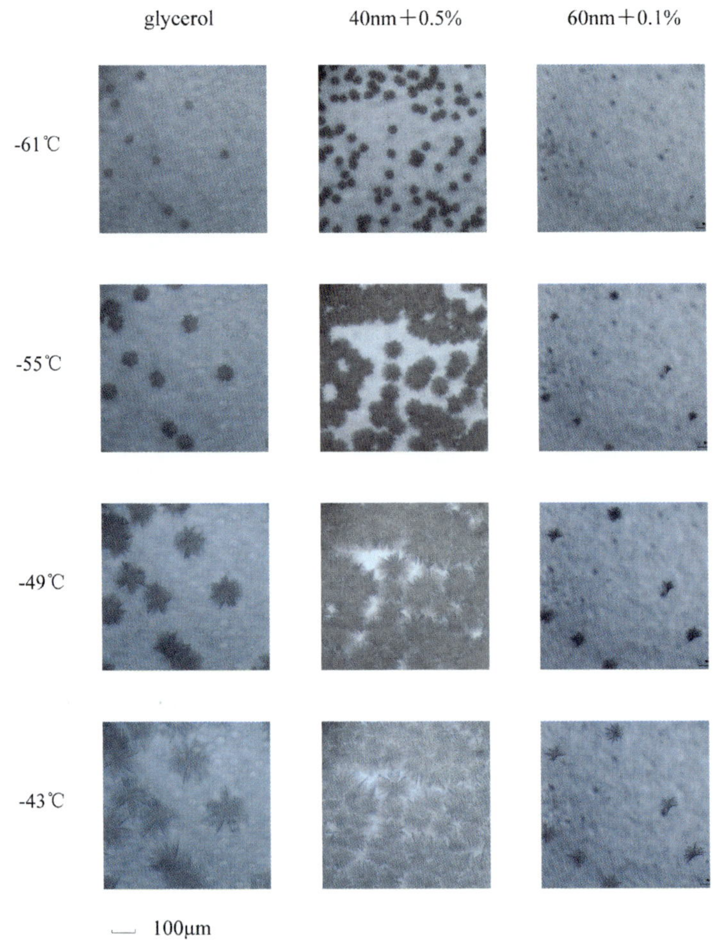

Figure 7. Microphotographs of dendritic ice crystal growth during warming in the quenched glycerol solutions (60% w/w) without nanoparticles, with 0.5% 40 nm and 0.1% 60 nm nanoparticles, at different temperatures between -61°C and –43°C. Solutions were quenched at 150°C/min to –160°C, and then warmed at 10°C/min. Bar = 100 μm. "40 nm + 0.5%" and "60 nm + 0.1%" mean the glycerol solutions contained 0.5% 40 nm and 0.1% 60 nm nanoparticles respectively.

The devitrification and recrystallization processes have normally the same probability in the homogeneous medium, and lead to the formation of symmetric ice crystals. The morphologies of ice crystals are dendritic in glycerol solutions compared with the spheric in PEG-600 solutions, and not changed with low content of nanoparticles. The morphology of ice crystals is mainly determined by the type of solution, and the data in the present study is consistent with that of Gonda and Sei [2005].

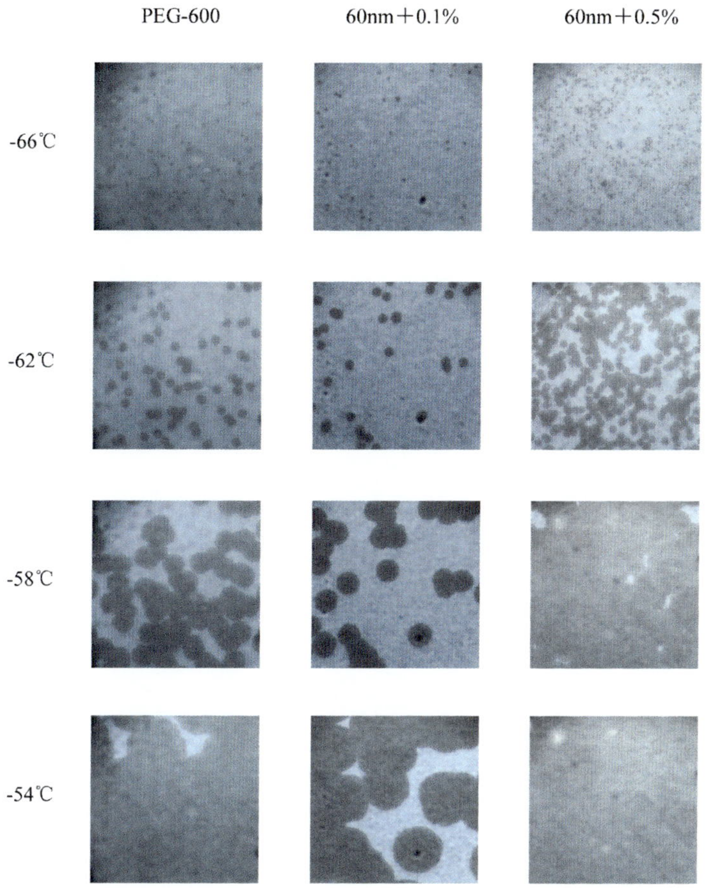

Figure 8. Microphotographs of spheric ice crystal growth during warming in the quenched PEG-600 solutions (50% w/w) without nanoparticles, with 0.1% 60 nm and 0.5% 60 nm nanoparticles, at different temperatures between -66°C and 54°C. Solutions were quenched at 150°C/min to –160°C, and then warmed at 10°C/min. "60 nm + 0.1%" and "60 nm + 0.5%" mean the PEG-600 solutions contained 0.1% 60 nm and 0.5% 60 nm nanoparticles respectively.

3.4. *Ice Fraction and Growth-Rate of Ice Crystal*

The investigation with the cryomicroscope exhibits that the effect of HA nanoparticles on ice crystal growth is very complex. It appears that there are complex interactions among solution type, nanoparticle size and content, which influence ice crystal growth or recrystallization in the quenched solutions upon warming. Figure 9 and Figure 10 show the fraction of ice crystals formed and the rate of ice crystal growth of nanoparticle-containing glycerol (60% w/w) and PEG-600 (50% w/w) solutions as a function of warming

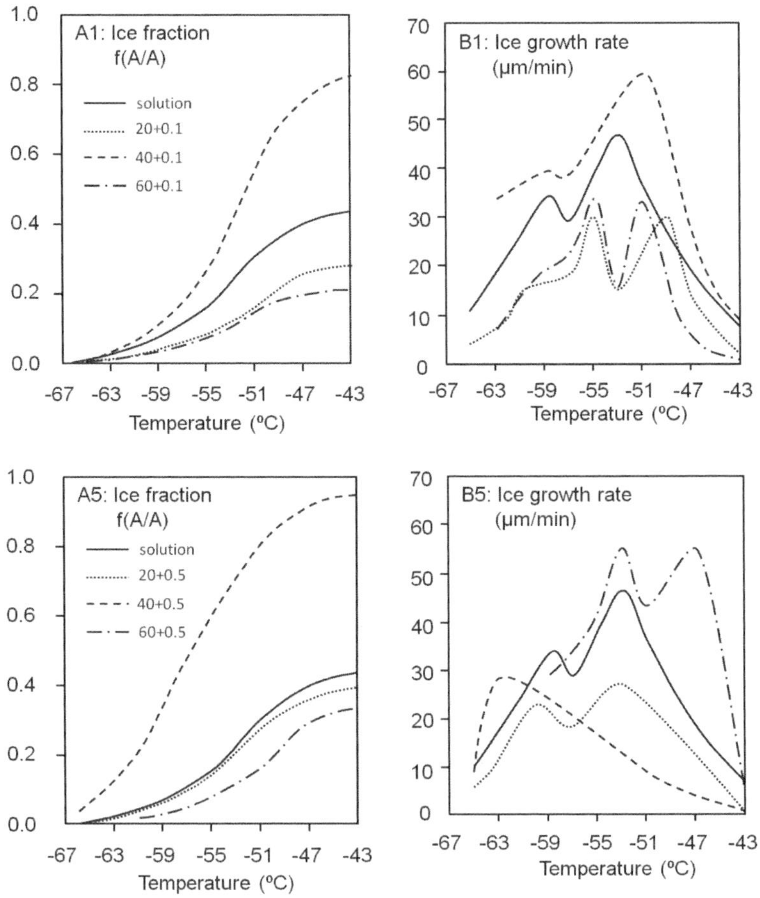

Figure 9. Fraction of ice crystals and the rate of ice crystal growth in quenched glycerol solutions (60% w/w) with and without HA nanoparticles. Solutions were quenched at 150°C/min to −160°C, and then warmed at 10°C/min. "20 nm + 0.1%" means the solution contained 0.1% 20 nm HA nanoparticles. The same meaning applies for other data. A and B represent the ice fraction and ice growth rate of glycerol solutions respectively, 1 and 5 means the contents of nanoparticles are 0.1% and 0.5% respectively.

temperature. As expected, the fraction of ice crystals increased as the solutions were warmed up. In the quenched glycerol solutions, 20 nm and 60 nm HA nanoparticles decreased, whereas 40 nm HA nanoparticles increased the fraction of ice crystals at both 0.1% and 0.5% (w/w) in comparison to the control solution, and the ice fraction increases with higher nanoparticles content for the same size. The rate of ice crystal growth is the highest with 0.1% 40 nm nanoparticles (60.0 μm/min), and the lowest with 0.5% 20 nm nanoparticles (26.8 μm/min). In the quenched PEG-600 solutions, ice fraction all

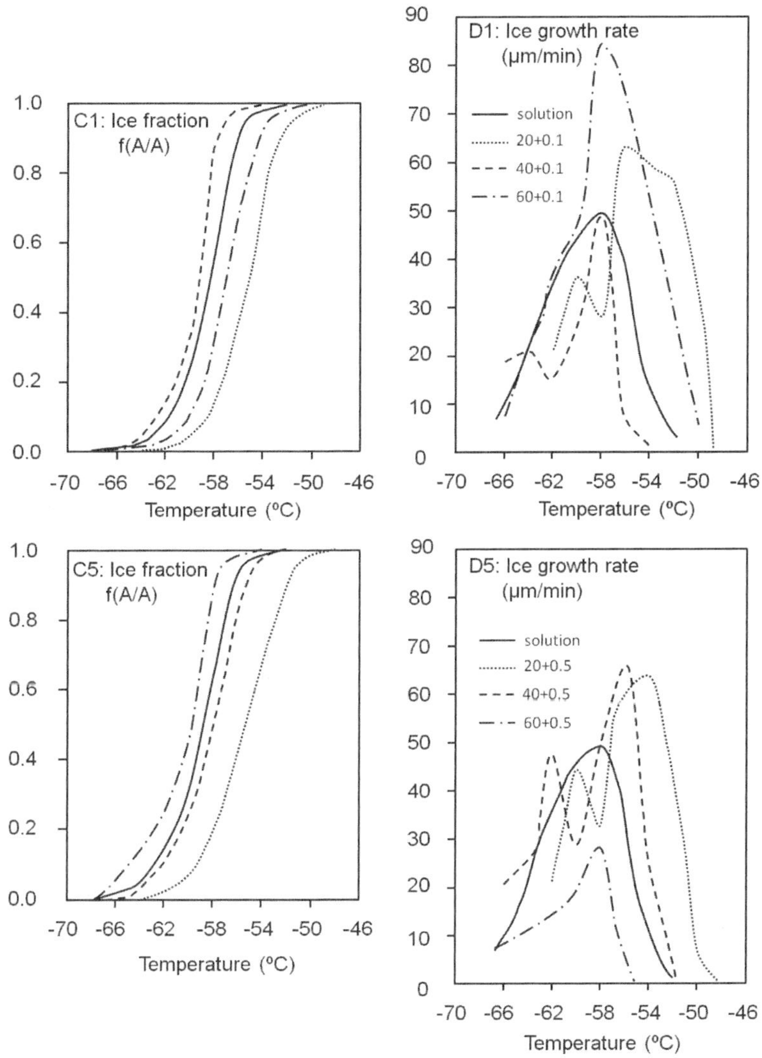

Figure 10. Fraction of ice crystals and the rate of ice crystal growth in quenched PEG-600 solutions (50% w/w) with and without HA nanoparticles. Solutions were quenched at 150°C/min to −160°C, and then warmed at 10°C/min. "20 nm + 0.1%" means the solution contained 0.1% 20 nm HA nanoparticles. The same meaning applies for other data. C and D represent the ice fraction and ice growth rate of PEG-600 solutions respectively, 1 and 5 means the contents of nanoparticles are 0.1% and 0.5% respectively.

increases to 1 by the end of recrystallization, that is to say, the ice crystal grows to the whole field of observation. 20 nm HA nanoparticles at both 0.1% and 0.5% (w/w) retards the ice crystal growth. The 40 nm HA nanoparticles accelerates the ice crystal growth at 0.1% (w/w), but inhibits the ice crystal growth at 0.5% (w/w). On the contrary, the largest 60 nm HA nanoparticles restricts the ice crystal growth at 0.1% (w/w), but promotes the ice crystal growth at 0.5% (w/w). The rate of ice crystal growth in PEG-600 solution is the greatest for 0.1% 60 nm nanoparticles (84.0 μm/min), and the lowest with 0.5% 60 nm nanoparticles (28.1 μm/min).

During warming, small crystals in the quenched solutions fuse together to form larger crystals. The recrystallization event has significant implications. It may enhance structural destruction induced by the preceding freezing. In general, the more and larger ice crystals form, the more structural destruction is generated. Recrystallization is one of the major causes of quality deterioration of frozen food during storage and distribution [Petzold and Aguilera, 2009; Hagiwara *et al.*, 2006]. The ice fraction of glycerol solution with 0.1% 60 nm HA nanoparticles diminished dramatically when compared to the control solution. This is favorable for cryopreservation because survival rate of biomaterial is decreased remarkably due to the osmosis of cell membrane, diminution of unfrozen water, toxicity of cryoprotectant solution, and mechanical injury with the formation of ice crystal.

As ice crystals grow during warming, a larger total contact surface in the ice-liquid interface results in the increment of ice growth rate. The surfaces of ice crystals make contact and extrude when the ice crystals grow to a certain diameter because of the large numbers of ice nuclei in the solution, rate-limiting process of the free individual ice crystal turns into surface kinetics dominant process; meanwhile, local increases in the temperature at the ice–solution interface due to the release of crystallization heat and thermal conductivity is limited, which result in the decrement of ice growth rate. The maximum of ice-crystal growth rate is generated as analyzed above. The rate of ice-crystal growth during heating is determined by the quantity of ice nucleus, diameter of ice crystal, interfacial kinetics, size and content of nanoparticle, characters of solution, warming rate, amongst other factors. The second maximum appears if the effect of speedup is strong than the inhibition.

As shown in Figure 9, there is only one maximum of ice-crystal growth rate in glycerol solution containing 0.5% 40 nm HA nanoparticles, which coincides with Figure 7 that shows there are so many ice nuclei in the solution. The growth rate of ice crystals appears as the first maximum as explained above, the growth rate keeps on decreasing due to the inhibition of interfacial kinetics being more than the speedup of enlarging contact surface all the time, resulting in only one maximum. Similarly, there is only one maximum in PEG-600 solutions with 60 nm HA and without nanoparticles as shown in Figure 10. Due to the nearly same temperature range for ice crystal growth in PEG-600 solutions, and the constant warming rate, we can deduce that the growth rate of ice crystal is inversely proportional to the number of ice nuclei. The ice growth rate is the lowest in PEG-600 solution containing 0.5% 60 nm HA nanoparticles owing to the most ice nuclei, as is depicted in Figure 10. Han *et al.* [2008] puts forth that the ice growth in EG solutions is prominently accelerated due to the nucleation improvement with the addition of diamond nanoparticles during freezing.

The liquid layering was generated on the surface of nanoparticles when we added nanoparticles into the solution, as explained by Kang and Kim [2006]. The liquid in the interfacial layer has a strong interaction with nanoparticles. The thickness of liquid layer varies with the content and size of nanoparticles, and decreases with the increase of size because of the smaller liquid layering. All of these changes the mass transport, heat transport and thermo-physical properties of the solution because of the different interactions between nanoparticles and liquid molecules, which result in the different behaviors of crystallization/vitrification, devitrification and recrystallization during freezing and rewarming. However, the exact mechanism seems to be unknown.

In conclusion, the presence of HA nanoparticles alters the behaviors of crystallization/vitrification, devitrification and recrystallization of quenched solutions upon freezing and rewarming. Data of the present study and earlier work [Han *et al.*, 2008; Hao *et al.*, 2009; Hao and Liu, 2011] demonstrate the complex interactions among solution type, nanoparticle size and the content of nanoparticles in the dispersion. The molecular nature of such interactions at subzero temperatures and the complexity of ice crystal growth kinetics in nanoparticle-containing dispersions are poorly understood, and further studies are needed to solve this problem.

4. HA Nanoparticles Improved Survival Rate of Vitrified Porcine Oocytes

Cryopreservation of oocyte is a potential technique for storing female germ cells for subsequent use in assisted reproduction. Among different cryopreservation strategies, vitrification is considered as the most effective, because it can solidify the solution in a glass-like amorphous state without the formation of ice crystals.

A number of factors affect the survival of oocytes after vitrification, such as vitrification solution formulation, freezing and thawing protocol. Seki and Mazur [2008] argued that ice formation can occur during the rewarming of a vitrified sample, even if it is prevented during cooling. Therefore, it is important to inhibit ice formation during thawing.

Since HA nanoparticles can modify the thermal physical properties and enhance the heat transfer of CPAs, they have been used to improve the survival rate of vitrified porcine oocytes [Li *et al.*, 2013a; Li *et al.*, 2013b; Zhou *et al.*, 2015].

4.1. *Materials and Methods*

4.1.1. *Equipment and devices*

BC-J160S CO_2 incubator was purchased from Shanghai Boxun Enterprise Co. Ltd, China. TS100 fluorescence inverted microscope system and Microscope BX50 were made by Nikon and Olympus Optical Co. Japan, respectively. Cryotops were purchased from Kitazato Biopharma (Fujinomiya, Japan).

4.1.2. *Chemicals*

All chemicals were purchased from Sigma Chemical Co. Ltd unless otherwise stated. TCM199, phosphate buffer solution (PBS) and fetal bovine serum (FBS) were purchased from Gibco. HA nanoparticles were purchased from Nanjing Nano Material Co., Ltd.

4.1.3. *Vitrification solution/Thawing solution*

The basic solution (BS) used for handling oocytes during vitrification and warming was TCM199 supplemented with 20% (v/v) FBS. The equilibrium solution (ES) was BS containing 8% (v/v) dimethylsulfoxide (Me2SO) and 8% ethylene glycol (EG). The vitrification solution (VS) was BS containing 15% (v/v) Me2SO, 15% (v/v) EG and 0.5 M sucrose. The nano-VS were prepared by suspending 60 nm HA nanoparticles at 0.01%, 0.02%, 0.05%, 0.1% or 0.5% (w/w) in vitrification solution, and constantly agitated for three hours by a JY92-2D ultrasonic vibrator at 200 w. This procedure resulted in a uniform distribution of nanoparticles.

4.1.4. *Collection of porcine oocytes*

Porcine ovaries recovered from a local slaughterhouse were placed in warm (37°C) 0.9% NaCl solution containing antibiotics (100 IU/ml potassium penicillin G and 100 mg/ml streptomycin sulfate) and transported to the laboratory within two hours after collection. The GV oocytes were aspirated from antral follicles (2–5 mm in diameter). Cumulus-oocyte complexes (COCs) with three or more layers of compact cumulus cells were chosen for the experiments and washed three times with TCM199 medium containing 10% FBS.

4.1.5. *Vitrification and thawing of oocytes*

Vitrification: Ten oocytes were transferred from basic solution to pre-equilibrium solution for three minutes. After equilibration, they were transferred to 50 μL VS to equilibrate for 30 seconds. Five other groups were transferred to 50 μL VS containing 0.01%, 0.02%, 0.05%, 0.1%, 0.5% HA nanoparticles for 30 seconds. Then, the oocytes were loaded onto the Cryotop carrier with mouth pipette. Before vitrification, the extra CPA was removed with capillary tube. Then, the Cryotop was submerged into liquid nitrogen quickly.

Thawing: The Cryotop with oocytes was taken out of liquid nitrogen quickly and inserted directly into TSA at room temperature for three minutes. The warmed oocytes were transferred to TSB for three minutes and then to TCM199 at 37°C.

4.1.6. *Viability of vitrified oocytes after warming*

Viability of oocytes derived from all experimental groups was morphologically evaluated immediately after treatment using an inverted microscope. Oocytes with structural alterations, such as faint cytoplasm or damaged zona pellucida and/or membrane, were recorded as not viable.

4.1.7. *Viability of vitrified oocytes after in vitro maturation*

After thawing, oocytes were cultured for 42 hours in MS, they were dyed with fluorescein diacetate (FDA) for three minutes and washed with BS three to four times, then they were observed with inverted microscope to see if they were dyed. Oocytes with fluorescence reaction were alive and those without were dead.

4.2. *Results*

4.2.1. *Effect of HA nanoparticles on vitrification of porcine GV oocytes*

HA nanoparticles of different concentrations (0.01%, 0.02%, 0.05%, 0.1%, 0.5%) were added into vitrification solutions as described in Section 2

Table 2. The effect of HA nanoparticles on vitrification of porcine GV oocytes.

Cryoprotectants	**Survival rate after warming (%)**	**Survival rate after *in vitro* maturation (%)**
VS	93.4 (140/150)[a]	14.7 (22/150)[a]
VS + 0.01%HA	98.1 (150/153)[a]	26.8 (41/153)[b]
VS + 0.02% HA	97.4 (143/147)[a]	27.2 (40/147)[b]
VS +0.05% HA	99.3 (147/148)[a]	30.4 (45/148)[b]
VS + 0.1% HA	97.2 (139/143)[a]	25.9 (37/143)[b]
VS + 0.5% HA	96.6 (142/147)[a]	17.7 (26/147)[a]

Note: Values with different letter superscripts [a,b] within columns are significantly different, $P < 0.05$.

(Preparation of CPAs with HA Nanoparticles). The oocytes were then vitrified and warmed as described above. Table 2 shows the viability of vitrified oocytes after warming and *in vitro* maturation. All the groups showed high survival rate immediately after warming, there are no significant difference between groups with or without HA nanoparticles. However, after 42 hr culture, the survival rate of oocytes vitrified with nanoparticles (0.01%, 0.02%, 0.05% and 0.1%) is significantly higher than the VS group, but there is no significant difference between the four groups. The survival rate of VS +0.05% HA nanoparticles group reached as high as 30.4%.

4.3. *Discussion*

Oocytes cryopreservation has numerous benefits on assisted reproduction as well as stock farming. However, the overall success of oocyte cryopreservation in terms of survival rate and developmental rate is not satisfactory. Even by the most efficient cryotop method, only 10% birth rate was reported. In this study, encouraging results were obtained which showed that HA nanoparticles at a certain concentration can increase the survival rate of vitrified oocytes. When 0.01%, 0.02%, 0.05% and 0.1% HA nanoparticles were added into vitrification solution, the survival rate of oocytes after *in vitro* maturation ranged from 25.9% to 35.4%, which is significantly higher than the control group without HA nanoparticles (14.7%). This is the first report about the positive effects of nano-CPA on cell cryopreservation.

Crystallization is lethal to cells during freezing, because a great number of crystals can puncture the wall and membrane of cells. Recrystallization is another important but unnoticed phenomenon during cryopreservation. Even if cells are vitrified without any crystals during freezing, they suffer the lethal damage after recrystallization. Seki and Mazur [2008] tried to inhibit the recrystallization by speeding up the warming rate, however, the warming rate is restricted by temperature that cells can bear. As discussed in Section 3.3, HA nanoparticles at a certain concentration can hinder the recrystallization of vitrification solution during rewarming. We speculate that this is the reason that nano-CPA promoted the survival rate of vitrified oocytes.

According to the Gibbs free energy theory, smaller ice crystals possess high Gibbs free energy and weak stability, while larger ice crystals possess low Gibbs free energy and strong stability. During warming, ice crystals obtain heat from outside and are metastable. Some small crystals begin to melt, while frozen crystals tend to absorb solutions and grow to larger ice crystals. When the growth interface of ice crystals comes into contact with HA nanoparticles, recrystallization becomes more complex. Since the melt solutions are isolated by nanoparticles and can't serve as the source of new ice crystals, the growth of recrystallized ice crystals are greatly constrained. Without large ice crystals, the survival rate of oocytes accordingly get promoted. However, when excessive addition of HA nanoparticles are added, severe heterogeneous nucleation occurs during rewarming. Therefore, the concentration of HA nanoparticles in nano-CPA is crucial to inhibit recrystallization.

Zhao *et al.* [2013] evaluated the toxicity of HA nanoparticles in catfish cells and zebrafish embryos, they concluded that nHA lower metabolic activity of the cells, cause axial deformation and hatching inhibition. In their experiments, they cultured the cells with nHA for 24 hr exposure, HA nanoparticles may enter into the cell and interact with embryo/emzyme during a long incubation time. However, in this experiment, the oocytes were exposed in nano-VS only for several minutes before freezing and after thawing, then they were washed away. The function of nHA is to inhibit the recrystallization, therefore, nHA have little biological effects on oocytes during vitrification.

Although nano-CPA showed primary advantages on oocyte vitrification, the effects of nanoparticles on microstructure and developmental potency of vitrified oocytes were not determined in this study. The mechanism by which nanoparticles benefit vitrification needs to be further investigated.

References

Boutron, P. (1986). Comparison with the theory of the kinetics and extent of ice crystallization and of the glass-forming tendency in aqueous cryoprotective solutions. *Cryobiology*, 4, pp. 88–102.

Choi, S. U. S. (1995). *Developments and Applications of Non-Newtonian Flows*, (FED, New York), Vol. 231, p. 99.

Eastman, J. A., S. U. S. Choi, S. Li, W. Yu and L. J. Thompson (2001). Anomalously increased effective thermal conductivities of ethylene glycol-based nanofluids containing copper nanoparticles. *Applied Physics Letters*, 78, pp. 718–720.

Edmund, A. and D. Marzio (1997). Configurational entropy approach to the kinetics of glasses. *Journal of Research of the National Institute of Standards and Technology*, 102, p. 135.

Gao, D. and J. K. Critser (2000). Mechanisms of cryoinjury in living cells. *Institute for Laboratory Animal Research Journal*, 41, pp. 187–196.

Gao, P. Y. and J. B. Li (2007). *Chemical Physics* (Science Press, Beijing).

Gao, Z. X., B. T. Hao, B. L. Liu *et al.* (2011). Effects of nanoparticles on specific heat of PVP cryoprotectant solutions. *Chinese Journal of Low Temperature Physics*, 33, pp. 36–39.

Gonda, T. and T. Sei (2005). The inhibitory growth mechanism of saccharides on the growth of ice crystals from aqueous solutions. *Progress in Crystal Growth and Characterization,* 51, pp. 70–80.

Hagiwara, T., R. W. Hartel and S. Matsukawa (2006). Relationship between recrystallization rate of ice crystals in sugar solutions and water mobility in freeze-concentrated matrix. *Food Biophysics*, 1, pp. 74–82.

Han, X., H. B. Ma, C. Wilson and J. K. Critser (2008). Effects of nanoparticles on the nucleation and devitrification temperatures of polyol cryoprotectant solutions. *Microfluid Nanofluid*, 4, pp. 357–361.

Hao, B.T. and B. L. Liu (2011). Thermal properties of PVP cryoprotectants with nanoparticles. *Journal of Nanotechnology in Engineering and Medicine*, 2, pp. 1–4.

Hao, B. T., B. L. Liu, S. J. Rong *et al.* (2009). Effect of HA nanoparticles on thermodynamic parameters of cryoprotective agent. *Cryobiology*, 2, pp. 409–410.

Kang, H. U. and S. H. Kim (2006). Estimation of thermal conductivity of nanofluid using experimental effective particle (volume). *Experimental Heat Transfer*, 19, pp. 181–191.

Lee, S., U. S. Choi and J. A. Eastman (1999). Measuring thermal conductivity of fluids containing oxide nanoparticles, *ASME Journal of Heat Transfer*, 121, pp. 280–289.

Li, W. J., X. L. Zhou, J. J. Dai *et al.* (2013a). Effect of hydroxyapatite nanoparticles on MII-stage porcine oocytes vitrification and the study of its mechanism. *Journal of Biomedical Engineering*, 4, pp. 789–793.

Li, W. J., X. L. Zhou, B. L. Liu *et al.* (2013b). The effect of nanoparticles on vitrification of porcine GV-stage oocytes. *Chinese Journal of Biomedical Engineering*, 32, pp. 601–605.

Liu, L. J., H. F. Xu, B. T. Hao *et al.* (2010). Effects of nanoparticles on specific heat of cryoprotectant solutions. *Journal of Refrigeration*, 33, pp. 55–60.

Lv, F. K., B. L. Liu, W. J. Li (2012). Effects of HA nanoparticles on devitrification crystallization of PEG-600 cryoprotectants. *Chinese Journal of Low Temperature Physics*, 4, pp. 315–320.

Lv, F. K., B. L. Liu, W. J. Li and K. Ganesh (2014a). Devitrification and recrystallization of nanoparticle-containing glycerol and PEG-600 solutions. *Cryobiology*, 1, pp. 84–90.

Lv, F. K., B. L. Liu, W. J. Li and X. Y. Song (2014b). Effects of nanoparticles on devitrification and recrystallization of aqueous glycerol and PEG-600 solutions, *SCIENCE CHINA Technological Sciences*, 2, pp. 264–269.

Maxwall, J. C. and A. C. Maxwell (1954). *A Treatise on Electricity and Magnetism*, 3rd ed. (Dover, New York), Vol. 1, p. 435.

Mazur, P. (1977). The role of intracellular freezing in the death of cells cooled at supraoptimal rates. *Cryobiology*, 3, pp. 251–272.

Mazur, P. (1984). Freezing of living cells: mechanisms and implications. *American Journal of Physiology*, 247, pp. 125–142.

Mazur, P. and S. Seki (2011). Survival of mouse oocytes after being cooled in a vitrification solution to -196 degrees c at 95 degrees to 70,000 degrees c/min and warmed at 610 degrees to 118,000 degrees c/min: a new paradigm for cryopreservation by vitrification. *Cryobiology*, 62, pp. 1–7.

Muldrew, K. and L. E. McGann (1990). Mechanisms of intracellular ice formation. *Biophysical Journal*, 57, pp. 525–532.

Muldrew, K., J. Rewcastle and B. J. Donnelly (2001). Flounder antifreeze peptides increase the efficacy of cryosurgery. *Cryobiology*, 42, pp. 182–189.

Pegg, D. E. (2010). The relevance of ice crystal formation for the cryopreservation of tissues and organs. *Cryobiology*, 60, pp. 536–544.

Petzold, G. and J. M. Aguilera (2009). Ice morphology: fundamentals and technological applications in foods. *Food Biophysics*, 4, pp. 378–396.

Seki, S. and P. Mazur (2008). Effect of warming rate on the survival of vitrified mouse oocytes and on the recrystallization of intracellular ice. *Biology of Reproduction*, 79, pp. 727–737.

Seki, S. and P. Mazur (2009). The dominance of warming rate over cooling rate in the survival of mouse oocytes subjected to a vitrification procedure. *Cryobiology*, 59(1), pp. 75–82.

Seki, S. and P. Mazur (2012). Ultra-rapid warming yields high survival of mouse oocytes cooled to -196°c in dilutions of a standard vitrification solution. *PloS one*. 7, pp. e36058–e36058.

Smith, M. J., G. W. Morris and R. Fowler (1988). Physical factors are involved in the destruction of embryos and oocytes during freezing and thawing procedures, *Human Reproduction*, 3, pp. 795–802.

Tang, L. L., H. F. Xu, B. T. Hao *et al.* (2010). Effects of nanoparticles on supercooling and hydration properties of polyalcohols solutions. *Refrigeration*, 8, pp. 60–63.

Wang, X., X. Xu and S. U. S. Choi (1999). Thermal conductivity of nanoparticle-fluid mixture. *Journal of Thermophysical Heat Transfer*, 13, pp. 474–480.

Xie, H., J. Wang, T. Xi, Y. Liu, F. Ai and Q. Wu (2002). Thermal conductivity enhancement of suspensions containing nanosized alumina particles. *Journal of Applied Physics*, 91, pp. 4568–4572.

Xu, H. F., B. T. Hao, L. J. Liu *et al.* (2010). Thermal properties of nano-cryoprotectant solutions during vitrification examination. *Chinese Journal of Refrigeration Technology*, 4, pp. 8–11.

Xu, H. F., B. L. Liu, B. T. Hao *et al.* (2011). Hydration and vitrification properties of nano-cryoprotectant solutions examined with DSC. *Chinese Journal of Low Temperature Physics*, 6, pp. 458–462.

Zhao, X. X., K. J. Ong, J. D. Ede *et al.* (2013). Evaluating the toxicity of hydroxyapatite nanoparticles in catfish cells and zebrafish embryos, *Small*, 9, pp. 1734–1741.

Zhou, X. L., W. J. Li, L. Fang *et al.* (2015). Hydroxyapatite nanoparticles improved survival rate of vitrified porcine oocytes and its mechanism. *CryoLetters*, 1, pp. 45–50.

Chapter 11

Use of FTIR Spectroscopy to Study Cells and Tissues during Cryopreservation Processing and Thermal Therapies

Willem F. Wolkers and Harriëtte Oldenhof

1. Introduction

Fourier transform infrared (FTIR) spectroscopy is emerging as a powerful technique to study intact cells and tissues during cryopreservation or during thermally based treatments of diseases. FTIR studies provide information on conformational and phase changes of endogenous biomolecules in cells and/or tissues [Crowe *et al.*, 1989b; Wolkers and Oldenhof, 2010]. By following the temperature dependence of characteristic peaks in FTIR spectra, temperature induced (thermotropic) and dehydration induced (lyotropic) membrane phase behavior and changes in protein secondary structure can be monitored simultaneously [Wolkers *et al.*, 2007; Wolkers and Hoekstra, 2003]. FTIR studies are minimally invasive and do not require labeling. Samples can be directly measured during drying or during freezing, while bands arising from water give information about the hydration state.

In an IR spectrum of a specimen, bands can be identified arising from characteristic molecular group vibrations. The intrinsic vibrational frequency of a molecular group primarily depends on the mass of the atoms. The primary band position is affected by both intra- and inter-molecular interactions. Phase and conformational changes, due for instance to exposure to different environmental conditions, freezing/heating or drying, result in a change of both intra- and inter-molecular interactions, which is visible as shifts in the position and/or shape of the absorption band.

The CH_2 stretching bands arising from lipid acyl chains can be used to detect phase transitions in liposomes, isolated biological membranes, as well as in whole cells [Mantsch and McElhaney, 1991; Crowe *et al.*, 1989b; Moore *et al.*, 1999]. Liposomes composed of pure lipids, display a defined cooperative gel-to-liquid crystalline phase transition at a specific temperature during heating [Caffrey and Hogan, 1992]. Membranes of mammalian cells typically exhibit a gradual increase in conformational disorder during heating with non-cooperative phase transitions. Membrane phase behavior of mammalian cells is influenced by the presence of membrane cholesterol. During freezing, mammalian cell membranes can undergo sharp fluid-to-gel phase transitions upon ice nucleation and concomitant cellular dehydration [Wolkers *et al.*, 2007].

The overall protein secondary structure can be studied using the amide bands, arising from group vibrations of the protein backbone [Surewicz *et al.*, 1993]. Most FTIR studies on isolated proteins rely on inspection of the amide-I (CO) band for protein secondary structure analysis [Haris and Chapman, 1995]. When cells or tissues are analyzed, the amide bands provide information on the overall protein secondary of all endogenous proteins. The H_2O scissoring vibrational band interferes with the amide-I band, which complicates analysis. Nevertheless, the amide-I band can be used to detect heat-induced protein denaturation in cells using a combination difference and second derivative spectral analysis (Wolkers *et al.* 2007). The different types of secondary structure are less well resolved in the amide-II (NH) band. The weaker amide-III band, predominantly arising from C-N groups, can also be used for protein analysis [Cai and Singh, 2004; Anderle and Mendelsohn, 1987].

In this work, we describe the application of FTIR to study cellular and tissue systems during cooling and heating. Approaches are described for studying membrane and protein conformational changes, water transport processes, and diffusion of protective agents, for applications in cryopreservation and thermal therapies.

2. Infrared Spectroscopy

The mid infrared region of the electromagnetic spectrum ranges from 400–4000 cm^{-1}. The energy of most molecular vibrations falls in the infrared region. However, a molecular vibration is only infrared-active if it is associated with a change in the electric dipole moment. If a diatomic molecule is considered as a simple harmonic oscillator, the vibrational frequency (f) is:

$$f = \frac{1}{2\pi}\sqrt{\frac{k}{\mu}} \tag{1}$$

where μ is the reduced mass of the two atoms ($\mu = (m_1 \times m_2) / (m_1 + m_2)$, where m_1 and m_2 represent the atomic masses), and k is the force constant determined by the strength of the bond. The potential energy (V) of the oscillator is given by:

$$V = \frac{1}{2}k\left(r - r_e\right)^2 \tag{2}$$

where r is the distance between the two vibrating atoms and r_e the equilibrium distance. In infrared spectroscopy the frequency is typically expressed as wavenumbers ($f = c\,(1/\lambda)$); this transforms Equation (1) into:

$$\frac{1}{\lambda} = \frac{1}{2\pi \times c}\sqrt{\frac{k}{\mu}} \tag{3}$$

where $1/\lambda$ represents wavenumber (ν) usually expressed in cm^{-1}.

The main experimental parameter that can be obtained from infrared spectra is the position of the maximum of an absorption band, in cm^{-1}, and band intensity (in arbitrary or relative units). For a macromolecule, there are many vibrational transitions, which can be assigned to particular bonds or groupings. Vibrational bands are shifted by intra- and intermolecular interactions. Formation of a hydrogen bond or other types of molecular interactions will cause any stretching vibration to broaden and shift to lower wavenumber.

3. FTIR Sampling Techniques

Transmission FTIR is widely used to study biomolecules in water. The strong absorbance of water in the CH stretching and amide-I region, however, complicates the analysis of samples in water. In order to record spectra in H_2O, short path length cells ($< 20\ \mu m$) need to be used. Consequently, a relatively high sample concentration is needed to obtain good quality spectra. D_2O does not have strong absorption bands in these regions and longer path length cells or more dilute samples can be used. For protein analysis, it is generally necessary to subtract overlapping absorption of the solvent and buffer components, from the spectrum of the sample [Goormaghtigh *et al.*, 1994].

Attenuated total reflection (ATR) is a sampling technique used in conjunction with infrared spectroscopy. Where transmission spectra are acquired after passage of IR light through a sample, ATR measurements rely on total internal reflection resulting in an evanescent wave extending into the sample. The penetration depth into the sample is determined by the wavelength of light, the angle of incidence and the indices of refraction of the ATR crystal and the medium. Most modern infrared spectrometers can easily be converted to study samples via ATR by mounting an ATR accessory in the sample compartment of the spectrometer. Using polarized light, ATR can be extended to study the orientation of components of lipid films and peptides in biomembranes. By using parallel and perpendicular polarized incident light, it is possible to detect changes in the orientation of a number of chemical bonds belonging to lipids and proteins [Goormaghtigh *et al.*, 1999]. ATR sample holders with temperature control can be used to study protein denaturation in materials which are less translucent such as tissues. Furthermore, ATR can be used to study diffusion of molecules though solid surfaces or tissues [Wang *et al.*, 2014].

A broad range of FTIR accessories is available for studying biological samples, including diffuse reflection, grazing angle specular reflection and FTIR microscopes. Coupling of optical microscopes to an FTIR spectrometer enables the study of very small quantities of biological samples and allows for biochemical mapping of tissues. Furthermore, new technical developments such as multichannel detectors or synchrotron IR microspectroscopy permit high-quality infrared microspectroscopic imaging of biomedical samples [Lasch and Naumann, 2006].

4. Use of FTIR to Study Biomolecules in Cells and Tissues

FTIR can be used to measure samples irrespective of the hydration state. This makes it a powerful to study cells during freezing, drying, or heating [Wolkers and Oldenhof, 2010]. On account of characteristic molecular vibrations, information can be derived on the conformation of biomolecules in their native environment.

Figures 1 and 2 show characteristic molecular vibrations that can be observed in infrared spectra of phospholipids and proteins. Figure 1 shows a spectrum of the glycerophospholipid POPC, 1-palmitoyl-2-oleoyl-sn-glycero-3-phosphocholine. Most notably, are the symmetric and asymmetric

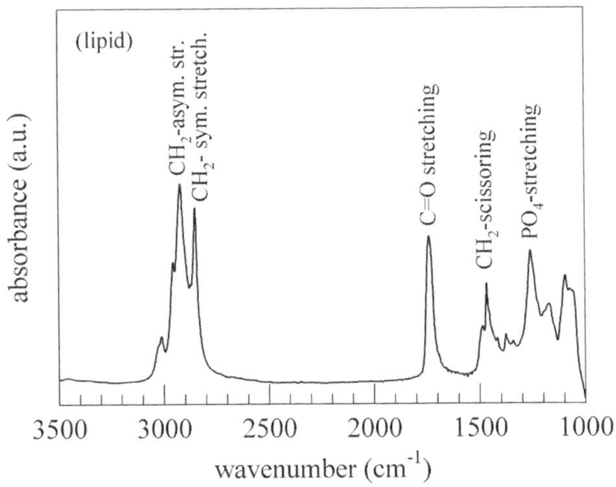

Figure 1. Infrared spectrum of the phospholipid POPC (1-palmitoyl-2-oleoyl-sn-glycero-3-phosphocholine). Characteristic bands of the lipid acyl chains (symmetric and asymmetric CH_2 stretching bands, and CH_2 scissoring band): and the lipid head group (C=O ester band, and PO_4 stretching band) are indicated in the figure.

CH_2 stretching bands at respectively 2850 and 2920 cm^{-1}, arising from the lipid acyl chains. The CH_2 scissoring vibrational mode is visible at 1470 cm^{-1}. Other characteristic bands are the C=O vibrational band, and the PO_4 stretching band arising from the lipid head group. Figure 2 depicts a spectrum of poly-L-lysine illustrating how the amide-I band, located between 1600–1700 cm^{-1}, can be used to detect changes in protein secondary structure after different drying treatments. In the presence of sucrose, the polypeptide adopts an α-helical structure, whereas it adopts a β-sheet structure when dried in the absence of sucrose.

Figure 3 shows in situ infrared spectra of fibroblast cells in hydrated and frozen state. Several bands are visible in the CH-stretching region located between 3000 and 2800 cm^{-1}, mainly arising from CH_3 and CH_2 groups of endogenous lipids and proteins. Characteristic bands arising from cellular proteins are the amide-I band, located between 1600–1700 cm^{-1}, the amide-II band, located between 1500–1600 cm^{-1}, and the amide-III band located between 1330–1200 cm^{-1}. In the region below 1500 cm^{-1}, a variety of characteristic IR group frequencies can be observed, but the bands in this region are difficult to assign. The shape of the spectra below 1500 cm^{-1}, however, is highly characteristic and can be used as a 'finger print' of a given cell type.

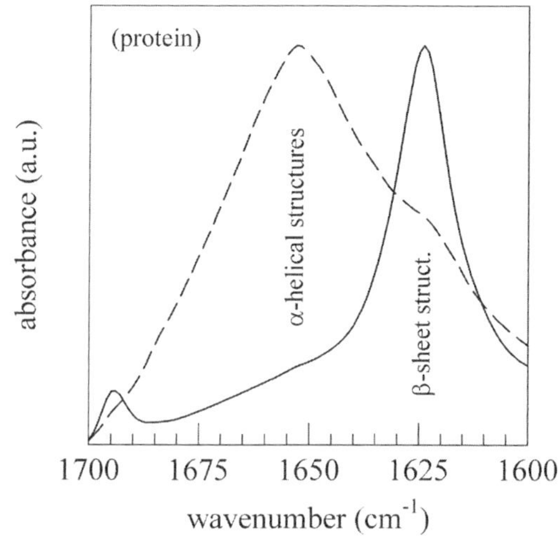

Figure 2. Infrared spectrum of the polypeptide poly-L-lysine in the amide-I region. The spectra reflect those of poly-L-lysine dried without protectants (solid line) and dried in the presence of sucrose (dashed line). Bands arising from α-helical and β-sheet structures are indicated in the figure.
Note: Data adapted from Oldenhof *et al.* [2005].

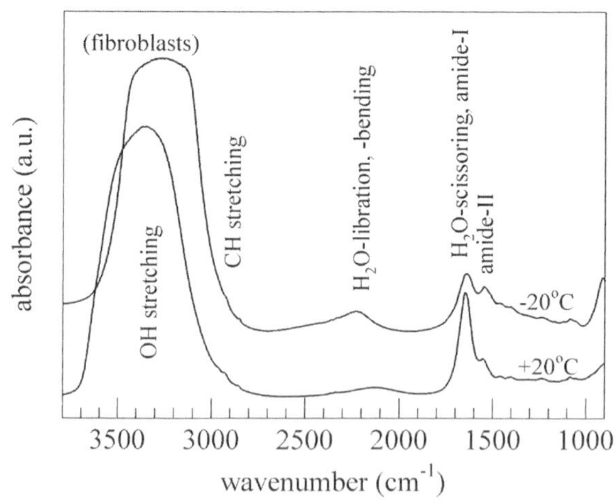

Figure 3. Infrared absorption spectra of fibroblasts, acquired at +20°C (lower trace) and −20°C (upper trace). The OH-stretching, H_2O-libration and bending combination band, and the H_2O scissoring band arising from water, the CH_2-stretching bands mainly arising from membrane lipids, and the protein amide bands are indicated.
Note: Data adapted from Akhoondi *et al.* [2011].

5. Temperature Dependent FTIR Analysis of Membrane Phase Behavior in Cells

FTIR has been widely applied to study conformational order of lipids in biological membranes [Mantsch and McElhaney, 1991; Moore *et al.*, 1995], by inspecting the CH-stretching modes arising from membrane acyl chains. Figure 4 depicts enlargements of the 3000–2800 cm^{-1} and the 2500–2000 cm^{-1} regions of IR spectra of fibroblast cells at 20°C and -20°C. The presence of ice is evident from inspection of the H_2O libration and bending combination band, which depicts a clear shift in band position and shape coinciding with the water-to-ice phase change. Second derivative analysis can be used to show the different bands in the CH-stretching region more clearly. The characteristic asymmetric and symmetric CH_2 stretching vibration bands (νCH_2as and νCH_2sy) are visible at around 2920 cm^{-1} and 2850 cm^{-1}, respectively. CH_3 group vibrations are visible at 2870 and 2950 cm^{-1}. The band positions of the CH_2 stretching vibrations shift to lower wavenumber upon freezing, indicating an increase in conformational order. Membrane phase behavior can be studied by constructing wavenumber versus temperature plots of the (symmetric or asymmetric) CH_2 stretching vibration band [Wolkers, 2009]. Deuterated lipids can be used to obtain information on the conformational status of specific lipid classes in membranes [Moore *et al.*, 1996].

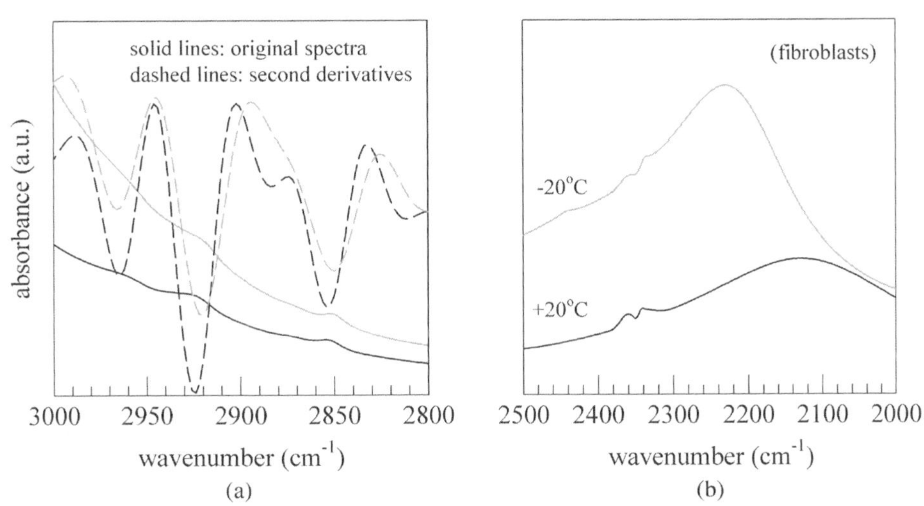

Figure 4. Infrared absorption spectra of fibroblasts at +20°C (black line) and –20°C (grey line) in the CH-stretching region between 3000–2800 cm^{-1} (a), and the 2500–2000 cm^{-1} region showing the H_2O libration and bending combination band (b). Original spectra (solid lines) are shown. Furthermore, second derivatives (dashed lines) were taken to resolve the different bands in the CH stretching region: the symmetric and asymmetric CH_2 stretching vibration bands at 2850 and 2920 cm^{-1}, and symmetric and asymmetric CH_3 stretching vibrations at 2870 and 2950 cm^{-1}.

Note: Data adapted from Akhoondi *et al.* [2011], Oldenhof *et al.* [2013].

6. Effects of Freezing on Membranes

Disruption of the plasma membrane is one of the primary causes of freezing injury. The mechanisms of injury include destabilization of the plasma membrane resulting from freezing-induced cell dehydration and lyotropic membrane phase transitions [Caffrey, 1987]. These include liquid crystalline-to-gel [Wolkers *et al.*, 2007] as well as liquid crystalline-to-hexagonal phase [Gordon-Kamm and Steponkus, 1984] transitions. Such phase transitions, occuring in different temperature regimes for different lipid classes, may lead to loss of membrane integrity and eventually to cell death [Steponkus and Lynch, 1989].

Figure 5 shows supra- and subzero membrane phase transitions for fibroblast cells, as are typical for most mammalian cells. A broad phase transition takes place at suprazero temperatures, but membranes remain relatively fluid. Under freezing conditions that cause cellular dehydration, membranes display a highly cooperative liquid crystalline-to-gel phase transition coinciding with ice formation in the system. The sharpness and extent of this phase transition primarily depends on the cooling rate and the ice nucleation temperature, which determine the extent of cellular dehydration and the incidence of

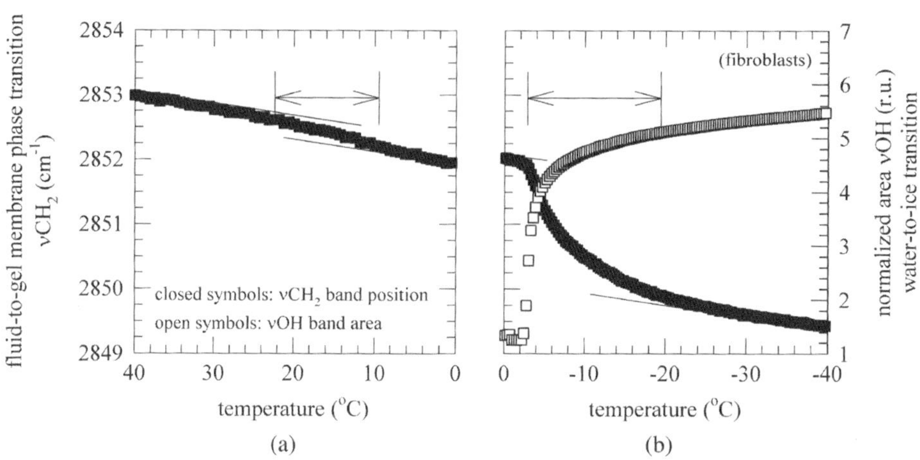

Figure 5. Membrane phase behavior of fibroblasts at suprazero (a) and subzero (b) temperatures, as well as ice formation; as determined using FTIR. Spectra were acquired during cooling from 40 down to –40°C at 1°C min^{-1}, and ice nucleation was induced at –3°C. The band position of the symmetric CH_2 stretching vibration band arising from endogenous lipids was determined and plotted as a function of the sample temperature (closed squares). Ice formation was followed by determining the band area of the H_2O libration and bending combination band versus the sample temperature (open squares). Ice formation is evident as an abrupt increase in band area.

Note: Data adapted from Akhoondi *et al.* [2011], Oldenhof *et al.* [2013].

intracellular ice formation [Wolkers *et al.*, 2007; Oldenhof *et al.*, 2010; Akhoondi *et al.*, 2011].

7. Use of FTIR to Study Water Transport Processes for Cryopreservation or Cryosurgery

The ice nucleation temperature dependency of the thermotropic response of the symmetric CH_2 stretching vibration is illustrated in Figure 6A. When the sample is nucleated at –3°C, membranes undergo a highly co-operative phase transition. Nucleation at –10°C, however, results in a much less co-operative phase transition. Intermediate membrane phase behavior is observed when the sample is nucleated at –6°C.

The so-called water transport model describes temperature-dependent cell volume responses during freezing. The response is dependent on cell intrinsic parameters as well as the cooling rate. In order to use FTIR data to predict cryobiological outcomes, membrane phase behavior data can be converted into cell volume data and used to solve the water transport model equations (Figure 6B). The reduction in cellular volume that occurs during freezing (dV/dT) is described by the following equation [Mazur, 1963; Levin et al., 1976; Toner et al., 1990]:

$$\frac{dV}{dt} = \frac{L_p ART}{v_w B}\left[\ln\left(\frac{V_o - V_b}{V_o - V_b + v_w v_s n_s}\right) - \frac{\Delta H_f}{R}\left(\frac{1}{T_R} - \frac{1}{T}\right)\right] \tag{4}$$

in this equation, the permeability of the plasma membrane to water (L_p in $ms^{-1}Pa^{-1}$: the membrane hydraulic permeability) is defined by:

$$L_p = L_{pg} \exp\left(-\frac{E_{Lp}}{R}\left(\frac{1}{T} - \frac{1}{T_R}\right)\right) \tag{5}$$

where L_{pg} is the reference membrane permeability at a reference temperature (T_R: 273.15 K, equals 0°C), E_{Lp} is the apparent activation energy (in J mol^{-1}) or the temperature dependence of the cell membrane permeability, R is the universal gas constant (8.314 $JK^{-1}mol^{-1}$ or $m^3PaK^{-1}mol^{-1}$), and temperatures (T) are given in Kelvin. V_o and V_b reflect the isotonic and osmotically inactive cell volume, respectively. B is the cooling rate. Furthermore, v_w is the molar volume of water, v_s is the dissociation constant of salt, n_s is the number of moles of salts in the cell, and ΔH_f is the latent heat of fusion of water.

In order to convert freezing-induced membrane dehydration data in corresponding cell volume responses during freezing it is assumed that the freezing-induced shift in νCH_2 is proportional to the reduction in cellular volume. Thus, the maximum shift in νCH_2 as attained with the highest possible subzero nucleation temperature corresponds to maximum cellular dehydration, and partial dehydration is scaled relative to this. This is described by the following equation:

$$\frac{V_o - V(T)}{V_o - V_b} = \frac{\nu CH_{2,o} - \nu CH_2(T)}{\nu CH_{2,o} - \nu CH_{2,\max}} \tag{6}$$

this equation can be rearranged, to translate the freezing-induced membrane phase behavior as measured using FTIR into the relative decrease in cell volume upon freezing:

$$\frac{V(T)}{V_o} = 1 - \left(\frac{\nu CH_{2,o} - \nu CH_2(T)}{\nu CH_{2,o} - \nu CH_{2,\max}}\right)\left(\frac{V_o - V_b}{V_o}\right) \tag{7}$$

V(T) represents the cell volume at a given temperature. Further: $\nu CH_{2,o}$ is νCH_2 just before freezing-induced dehydration (at *Tn*), $\nu CH_2(T)$ is νCH_2 at a given temperature, and $\nu CH_{2,max}$ is the νCH_2 as attained at –30°C for the highest subzero nucleation temperature possible (condition that results in the maximum shift in νCH_2. In order to determine the shift in νCH_2 that is caused by the dehydrating effect of ice formation and not by the thermotropic decrease in νCH_2 with decreasing temperatures, a baseline correction needs to be performed using the linear regression below –25°C, where νCH_2 only exhibits a thermotropic decrease.

The membrane hydraulic permeability parameters, L_p, E_{Lp} and L_{pg}, can be derived from cell volume-normalized νCH_2 versus temperature plots. For a particular nucleation temperature, the slope $d(V/V_o)/dT$ at nucleation is determined from the cell volume versus temperature plot and used to determine L_p using Equation (4). L_p values can then be plotted versus Tn, in an Arrhenius plot (Figure 6C), to derive E_{Lp} and L_p at 0°C (L_{pg}) using a linearized version of Equation (5).

With experimentally determined cell specific volume and hydraulic permeability parameters, simulations of the cell volume response during freezing at different cooling rates can be performed [Toner *et al.*, 1990]. The optimal cooling rate is considered to correspond to the cooling rate at which 90 to 95% of the maximum cellular dehydration has taken place at –30°C [Devireddy *et al.*, 2002; Mazur *et al.*, 1984].

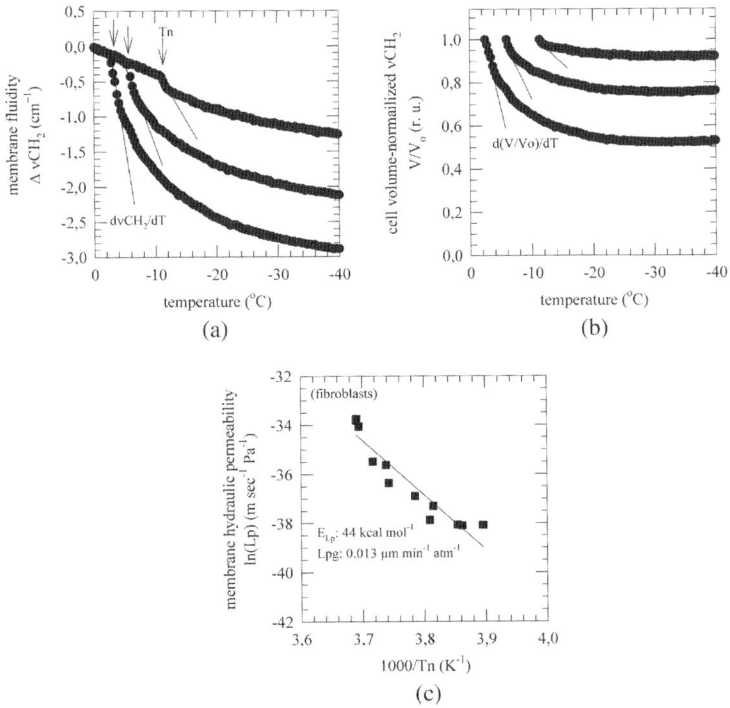

Figure 6. Membrane and cellular dehydration behavior of fibroblasts that were cooled from 20 down to −40°C at 1°C min^{-1}, and nucleated at various temperatures. The data points reflect the relative shift in band position of the symmetric CH_2 stretching band ($\Delta\nu CH_2$) from 0 to −30°C (a). Freezing-induced membrane dehydration (slopes) as well as the ice nucleation temperature (arrows) are indicated. These data were scaled (b) assuming that the measured freezing-induced shift in νCH_2 is proportional to cellular dehydration, and maximum dehydration was equivalent to the isotonic inactive volume (V_b: 52% of the isotonic volume V_o). The initial slope, dV/dT was determined from these plots and used to derive the membrane hydraulic permeability (L_p). L_p was plotted as a function of Tn in an Arrhenius plot (c), which was used to derive the activation energy for subzero water transport (E_{Lp}) and permeability of the membrane for water at 0°C (L_{pg}).
Note: Data adapted from Akhoondi *et al.* [2011], Oldenhof *et al.* [2013].

8. Diffusion of Protectants in Tissues

Whereas for cryopreservation of tissues, permeating protectants like glycerol or DMSO can be used, freeze-drying requires protectants like sucrose or trehalose that form a glassy state at room temperature [Crowe *et al.*, 1989a]. In all cases, complete penetration of protectants in the tissue is needed prior to preservation.

Kinetics of penetration of molecules into tissues can be studied using ATR-FTIR (Figure 7). For this, samples of a defined geometry need to be

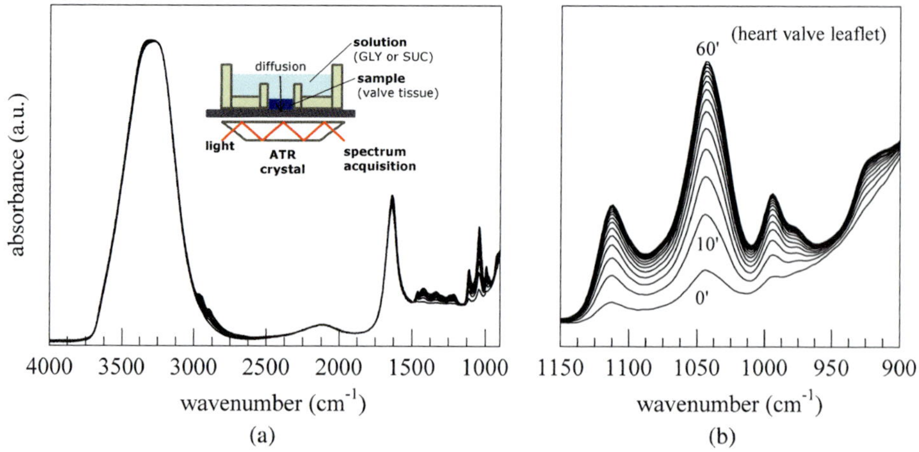

Figure 7. Infrared spectra of leaflet tissue from decellularized heart valves as collected during diffusion of glycerol through the tissue. The inset in panel A depicts a schematic presentation of the setup. The tissue was mounted on the ATR–FTIR sample holder, glycerol was added on top and spectra were recorded every 5 min up to 1 h. Full spectra are shown (a) as well as an enlargement of the spectral region between 1200 and 900 cm^{-1} (b). In this region, glycerol specific bands that can be recognized were used for further analysis of diffusion kinetics.
Note: Data adapted from Wang *et al.* [2014].

mounted such that one side is in contact with the ATR accessory crystal while protectant solution can be added on top of the sample. During diffusion of protectant solution through the sample, infrared spectra need to be recorded at distinct time intervals. Diffusion of protectant through the tissue sample, reaching the proximity of the ATR crystal, is evident from an increase in the intensity of protectant specific bands. Figure 7 shows a typical diffusion measurement of glycerol permeation in decellularized heart valve tissue. The intensity of characteristic glycerol bands visible in the 1200–900 cm^{-1} region, increases over time. To reveal the appearance of protectant specific bands more clearly, difference spectra can be calculated by subtracting the first recorded spectrum from that at a given time point. After this, the area of the glycerol bands between 1150 and 960 cm^{-1} can be determined and plotted as a function of the incubation time to quantify diffusion kinetics.

Fitting of ATR-FTIR data on diffusion of penetrating molecules into tissues can be done as previously described [Fieldson and Barbari, 1995]. Combining Fick's laws on *diffusion* and the continuity equation for the penetrating molecule gives:

$$\frac{\partial c}{\partial t} = D\frac{\partial^2 c}{\partial z^2} \tag{8}$$

where C is the concentration of the penetrating molecule (in mol m^{-3}), and D the diffusion coefficient. ATR-FTIR data on the absorbance of the penetrating molecule at a given time (A_t) to that at equilibrium (A_∞) is given by:

$$\frac{At}{A\infty} = 1 - \frac{8\gamma}{\pi(1-\exp(-2\gamma L))}$$

$$\times \sum_{n=0}^{\infty} \frac{\exp\left(\frac{-D(2n+1)^2\pi^2 t}{4L^2}\right)\left[(-1)^n 2\gamma + \frac{(2n+1)\pi}{2L}\exp(-2\gamma L)\right]}{(2n+1)\left[4\gamma^2 + \left(\frac{(2n+1)\pi^2}{2L}\right)\right]} \tag{9}$$

where L is the tissue thickness in m, and γ the evanescent field decay coefficient, which corresponds to the reciprocal of the depth of penetration, d_p (typically ~10 μm), defined as:

$$\gamma = \frac{1}{d_p} = \frac{2n_2\pi\sqrt{(\sin\theta)^2 - \left(\frac{n_1}{n2}\right)^2}}{\lambda} \tag{10}$$

here n_1 represents the refractive index of tissue, n_2 the refractive index of the ATR crystal, q the incident angle of the IR rays, λ the inverse of the wavenumber position with maximum absorbance (ν_{max}; 1044 cm^{-1} for glycerol). The glycerol band area versus time data can be fitted with a least-squares minimization algorithm to solve Equation (9) and to derive the diffusion coefficient (Figure 8). By extrapolating the experimental data, the times needed to reach saturation with glycerol can be revealed. It should be noted that *diffusion* is measured in one direction, so the actual loading times needed for complete infiltration are actually shorter.

9. Temperature Dependent FTIR Analysis of Protein Denaturation in Cells

FTIR can be used to determine protein denaturation in cells and tissues, by analyzing one of the protein amide bands as a function of temperature. If the amide-I band is used it should be noted that the amide-I band and the H_2O-scissoring band overlap, which complicates analysis. In order to analyze changes in the amide-I band during thermal denaturation, difference spectra can be calculated, with respect to the first recorded spectrum. Then, to reveal contributions of different conformational structures, second derivative spectra can be computed from such difference spectra (Figure 9).

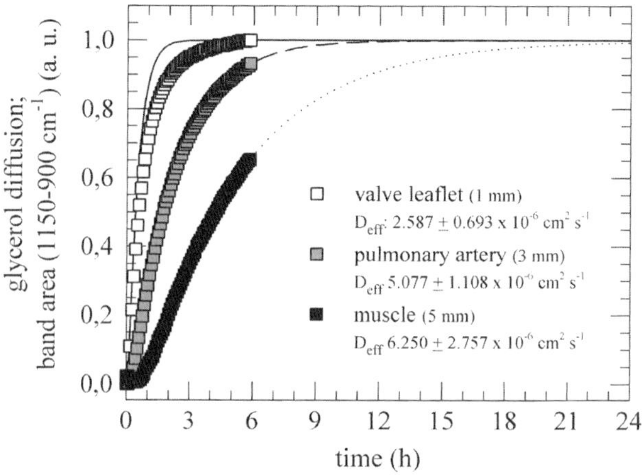

Figure 8. Diffusion of glycerol in decellularized heart valve tissues, as determined from ATR-FTIR spectra (symbols) and simulated by using derived diffusion coefficients (lines). Tissues included: valve leaflets (open squares, solid line), pulmonary artery (grey filled squares, dashed line) and muscle tissue (black filled squares, dotted line).
Note: Data adapted from Wang *et al.* [2014].

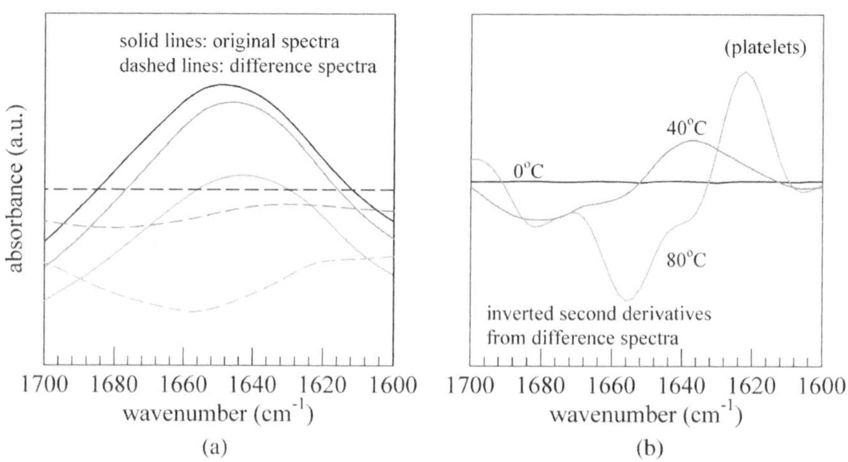

Figure 9. Original spectra (solid lines) and difference spectra (dashed lines) in the amide-I region, as acquired during heating of porcine platelets at 0, 40 and 80°C (a). Difference spectra were calculated by subtracting of the spectrum recorded at 0°C. Panel (b) shows second derivative spectra of the difference spectra. Denaturation coincides with a decrease in the band at 1625 cm^{-1} and a decrease in the band at 1655 cm^{-1}.
Note: Data adapted from Wolkers and Oldenhof [2010].

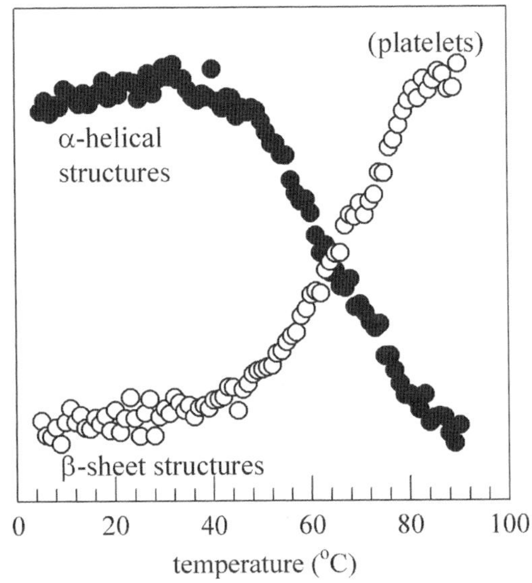

Figure 10. Protein denaturation of porcine platelets during heating from 0 up to 90°C, at 1°C min^{-1}. FTIR spectra were acquired every 1 min, from which the amide-I band was selected and analyzed. The data points reflect the relative area of the band at 1625 and 1655 cm^{-1}, assigned to β-sheet and α-helical structures, respectively.
Note: Data adapted from Wolkers and Oldenhof [2010].

In panel A, original as well as difference spectra of porcine platelets at different temperatures are shown. Panel B shows (inverted) second derivatives of such difference spectra. By using this approach, protein denaturation becomes visible as an increase of a band at 1625 cm^{-1} representing extended β-sheet structures, and a concomitant decrease of a band at 1655 cm^{-1} representing α-helical structures. The area under these bands can be computed and plotted against the sample temperature to reveal the protein denaturation profile (Figure 10).

10. Conclusion

FTIR is a powerful tool to study biomolecules in cells and tissues during treatment steps that are involved in preservation processing or thermal therapies. Membrane phase behavior of cells can be studied by following the temperature dependence of the symmetric CH_2 stretching vibration. Membrane phase behavior studies during freezing can be used to predict cryobiological outcomes after converting membrane phase behavior data into cell volume data. Cell volume data can be used to derive cell-specific membrane hydraulic permeability parameters that are needed to predict optimal cooling rates for cryopreservation. ATR-FTIR can be used to study diffusion kinetics of protective molecules in tissues, which can be used to estimate loading times needed for complete infiltration with protective molecules. Protein denaturation can be studied using the amide-I band. Data analysis requires a combination of difference and second derivative analysis.

Acknowledgments

This work was supported by funding from the German Research Foundation (DFG, Deutsche Forschungsgemeinschaft) via the Cluster of Excellence 'From Regenerative Biology to Reconstructive Therapy' (REBIRTH) and the CORTISS foundation.

References

Akhoondi, M., H. Oldenhof, C. Stoll, H. Sieme and W. F. Wolkers (2011). Membrane hydraulic permeability changes during cooling of mammalian cells. *Biochimica et Biophysica Acta*, 1808, pp. 642–648.

Anderle, G. and R. Mendelsohn (1987). Thermal denaturation of globular proteins. Fourier transform-infrared studies of the amide III spectral region. *Biophysics Journal*, 52, pp. 69–74.

Caffrey, M. (1987). The combined and separate effects of low temperature and freezing on membrane lipid mesomorphic phase behavior: relevance to cryobiology. *Biochimica et Biophysica Acta*, 896, pp. 123–127.

Caffrey, M. and J. Hogan (1992). LIPIDAT: A database of lipid phase transition ry and temperatures and enthalpy changes. DMPC data subset analysis. *Chemistry and Physics of Lipids*, 61, pp. 1–109.

Cai, S. and B. R. Singh (2004). A distinct utility of the amide III infrared band for secondary structure estimation of aqueous protein solutions using partial least squares methods. *Biochemistry*, 43, pp. 2541–2549.

Crowe, J. H., J. F. Carpenter and L. M. Crowe (1989a). The role of vitrification in anhydrobiosis. *Annual Review of Physiology*, 60, pp. 73–103.

Crowe, J. H., F. A. Hoekstra, L. M. Crowe, T. J. Anchordoguy and E. Drobnis (1989b). Lipid phase transitions measured in intact cells with Fourier transform infrared spectroscopy. *Cryobiology*, 26, pp. 76–84.

Devireddy, R. V., D. J. Swanlund, T. Olin, W. Vincente, M. H. T. Troedsson, J. C. Bischof and K. P. Roberts (2002). Cryopreservation of equine sperm: optimal cooling rates in the presence and absence of cryoprotective agents determined using differential scanning calorimetry. *Biological Reproduction*, 66, pp. 222–231.

Fieldson, G. T. and T. A. Barbari (1995). Analysis of diffusion in polymers using evanescent field spectroscopy. *AIChE Journal*, 41, pp. 795–804.

Goormaghtigh, E., V. Cabiaux and J. M. Ruysschaert (1994). Determination of soluble and membrane protein structure by Fourier transform infrared spectroscopy. I. Assignments and model compounds. *Subcellular Biochemistry*, 23, pp. 329–362.

Goormaghtigh, E., V. Raussens and J. M. Ruysschaert (1999). Attenuated total reflection infrared spectroscopy of proteins and lipids in biological membranes. *Biochimica et Biophysica Acta,* 1422, pp. 105–185.

Gordon-Kamm, W. J. and P. L. Steponkus (1984). Lamellar-to-hexagonalII phase transitions in the plasma membrane of isolated protoplasts after freeze-induced dehydration. *Proceedings of the National Academy of Sciences of the United States of America,* 81, pp. 6373–6377.

Haris, P.I. and D. Chapman (1995). The conformational analysis of peptides using Fourier transform IR spectroscopy. *Biopolymers,* 37, pp. 251–263.

Lasch, P. and D. Naumann (2006). Spatial resolution in infrared microspectroscopic imaging of tissues. *Biochimica et Biophysica Acta,* 1758, pp. 814–829.

Levin, R. L., E. G. Cravalho and C. E. Huggins (1976). A membrane model describing the effect of temperature on the water conductivity of erythrocyte membranes at subzero temperatures. *Cryobiology,* 13, pp. 415–429.

Mantsch, H. H. and R. N. McElhaney (1991). Phospholipid phase transitions in model and biological membranes as studied by infrared spectroscopy. *Chemistry and Physics of Lipids,* 57, pp. 213–226.

Mazur, P. (1963). Kinetics of water loss from cells at subzero temperatures and the likelihood of intracellular freezing. *Journal of General Physiology,* 47, pp. 347–369.

Mazur, P., W. F. Rall and S. P. Leibo (1984). Kinetics of water loss and the likelihood of intracellular freezing in mouse ova. Influence of the method of calculating the temperature dependence of water permeability. *Cell Biophysics,* 6, pp. 197–213.

Moore, D. J., S. Gioioso, R. H. Sills and R. Mendelsohn (1999). Some relationships between membrane phospholipids domains, conformational order, and cell shape in intact human erythrocytes. *Biochimica et Biophysica Acta,* 1415, pp. 342–348.

Moore, D. J., R. H. Sills and R. Mendelsohn (1995). Peroxidation of erythrocytes—FTIR spectroscopy studies of extracted lipids, isolated membranes, and intact cells. *Biospectroscopy,* 1, pp. 133–140.

Moore, D. J., R. H. Sills, N. Patel and R. Mendelsohn (1996). Conformational order of phospholipids incorporated into human erythrocytes: an FTIR spectroscopy study. *Biochemistry,* 35, pp. 229–235.

Oldenhof, H., K. Friedel, H. Sieme, B. Glasmacher and W. F. Wolkers (2010). Membrane permeability parameters for freezing of stallion sperm as determined by Fourier transform infrared spectroscopy. *Cryobiology,* 61, pp. 115–122.

Oldenhof, H., W. F. Wolkers, F. Fonseca, S. Passot and M. Marin (2005). Effect of sucrose and maltodextrin on the physical properties and survival of air-dried Lactobacillus bulgaricus: an in situ Fourier transform infrared spectroscopy study. *Biotechnology Progress,* 21, pp. 885–892.

Steponkus, P. L. and D. V. Lynch (1989). Freeze/thaw-induced destabilization of the plasma membrane and the effect of cold acclimation. *Journal of Bioenergetics and Biomembranes,* 21, pp. 21–41.

Surewicz, W. K., H. H. Mantsch and D. Chapman (1993). Determination of protein secondary structure by Fourier transform infrared spectroscopy: a critical assessment, *Biochemistry,* 32, pp. 389–394.

Toner, M., E. G. Cravalho and D. R. Armant (1990). Water transport and estimated transmembrane potential during freezing of mouse oocytes. *The Journal of Membrane Biology,* 115, pp. 261–272.

Wang, S., H. Oldenhof, X. Dai, A. Haverich, A. Hilfiker, M. Harder and W. F. Wolkers, (2014). Protein stability in stored decellularized heart valve scaffolds and diffusion kinetics of protective molecules. *Biochimica et Biophysica Acta,* 1844, pp. 430–438.

Wolkers, W. F. (2009). Biomedical FTIR spectroscopy of lipids. In A. Barth and P. I. Haris (eds.) *Biological and Biomedical Infrared Spectroscopy*, pp. 272–287. ISBN: 978-1-60750-045-2.

Wolkers, W. F., S. K. Balasubramanian, E. L. Ongstad, H. C. Zec and J. C. Bischof (2007). Effects of freezing on membranes and proteins in LNCaP prostate tumor cells. *Biochimica et Biophysica Acta,* 1768, pp. 728–736.

Wolkers, W. F. and F. A. Hoekstra (2003). *In situ* FTIR assessment of desiccation-tolerant tissues. *Spectroscopy*, 17, pp. 297–313.

Wolkers, W. F. and H. Oldenhof (2010). In situ FTIR studies on mammalian cells. *Spectroscopy,* 24, pp. 525–534.

Index